Russell's Hidden Substitutional Theory

Bagley Wood. Jan. 22. '07

Dear Hawtrey

I forgot to send you the paradox which pilled the substitution-theory. Here it is. Put

$$p_0 . = : (\exists p, a): a_0 . = . p\frac{b}{a}!q : \sim(p \frac{a_0}{a})$$

[where "$p\frac{b}{a}!q$" means "p becomes q by substituting b for a"]

Then

$$p_0 \frac{p_0 \frac{b}{a}!q}{a_0} . = : (\exists p, a): p_0 \frac{b}{a_0}!q . = . p\frac{b}{a}!q : \sim(p \frac{p_0 \frac{b}{a_0}!q}{a}) ?$$

Hence $p_0 \frac{p_0 \frac{b}{a}!q}{a_0} . \supset : (\exists p, a): p_0 \frac{b}{a_0}!q = p\frac{b}{a}!q : p_0 \frac{p_0 \frac{b}{a_0}!q}{a_0} . \sim(p \frac{p_0 \frac{b}{a_0}!q}{a})$ (1

$$\sim(p_0 \frac{p_0 \frac{b}{a_0}!q}{a_0}) . \supset :: p_0 \frac{b}{a_0}!q . = . p\frac{b}{a}!q : \supset_{p,a} p \frac{p_0 \frac{b}{a_0}!q}{a} ::$$

$$\supset :: p_0 \frac{p_0 \frac{b}{a_0}!q}{a_0} \quad (2)$$

$\vdash$ (1). (2). $\supset$ ~~[struck out]~~ $:: p_0 \frac{p_0 \frac{b}{a_0}!q}{a_0} :: (\exists p, a): p_0 \frac{b}{a_0}!q . = . p\frac{b}{a}!q : \sim(p \frac{p_0 \frac{b}{a_0}!q}{a})$ (3

But if $p_0 \frac{b}{a_0}!q$ is the same prop as $p\frac{b}{a}!q$, it seems plain we must have $p = p_0 . a = a_0$, whence $p \frac{p_0 \frac{b}{a_0}!q}{a} . \equiv . p_0 \frac{p_0 \frac{b}{a_0}!q}{a_0}$. Thus it is impossible that $p \frac{p_0 \frac{b}{a_0}!q}{a}$ should be false while $p_0 \frac{p_0 \frac{b}{a_0}!q}{a_0}$ is true, which, by (3), is shown to be involved.

In trying to avoid this paradox, I modified the substitution-theory in various ways, but the paradox always reappeared in more & more complicated forms.

Yrs ever B. Russell

Letter from Russell to Hawtrey, dated January 22, 1907. By permission of the Bertrand Russell Archives, McMaster University.

Russell's Hidden Substitutional Theory

GREGORY LANDINI

New York   Oxford
Oxford University Press
1998

Oxford University Press

Oxford New York
Athens Auckland Bangkok Bogota Bombay
Buenos Aires Calcutta Cape Town Dar es Salaam
Delhi Florence Hong Kong Istanbul Karachi
Kuala Lumpur Madras Madrid Melbourne
Mexico City Nairobi Paris Singapore
Taipei Tokyo Toronto Warsaw

and associated companies in
Berlin Ibadan

Published by Oxford University Press, Inc.
198 Madison Avenue, New York, New York 10016

Library of Congress Cataloging-in-Publication Data
Landini, Gregory.
Russell's hidden substitutional theory / Gregory Landini.
p. cm.
Includes bibliographical references and index.
ISBN 0-19-511683-6
1. Russell, Bertrand, 1872–1970. 2. Proposition (Logic—History—20th century. 3. Logic, Symbolic and mathematical—History—20th century. I. Title.
B1649.R94L36 1998
160′.92—dc21 97-22750

9 8 7 6 5 4 3 2 1

Printed in the United States of America
on acid-free paper

Preface

It is not well known that Russell solved the paradoxes that bear his name. The solution lay unnoticed, buried in unpublished manuscripts, for nearly sixty-seven years. It resurfaced anew in the early 1970s as scholars uncovered Russell's substitutional theory from his archival papers. The initial splash was large; perhaps the theory can fill the many gaps in the present understanding of the historical development of ramified type theory. Grattan-Guinness filled one. Substitution, he says, is the "missing link" that explains Russell's frequent recollections that the theory of descriptions "made it possible to see, in a general way how a solution of the contradictions might be possible" (*MMD*, p. 14). In the wake of further examination, however, it has become clear that the theory is out of sorts with the tradition of scholarship that views Russell's logic as a theory of "propositional functions." The substitutional theory can only have been one of many atrophied branches of the decussate evolutionary tree that generated the ramified type theory of *Principia Mathematica*.

When I first came across substitution I was enthralled by Russell's genius and animated by the thought that Logicism cannot be dead after all. A formulation of Logicism via substitution may yet be viable. Why, then, did Russell abandon it? While I was working on this problem and fitting together the meager fossilized remains of substitution, it slowly became clear that the entire tradition was mistaken—that the substitutional theory must be at the center of any historically accurate account of the development of Russell's ramified type theory. Collaborating with Whitehead, Russell originally set out to write a second volume of his 1903 *Principles of Mathematics*. This volume would formally demonstrate the logicist foundations of arithmetic, analysis, rational dynamics, and geometry, foundations which were informally set out in *Principles*. The second volume needed a solution of the paradoxes embodying the fundamental doctrines of *Principles*. That solution was precisely the substitutional theory. For this reason, substitution offers keen new insights into the logic of *Principles*. Through it, Russell's formal systems up to *Principia* can be viewed as calculi for the logic of *propositions* rather than as confused and archaic attempts at (higher order) predicate calculi. Moreover, although substitution was abandoned with *Principia*'s "no-propositions" theory, it affords a wholly new understanding of the formal system of *Principia* and its philosophical motivations. It is the tradition that Russell's logic is a theory of propositional functions (in the ontological sense) that has to go, not substitution's central role in the historical development of ramified type theory.

Uncovering the substitutional theory from the archival manuscripts is no simple matter. The calculus of propositions upon which substitution is based is not a modern predicate calculus. The reader is without a mooring from the onset, and there is a strong inclination to give up, assessing the theory as a jumble of use–mention confusions. One must resist such inclinations. This book shows how. I presuppose only a knowledge of predicate logic. Nevertheless, the reader may find some of the proofs of Chapters 4–7 dauntingly tedious. I felt it necessary to include them because of the uniquely nonstandard nature of the calculus of substitution. It adopts only one style of entity ("individual") variable, and yet Russell promised that it could proxy a type-regimented calculus that allows bindable predicate variables in subject as well as predicate positions. Against a tradition of naysaying, I show that Russell made good on his promise (and this demanded the proofs). I have softened this a bit for readers interested more in history. One can learn the basics of the calculus of propositions from Chapters 4 and 5. The more demanding Chapters 6 and 7 can be skipped without loss of continuity. It is my hope, however, that my work will kindle new research on substitution. In Russell's words, we have a new "beast for our zoo"; the calculus of substitution seems neither a "first-order" nor a "higher-order" calculus!

I came upon substitution in 1984 when I was a graduate student of Nino Cocchiarella at Indiana University. This book was inspired by his work on Russell and, in particular, by his discovery that the substitutional theory conflicts with Cantor's power-class theorem. I wondered how Russell could have missed the conflict. Later searching the archival manuscripts, I uncovered what I called the "p_0/a_0" paradox. Russell had not missed the conflict at all. This at once turned on the flood lights. I knew then that the liar paradox (and "semantic" paradoxes such as the Richard paradox) did not lead to ramification, and that substitution is central to its historical development.

I have benefited from many discussions with I. Grattan-Guinness whose *Dear Russell – Dear Jourdain* is essential reading. Nicholas Griffin has been a valuable resource, especially where the multiple-relation theory of judgment is concerned. My thanks to Ken Blackwell at the Russell Archives (and to Sheila Turcon as well). Special thanks to Russell Wahl and to Francisco Rodriguez-Consuegra for helpful critical comments along the way. Permission to use quotations has been obtained from the William Ready

Division of Archives and Research Collections, McMaster University Library, Hamilton, Ontario, Canada.

The book is dedicated to my mother, whom I miss very much; and it is written in admiration of my father, who first brought *Dear Russell – Dear Jourdain* to my attention.

Iowa City, Iowa
October 1997

G.L.

Contents

Russell's Hidden Substitutional Theory

Introduction

Quodlibet Ens Est Unum

This book concerns the historical development of Bertrand Russell's ramified type theory from the 1903 *Principles of Mathematics* to the 1910 *Principia Mathematica*. Its central theme is to show that one doctrine dominated Russell's thinking throughout the history. The doctrine is *quodlibet ens est unum*. "Whatever is, is one," according to Russell, and by this he meant that any calculus for pure logic must treat all entities alike; it must adopt only one style of pure and unrestricted "entity" variable which regards as one whatsoever is.

Russell may well have first encountered the doctrine in his study of Leibniz. It was through this doctrine that the ancient and timeless conundrum of the One and the Many seized him by the throat. In Russell's *Philosophy of Leibniz*, we find (*L*, p. 71):

> "Where there are only beings by aggregation," Leibniz says, "there are not even real beings. For every being by aggregation presupposes beings endowed with a true unity, since it only derives its reality from that of those which it is composed, so that it will have none at all if every component is again a being by aggregation. . . ." What is not truly one being, is not truly a being.[1]

To Russell, a class seems to be an aggregate so that its being is in its members. Still, the theory of classes requires that a class can be treated as one—for it can be one among members of a class (e.g., its singleton). All seems well if the

1. Thanks to Jan Dejnožka for pointing this out to me.

members of any class, or the members of its members, or the members of the members of its members, or... (in a finite train) are not themselves classes (or aggregates). Not so. There is the universal class V whose being lies in its members notwithstanding the fact that its members include V itself!

What is the solution of the puzzle of the One and the Many? Frege had a ready solution. A class (*Wertverlaufe*) does not have its being in its members, but rather in the concept of which it is the extension. Frege's notion of a class fell to incoherence with the discovery of Russell's paradox, and this sealed Russell's fate. Adhering to Leibniz's doctrine, he would seek a solution to the problem of the One and the Many.

This book will argue that Russell's allegiance to the doctrine never wavered. As a major part of the demonstration, a good deal of time will be spent with a theory Russell developed extensively but which, nonetheless, remained largely buried in manuscripts until the early 1970s. This theory has come to be called Russell's "substitutional theory." Russell held it between 1905 and 1907. It is of central importance to us because Russell believed that it was his best effort at preserving the conception of logic espoused in *Principles*. It is also of central importance to a proper understanding of the mature ramified type theory of the 1910 *Principia Mathematica*. In fact, we shall see that the substitutional theory is the conceptual "linchpin" for a unified account of Russell's (and Whitehead's) work on logicism from *Principles* to *Principia*.

Grattan-Guinness deserves credit for glimpsing this. As he put it (Grattan-Guinness 1974, p. 94):

> The substitutional theory is the missing link between Russell's theory of denoting and *Principia Mathematica*. But it fell soon after it rose, and remained only as a vague and regrettable episode in his memory when he came to think back years later about the period of his logical studies.

Grattan-Guinness recognizes that the substitutional theory is of considerable importance for the development of his thought. Unfortunately, he views the theory as essentially fouled by use–mention confusions. He considers it a valient last-ditch effort at salvaging the fundamental ideas of *Principles*—an effort which Russell came correctly to see was "all nonsense."[2] Abandoning *quodlibet ens est anum*, Russell is said to have adopted a theory of types with restricted individual and predicate variables. Quite the contrary is the case, as I hope to reveal in this book.

Most of the literature on the historical development of type theory finds Russell with many "changes of mind" in his search to find a solution to the paradoxes plaguing logicism. The 1907 "Mathematical Logic as Based on the Theory of Types" is said to mark a pivotal turning point in the historical development of the theory of logical types. But for the brief and rudimentary sketch in *Principles* (Appendix B), the theory of types is held to have begun in this article. *Principia*, the view continues, is (in its formal aspects) by and large

2. In his autobiography, Russell refers to some of his work of 1906 as "all nonsense" and Grattan-Guinness surmises that he must have meant his work on substitution.

the 1907 system put to work. Quine has been a steadfast proponent of this interpretation (Quine 1967). There are stable periods of equilibrium in which a given theory is worked out, but then, with a flash of inspiration, a rapid change of mind occurs which settles into another relatively stable period.

Russell's own remarks confound this interpretation, however. In "Mathematical Logic" he persists in claiming that the apparent variables are unrestricted. Apparent variables are to have "limitations" given by the constraints of meaningfulness. Moreover, Russell explicitly says that he uses type indexed predicate variables (i.e., "propositional function symbols") only for "technical convenience"—a convenience replaceable through the method of substitution (*ML*, p. 77). He does not intend a formal assumption of types of entities in the work.

In fairness to Quine, we should note that for some time it has been a mystery just what substitutional theory Russell could have had in mind. Without any alternative, it appeared that Russell simply could not make good on his claim of "technical convenience." To make sense of the use of predicate variables, one would have to assume a Platonic realm of attributes (propositional functions).

The reason for this situation is that many of Russell's writings on substitution were hidden in unpublished manuscripts. When these manuscripts come to light, an entirely new perspective emerges. We can see that many of Russell's published writings on substitution have been unduly neglected, and, in many cases, entirely misunderstood. For instance, a brief discussion of substitution occurs in Russell's 1905 paper "On Some Difficulties in the Theory of Transfinite Numbers and Order Types." Quine writes that "the theory of types is not even mentioned in this paper" (Quine 1967, p. 150). But contrast the following passage from Russell's 1906 "Les paradoxes de la logique" (*InS*, p. 193):

> *Technically*, the theory of types as suggested in Appendix B [of *Principles*] differs little from the no-classes theory. The only thing that induced me at that time to retain classes was the technical difficulty of stating the propositions of elementary arithmetic without them—a difficulty which then seemed to me insuperable.

The "no-classes" theory is just the substitutional theory. Classes (and propositional functions) are not assumed as single entities. Nonetheless, the apparatus of substitution proxies a theory of propositional functions (and through them a theory of classes and relations-in-extension). It achieves this in such a way that types become part of the logical grammar of a type-free logic of propositions. In his way, the purely formal dodge—the theory of types—became philosophically viable. Contrary to Quine, the substitutional theory was Russell's development of the theory of types!

It is revealing in this regard to return to the following recollection in Russell's autobiography. "In 1906," he says, "I discovered the theory of types. After this it only remained to write the book [*Principia*] out" (*A*, p. 299). Most take Russell to have been referring to the theory of types in

"Mathematical Logic." But the 1906 "theory of types" and the theory behind the technical conveniences of "Mathematical Logic" is the substitutional theory. To be sure, we shall see that much more transpired than merely "writing the book out." But in understanding what transpired we must be prepared to see that "Mathematical Logic" does not embody an abrupt change of mind, a radical departure from the doctrine of the unrestricted variable.

Neither, so I shall argue, does *Principia*. If *Principia* adopts genuinely restricted variables then almost its entire philosophical introduction becomes a jumble of views in tension with one another and with the body of the work. This has come to be the accepted position. But this position forces one to dismiss the philosophical justification, motivation, and nature of ramification offered in *Principia*. The requirement that functions can only take arguments of like order and type becomes quixotic. (Are types within orders or orders within types? And what favors one over the other?) Church is able to find two distinct accounts of "predicativity" in the work, and regards the "multiple-relation" theory as an epistemological add-on poorly integrated into the formal system. Echoing Church's point, Goldfarb writes that the multiple-relation theory would seem to "play no role in Russell's explanation of his logical system" (Goldfarb 1989, p. 34).

This whole mess is produced by the view that the unrestricted variable was abandoned in *Principia*. It is avoided if we look more carefully at the substitutional theory. Once cleansed of Russell's characteristically sloppy use–mention errors, the substitutional theory can be seen to be as adequate for pure mathematics as is simple type theory and requires no infinity axiom to boot. The language of substitution employs only one style of variable—the entity variable—and is wholly type free. In this way, the non-propositional paradoxes are solved in consort with the conception of pure logic in *Principles*. Within substitution, types become viable. Types of propositional functions, and thereby classes and relations-in-extension, are proxied in a type-free grammar. The repetition of arithmetic within types is a natural consequence of the proxy, and not an odious feature of an ad hoc dodge of the paradoxes. Within substitution, types are not ad hoc; they are part of a genuine solution found in a "no-classes," "no-relations-in-extension," and "no-propositional functions" theory. With all this success, Russell would have to be pressed hard to abandon substitution for a system which embraces genuine restricted variables.

To be sure, substitution was abandoned in *Principia*. The theory encounters paradoxes of propositions (among which is the "liar"). But Russell knew well enough that substitution can survive the paradoxes of propositions so long as restricted propositional variables are introduced. In fact (as I shall argue) this is precisely the system of substitution lurking behind the "technical convenience" of predicate variables in "Mathematical Logic." Substitution was indeed abandoned in *Principia*. But this was because Russell favored a correspondence theory of truth involving a "no-propositions" or multiple-relation theory of judgment. And this was favored because it *preserves* the doctrine of the unrestricted variable—a doctrine which Russell thought could not be preserved within substitution. If *Principia* has restricted variables, then Russell has no reason[3] (aside from notational convenience) to abandon substitution! If *Principia* retained the unrestricted variable, accepting only "limitations given by constraints of significance," then we get a smooth and consistent understanding of its philosophical introduction and the central role the multiple-relation theory has in it.

There is, then, a much greater unity in Russell's early philosophy than has been heretofore thought possible. The central tenet of this book is that if we consider the views that Russell advanced,[4] the historical development of the theory of types is not one of "changes of mind" or "punctuated equilibria" (to borrow a phrase from Gould and Eldredge). It is one of gradual evolutionary change on a central doctrine—the unrestricted variable.

Overview

The book divides into three parts. Part I discusses Russell's logicism and its formulation in terms of the doctrine of the unrestricted variable. Russell's conception of logic as the science of propositions is taken up, and the account of the constituents of propositions in *The Principles of Mathematics* is explained. Part II takes up Russell's substitutional theory, showing how the logic of propositions was to be extended to proxy a type-free theory of attributes ("propositional functions" in an ontological sense) and through them a theory of classes and relations-in-extension which embraces the doctrine of the unrestricted variable. Part III examines the paradoxes of propositions plaguing the substitutional theory, Russell's attempt to avoid them, and his eventual abandonment of the theory in favor of the system of *Principia Mathematica*.

Part I breaks down into three chapters. Chapter 1 examines the difference between Russell and Frege's logicist programs, explaining what Russell meant by "pure mathematics" and why he included geometry (Euclidian and non-Euclidian) in his logicism. The chapter explains Russell's conception of "necessity" in mathematics and logic. Chapter 2 examines Russell's *The Principles of Mathematics*. Russell's conception of a "logical subject" of a proposition will be the focus here. This conception is the philosophical foundation of Russell's view that any calculus for logic must embrace only one style of genuine entity/individual variables. The chapter shows how Russell used a theory of denoting concepts to explain the constituents of general propositions named by nominalized formulas containing variables.

Chapter 3 witnesses the collapse of the theory of denoting concepts in *Principles*, the origins of Russell's new theory of the variable, and the invention of the 1905 theory of definite descriptions. Russell shelves the problem as to what are the constituents of general propositions. The variables of quantifica-

3. No reason indeed, for even with order indices substitution can preserve an infinity theorem. See Appendix A.

4. I do not mean to deny that Russell tried all sorts of experiments in his efforts to solve the paradoxes besetting logicism. I am considering his settled views and their changes.

tion are taken as single letters whose "assignment" at this or that value is primitive and unanalyzable—a view not foreign to the modern view which makes the "assignment" of a variable part of the semantic interpretation of a formal language.

Chapter 4 begins Part II. It sets forth the calculus Russell envisioned for his substitutional logic of propositions. The calculus has only one style of variables. Chapter 5 shows how a proxy for a type regimented "second-order calculus with nominalized predicates" can be formulated from within that calculus. Chapters 6 and 7 set out the substitutional treatment of class and relation-in-extension symbols, comparing the substitutional treatment to that of *Principia*. In Appendix A the Peano postulates are proved substitutionally.

In Part III we take up Russell's attempts to solve the paradoxes of propositions. Chapter 8 examines Russell's discussion of these paradoxes in his manuscripts of 1906. It goes on to explain Russell's attempt at a solution in his 1906 "Les paradoxes de la logique." The solution was the abandonment of general propositions, and it collapsed at the hands of a hitherto unknown paradox esoteric to substitution. I call this the "p_0/a_0" paradox. It will be shown that it was this paradox, not the *liar*, or the paradox of propositions of Appendix B (of *Principles*) which led Russell to imagine grafting orders of propositions on to substitution. Chapter 9 takes up this theory, maintaining that it was behind the 1907/8 system set out in Russell's "Mathematical Logic as Based on the Theory of Types."

Chapter 10 offers an entirely new formulation of the formal system of *Principia Mathematica*. Contrary to much of the literature, circumflexion is not taken as a predicate term forming operator and reducibilities serve as the only comprehension principles for the work. The chapter argues that Russell abandoned the substitutional theory in favor of *Principia* because of his desire to *preserve* the view that his *genuine* variables are unrestricted.

The substitutional calculus offers a syntactic solution to the paradoxes. *Principia* offers a semantic solution. Predicate variables (with order/type indices) are allowed, but they are not genuine. They are introduced in such a way as to be "internally limited by their conditions of significance." This internal limitation is philosophically justified by an informal semantics, the centerpiece of which is a new "no-propositions" correspondence definition of "truth" whose foundation was Russell's "multiple-relation theory of judgment."

In *Principia* the only genuine variables are the individual/entity variables; predicate variables are introduced with a nominalistic semantics. An objectual interpretation of *Principia*'s predicate variable is, of course, possible. The calculus can be interpreted as a realist theory of order/types of attributes ("propositional functions") philosophically founded upon Poincaré's "vicious circle principle" (VCP). This is the common wisdom. But it is open to the Ramsey-Gödel-Quine criticism that realism undermines any application of the VCP. A nominalistic semantics for predicate variables, on the other hand, avoids this criticism. The VCP is thereby demoted to a regulative principle for the formation of a solution of the paradoxes. *Principia*'s philosophical justification of order and order/type indices on predicate variables rests on their semantics, not on the VCP.

Russell's view in *Principles* that logic is a universal synthetic a priori science of structure is shown to persist in *Principia*. The structures, however, are no longer reified as propositions but are "logical forms." The viability of this view of logic within the framework of *Principia*'s ramified type theory is assessed. The tractarian conception of logic is contrasted with Russell's and shown to collapse with Wittgenstein's "doctrine of showing." Finally, it is argued that there are yet unexplored prospects for a resurrection of logicism by returning to substitution.

PART I

The Unrestricted Variable

CHAPTER ONE

Russell's Logicist Program

1.0 It is well known that Russell's original conception of logicism differed from that of Frege. Exactly how they differed is, however, a thorny subject. Frege's logicism proclaimed that arithmetic is logic. Geometry was excluded, and as Grattan-Guinness points out, it is not even clear how much of analysis he wanted. Russell's logicism, on the contrary, was that all of "pure mathematics" is logic. Not a few regard Russell's conception of "pure mathematics" as confused. If so, the whole of Russell's work is marred by this original sin. It is important, therefore, to understand what Russell meant in speaking of "pure mathematics." It is no less important to properly grasp Frege and Russell's conceptions of logic. This is the focus of the present chapter.

Two Conceptions of Logicism: Frege and Russell

1.1 In Kant's day logic was still largely confined to the Aristotelian tradition of the laws of categorical and syllogistic reasoning. From Aristotle's *Organon*, the study of the syllogism and of categoricals was developed in the medieval theories of *suppositio* and the rules of valid inference found their way to doctrines such that of the "illicit process" and "undistributed middle" (Geach 1962, p. 90). Logical validity, so far as this tradition goes, was understood as the "containment" of the meaning of the conclusion in the premises' meaning. In an "analytically true" categorical, the grammatical subject expression—a common noun (or noun clause) to which one of the applicative[1] words "all," "every," "some," "any," "a," "no," "most," and the like, are appended—

1. The expression "applicative" is from W. E. Johnson (1922). See Geach (1962), p. 73.

contains in its meaning the meaning of predicate phrase. Because the meaning is contained, the objects referred to by the predicate would of necessity have to be among those referred to by the subject expression. Analytic truth is explained as the necessary truth of such a categorical.

Of course, the necessity of the theorems of geometry and arithmetic could not be explained by appeal to analytic containment. Empirical and synthetic intuition concerning the kinds of entities appropriate to the branch of mathematics seemed required. In arithmetic there seemed to be the primitive intuition of "sequence" or "succession" underlying mathematical induction. Grounding this, Kant's transcendental aesthetic posited an internal intuition of temporal sequence—the ideality of time. In the case of geometry, Kant adopts the view that there are semantically privileged terms, for example, "point," "between," "continuous," and "distance." Axioms, grounded in intuitions concerning the primitives, are then introduced. The proof of theorems, however, frequently involved further appeals to intuition bolstered by perceptions of figures in space. For example, there was the intuition of *uniformity* of space (underlying congruence), the intuition of the *infinity* of a line in space, the intuition of the *continuity* of points on a line, and intuition of their *order* (e.g., Pasch's axiom).[2] As Russell pointed out, mathematicians could not have hoped to prove Euclid's seventh proposition from his axioms without appeal to such intuitions. Accordingly, if geometry is to yield necessary truths a priori, Kant held that these pure empirical intuitions would have to be grounded in the ideality of space.

In his *Begriffsschrift* (1879) Frege dissented from Kant where arithmetic is concerned. He explained that arithmetic relies on no intuition of temporal sequence. He wrote (p. 55):

> Pure thought, irrespective of any content given by the senses or even by an intuition a priori can, solely from the content that results from its own constitution, bring forth judgments that at first sight appear to be possible only on the basis of some intuition. . . . The propositions about sequences developed in what follows far surpass in generality all those that can be derived from any intuition of sequences.

The point is even more vivid in a letter to Anton Marty of 29 August 1882 (Frege 1980, p. 100):

> I regard it as one of Kant's great merits to have recognized the propositions of geometry as synthetic judgments, but I cannot allow him the same in the case of arithmetic. The two cases are anyway quite different. The field of geometry is the field of possible spatial intuition; arithmetic recog-

2. Hintikka (1969) hopes to salvage the Kantian view by abandoning the transcendental aesthetic and distilling out the essential (and in his view non-aesthetic) elements of Kant's notion of "intuition." I find the view difficult. Surely geometers did actually employ intuitions of infinity, continuity, and order, essentially in their proofs. Kant knew this and grounded these intuitions in the transcendental aesthetic.

> nizes no such limitation. The area of the enumerable is as wide as that of conceptual thought and a source of knowledge more restricted in scope, like spatial intuition or sense perception, would not suffice to guarantee the general validity of arithmetical propositions.

The enumerable is a part of conceptual thought itself and is much wider in its scope than that of geometry. Geometry governs only our spatial intuitions (real or imagined) and so its axioms can be denied (as, e.g., Boyali denied the parallel postulate) without contradiction for the purposes of "conceptual thought." But one cannot deny the principles of arithmetic for they involve the laws of conceptual thought itself. Frege maintains that logic itself, with only the formal content of the logical connectives, can reach arithmetic notions such as that of "sequence" without any appeal to Kantian intuition.

In his *Grundlagen der Arithmetik* (1884) Frege went on to rough out the general outline of his logicism. Appeals to intuitions of sequence in arithmetic (i.e., number theory including fractions, irrational numbers, complex numbers, and analysis in general) are both unjustified and unnecessary (Frege 1980, p. 113). The work provided a definition of "natural number" and "cardinal number' generally. Taking the predicative use of concept expressions as fundamental the nominative use of natural number words (e.g., "twelve is a whole number") is captured through Frege's theory of the extensions of concepts and relations. The theory of extensions later matured into the theory of *Grundgesetze der Arithmetik* (1893), viz., Frege's theory of *Werthverlaüf* (classes and ranges of values). Thus, from the logic of predication together with the assumption that the logic could be projected into a theory of extension, Frege claimed logicism to be true: arithmetic is logic.

1.2 Unlike Frege, Russell put his logicism as the thesis that "pure mathematics is logic." This is a much wider doctrine. What exactly is meant by "pure mathematics"? In "Mathematics and the Metaphysicians" (1901) Russell explains (*MM*, p. 60):

> It is common to start any branch of mathematics—for instance, Geometry—with a certain number of primitive ideas, supposed incapable of definition, and a certain number of primitive propositions or axioms, supposed incapable of proof. Now the fact is that, though there are indefinable and indemonstrables in every branch of applied mathematics, there are none in pure mathematics except such as belong to general logic. Logic, broadly speaking, is distinguished by the fact that its propositions can be put into a form in which they apply to anything whatsoever. All pure mathematics—Arithmetic, Analysis, and Geometry—is built up by combinations of the primitive ideas of logic.

The truths of pure mathematics are universal generalizations ("formal implications") asserting that if such and such a proposition is true of *anything*, then such and such another proposition is true of that thing. In *Principles* Russell puts it as follows (*PoM*, p. 3):

> Pure mathematics is the class of all propositions of the form "*p* implies *q*" where *p* and *q* are propositions containing one or more variables, the same in the two propositions, and neither *p* nor *q* contains any constants except logical constants.

It is assumed, of course, that pure mathematics is true—so that we have a class of true propositions. Observe as well that "all propositions of this form" gives us full universal generality. Thus, pure mathematics contains only true universal generalizations all of whose constants are logical constants.

Some care is needed in understanding this. After all, to characterize the sort of mathematics one wants as a class of true fully general formal implications and then to define logic as whatever one needs for pure mathematics would be circular.[3] Russell's intent must not be to literally *define* the mathematics he wants as "pure" mathematics; his intent is to *show* that all of non-applied mathematics can properly be understood to be "pure."

Understood in this way, it is misleading to characterize logicism as if it requires that all the theorems of mathematics are deducible from the axioms and rules of an appropriate calculus for logic. Russell did think that all logical truths (and thus the truths of pure mathematics) are deducible from within the proper formal calculus for logic. But this is incidental. The point is about intuition. In Russell's view, the appearance that there are uniquely mathematical intuitions is produced, from the one side, by logic's having been underdeveloped, and from the other, by a lack of rigor in the construction of proofs in some branches of mathematics (e.g., geometry). The purpose of deduction of mathematical formulas within the formal calculus for logic is to demonstrate that intuitions which were thought to be uniquely mathematical are, in fact, logical.

Consider, for example, mathematical induction. This seems to be grounded on a uniquely mathematical and non-logical intuition of the sequence of the numbers. Russell intends to show that the intuition is logical after all. This is to be accomplished by demonstrating that mathematical induction can be derived from logical principles alone. This does not deprive the mathematician of her intuitions—as if mathematical knowledge requires only knowledge of a few logical primitives. The claim is that the intuitions of mathematics are just those of logic and our knowledge of logic is synthetic a priori and rich with intuition. Russell remarks that Kant "rightly perceived that mathematics is synthetic but that he went wrong in his view of logic," for "it has since appeared that logic is just as synthetic" (*PoM*, p. 457). There is a genuine intuition of the sequence of the natural numbers underlying mathematical induction, but the logicist finds the intuition to be a purely logical intuition revealed by a logical analysis of the nature of number.

3. See Grattan-Guinness (1987), p. 237.

Arithmetization

1.3 Russell's logicism, then, is the claim that (non-applied) mathematics can be shown to be "pure" (pure mathematics)—that is, its propositions are formal implications which are true and fully general. Now one might argue, against logicism, that non-applied mathematics is not pure. It is rife with existential posits and uniquely mathematical intuitions governing them. How can Russell have believed otherwise? In his *The Principles of Mathematics* (1903), we find (*PoM*, p. 158):

> Weierstrass, Dedekind and Cantor, and their followers, have pointed out that, if irrational numbers are to be significantly employed as measures of quantitative fractions, they must be defined without reference to quantity. . . . In this way, during the last thirty or forty years, a new subject, which has added quite immeasurably to theoretical correctness, has been created, which may legitimately be called Arithmetic; for starting with integers, it succeeds in defining whatever else it requires—rationals, limits, irrationals, continuity, and so on. It results that, for all Algebra and Analysis, it is unnecessary to assume any material beyond the integers, which as we have seen, can themselves be defined in logical terms. . . . Continuity and irrationals were formerly the strongholds of the school who may be called intuitionists, but these strongholds are theirs no longer. Arithmetic has grown so as to include all that can be called pure in the traditional mathematics.

Russell tells us that symbolic logic is essentially concerned with inference in general, and is distinguished from various special branches of (non-applied) mathematics mainly by its generality (*PoM*, p. 11). In this the view is similar to Frege. But identifying non-applied mathematics with the new arithmetic, Russell felt able to include not only analysis, but also geometry (Euclidean and non-Euclidean) and the non-applied theory of space-time and motion (rational dynamics) (*ibid.*, p. 112). This is the key. In Russell's view, non-applied mathematics can be shown to be pure by exploiting the view that it is, in essence, nothing more than arithmetic! To grasp how it was that Russell came to this arithmetization of all of non-applied mathematics we need a bit of history.

Spatial and physical intuitions had been injected into analysis through the problem of continuity. The discovery of the irrational had been thought to place some physical magnitudes out of reach of number theory. (Numbers are discrete units.) Physical space is continuous, and continuity cannot be reached number theoretically—or so it seemed. Certain problems could not be given algebraic solutions, and geometric constructions had to be sought—for example, finding a square whose area is that of a given circle. Further, number theoretic approaches to the continuum led to the paradoxes and antinomies of the infinite—antinomies many of which were already heralded by Zeno.

The search for rigor in the calculus eventually forced geometric and spatial intuitions out of analysis. Newton (in earlier works)[4] and Leibniz both appealed to infinitesimals; and both felt the sting of the difficulties they carry. Newton's later emphasis on his "fluxion" calculus was to side-step difficulties (and emphasize his distinctness from Leibniz's methods of differentiation) by appealing to kinematic notions such as "velocity" in the explanation. The fundamental idea in the calculus is that of a "limit of an infinite sequence of elements" and both attempted to explain it by appeal (albeit sometimes tacit) to intuitions of continuity—Newton tacitly hiding part of the explanation in what he took to be the comforting notion of continuous motion. But the pursuit of a rigorous formulation of the calculus in the hands of Bolzano, Cauchy, Weierstrass, Cantor, Dedekind, and Russell, eventuated in a purely mathematical solution of the nature of the continuum, a solution which made no appeals to spatial, geometric, or kinematic intuitions.

Cauchy showed that the notion of a "limit" is involved in differentiation and integration, in the notion of a "continuous function," and in the definition of a "convergent series" (Boyer 1949, p. 280). Cauchy's theorem providing a necessary and sufficient condition for a sequence to converge to a limit, however, required a definition of the irrational numbers which is independent of the notion of a "limit." Unfortunately, geometric intuitions crept into Cauchy's notion of irrational numbers and this lead him to believe that the continuity of a function was sufficient for its geometric representation and for the existence of a derivative; and similar intuitions had led Ampère to try to prove that every continuous function has a derivative, except for certain isolated values in the interval (*ibid.*, p. 282). Bolzano knew this was false (in manuscript) and Weierstrass made it well known. Finding a definition of "irrational number" independent of the concept of the "limit," he would construct a purely formal arithmetical basis for analysis and the calculus (*ibid.*, p. 284). In the hands of Weierstrass, the "limit" does not involve intuition of continuous motion; allusions to "approaching but never reaching and never exceeding a value" are vanquished. Cantor's discovery of non-denumerable infinities (infinities not in one-one correspondence with the natural numbers) and Dedekind's work on irrational numbers revealed that the continuity of a geometric magnitude does not lie in its inability to be "numbered" by discrete rationals. Its continuity lies in the fact that some sets of discrete units form a dense, compact set. The new definitions of number allowed for the construction of real numbers from rational, and rational from natural. The new constructions arithmetizing analysis revealed that it is order, not magnitude, that is basic to continuity. The derivative and the integral became, through the definition of "number" and "limit," not quantitative but ordinal concepts. Ordinal concepts are relational. The calculus is not a branch of the science of quantity, but of the logic of relations (Boyer 1949, p. 294).

4. See his *De Analysi per Aequationes Numero Terminorum Infinitas* (written in 1669 and published in 1711). For a nice discussion, see Boyer (1949).

As Russell sees it, this development shows that it is the logic of relations that is all that is fundamental to non-applied mathematics. Analysis, for example, takes up the investigation of what inferences are possible in respect of continuous relations (i.e., relations generating continuous series). The only difference between logic and analysis is the degree of specialization of the sort of relations studied. The same view is to apply to other branches of mathematics. Russell believed that Kant's antinomies of the infinite had been shattered by this new arithmetization of analysis, and that paradoxes of space, time, and motion (Zeno's dynamical paradoxes) were solved by the new arithmetic theory of continuity. Rational dynamics succumbs to logic. With the notions of "continuity" and "infinity" captured, Russell believed that pure geometry (Euclidean and non-Euclidean) would yield as well to the logic of relations.

As Russell envisioned it in *Principles*, the notion of a "point" in projective (as well as descriptive and metric) geometry can be arithmetized. In his chapter on dimensions and complex numbers, Russell explains (*PoM*, p. 372):

> At present I shall set up by anticipation the following definition: *Geometry is the study of series of two or more dimensions*. This definition, it will be seen, causes complex numbers to form the subject matter of geometry, since they constitute a two dimensional series.

Projective geometry sets out the axioms defining the relationships of the "points" of a projective space. From the axioms, the theorems of projective geometry will be derived. Russell decides that he need not give "point" a fixed meaning any more than he needs a fixed meaning for "irrational number." He gives an existence theorem for a projective space by constructing a projective space out of complex numbers previously defined in a purely arithmetical sense. Russell writes (*ibid*., p. 430):

> The so-called axioms of Geometry, for example, when Geometry is considered as a branch of pure mathematics, are merely the protasis in the hypotheticals which constitute the science.

Geometry, it was once thought, must assert the truth of axioms containing primitive notions such as that of a "point." Russell hopes to accommodate both Euclidean and non-Euclidean geometries by taking non-applied geometry to assert a conditional relationship between its axioms and theorems (*PoM*, p. 373).

Arithmetization reveals the essence of non-applied mathematics, and the essence lies in the notion of order. But in Russell's view ordinal numbers (and the notions of order in general) are part of the logic of relations, so that non-applied mathematics is but a development of the logic of relations.

1.4 As we see, arithmetization is at the center of Russell's logicism. Arithmetization renders the conditionalization of the branches of non-applied mathematics. But, it may be objected, such conditionalization is empty. Russell's logicism is just the simple claim that we are to axiomatize so that the mathematical "primitives" of the various branches of mathematics can be made into variables and then claim that mathematics is "pure" because its sentences are true conditionals. Coffa (1981, for instance, sees matters this way and criticizes Russell's view as a form of what Putnam calls "if-thenism." He writes (*ibid.*, p. 250):

> [Russell] . . . viewed each geometry as asserting not its theorems but the implications from its axioms to each theorem. In a similar vein group theory, topology, measure theory, and most mathematical theories (but not arithmetic) are naturally seen as asserting (formal) implications of the form $A \supset T$, where T is a theorem of the corresponding discipline and A the conjunction of the appropriate axioms. "If-thenism," as Putnam has designated this aspect of Russell's doctrine, is a generalization of Russell's conception of geometry to the whole of mathematics.

Coffa recognizes that Russell did not accept conditionalization for arithmetic. But he rejects Russell's plan to capture the other branches of non-applied mathematics on the grounds that conditionalization is empty. Conditionalization trivializes the thesis of logicism. Wherever intuitions (empirical or otherwise) governing the subject matter of a given field seem needed for inferences, simply render a first-order axiomatization of the field and then all the inferences will thereby be "reducible" to logical inference. We have pure physics, pure biology, pure geography!

This is a misrepresentation. Russell's approach to the branches of non-applied mathematics other than arithmetic was one of arithmetization. Arithmetization is not mere conditionalization ("if-thenism'). What arithmetization did for analysis (in the hands of Weierstrass) obviously could not have been accomplished by the simplistic idea of conditionalization. The work of Weierstrass, Dedekind, and Cantor uncovered the fact that ordinal notions (characterizable by means of the logic of relations) are what is essential to the proper notion of continuity. Arithmetization does not apply everywhere, and is certainly not realized by the Lowenheim completeness result, that one can find a model in the natural numbers for any first-order theory! A necessary condition for arithmetization is a *conceptual analysis* which shows that it is order, not magnitude, that is proper to many mathematical concepts.

In fact, there is ample evidence that Russell was well aware of the difference between mere conditionalization and arithmetization, and that he rejected conditionalization. In Russell's view, there will be within the theory of relations different possible conceptual analyses of (and so constructions of) fractions, irrational numbers, positive and negative integers, complex numbers, and so on. This is, however, acceptable. There is no need to fix constant meanings here. One need only characterize the formal relationships that, say, irrational numbers must have. To be sure, this conditionalizes the theory of the fractions, irrationals, etc. But Russell glosses over this by maintaining that one must fix meanings when it comes to the basis of these numeric notions—that is, when it comes to the notions of "cardinal number" (and "ordinal number"). To see this, let us turn to Russell's discussion of arithmetic itself. We find (*PoM*, p. 430):

> We defined the class of classes called *progressions* by certain logical characteristics of the internal relations of terms of any class which is a progression, and we found that Arithmetic, in so far as it deals with numbers, and not with the terms or classes of which numbers can be asserted, applies equally to all progressions.

Arithmetic is the study of progressions. All progressions are characterized (up to isomorphism)[5] by the Dedekind–Peano postulates and here only the notions of "zero," "number," and "successor" are descriptive constants. We can make these three notions variable and then we have a purely logical theory of trios satisfying these axioms (*PoM*, p. 126). We can then deduce by logic alone the theorems that follow of anything which is such a trio. But, Russell warns, this is not logicism. The natural numbers are not to be just "whatever satisfies the formal structure of the Dedekind–Peano postulates." For arithmetic, logicism must yield a conceptual analysis of the notion of "natural number," and for analysis a conceptual analysis of "ordinal number." As we shall see, Russell's logical analysis of the notion of natural number—an analysis arrived at by use of his infamous "principle of abstraction"—enabled him to fix on a meaning that Frege had arrived at some eighteen years earlier. The natural number 2 becomes the class of all classes similar to a two-membered class. The finite ordinal number 2_r becomes the class of all relations similar to a well-ordering relation with a field of exactly two members.

Russell's Principle of Abstraction

1.5 To understand how Russell arrived at his conceptual analysis of natural number we must examine his "principle of abstraction." Unfortunately, Russell's use of the principle has been widely misunderstood. To set matters straight,we need to turn to the relationship between Russell and Peano.

In 1900 Russell and Whitehead attended the International Congress on Philosophy in Paris. It was, as Russell later recalled in "My Mental Development," "the most important year of my intellectual life." Russell was struck by the techniques and logical apparatus demonstrated at the conference by Peano and his disciples. He found the logical precision affordable by Peano's notation of inestimable value; and he was very much influenced by the philosophy of mathematics espoused by Peano and his school.

It is often said that Russell learned only a symbolic method from Peano and his school. This is but a myth. Rodriguez-Consuegra (1991) has argued convincingly that the early tomes of the annual series *Formulaire de Mathématique* (1894–1905) played a central role not only in bringing Russell to the logicist thesis but in establishing the main threads of his work. The work of Peano's school, for example, that of Pieri, Burali-Forti, and Padoa, and the discussion of such figures as Cantor, Dedekind, and Weierstrass expressed in the

5. This holds if it is set within an adequate logic of relations (a theory of classes and relations-in-extension).

Formulaire, presented to Russell the most contemporary work on the arithmetization of the various branches of non-applied mathematics, including real analysis and geometry. In fact, it was the Italian school that provided Russell's *Principles of Mathematics* with its global philosophical view of the mathematical sciences (Rodriguez-Consuegra 1991).

Peano and his disciples stopped short of carrying their methods to logicism, however. In Russell's words, "they hold that the various branches of mathematics have various indefinables by means of which the remaining ideas of the said subjects are defined" (*PoM*, p. 112). In spite of their great successes, Peano's school did not hold that their method grounds the unity of all non-applied mathematics. Russell proffers an explanation in "Mathematics and the Metaphysicians," illustrating by means of the Peano postulates. The notions of "zero," "natural number," and "successor" that Peano takes as mathematical primitives, he says, "can be explained by means of the notion *relation* and *class*; but this requires the logic of relations, which Professor Peano has never taken up" (*MM*, p. 62).

Peano had admitted only the relation of membership and class inclusion, and although the notion of "couple" was adopted as primitive, this led to some errors in the definition of dyadic functions and in other distinctly relational notions. The reliance on primitive mathematical notions in Peano's *Formulaire* is, in fact, rather greater that it might at first seem. Nominal definitions clearly are non-creative and allow the elimination of the defined symbol in all contexts. Peano regarded nominal definitions as desirable. But he could not always attain them and so allowed "definition by postulates," "inductive definition," and "definition by abstraction." All these pose difficulties if arithmetic primitives are to be avoided.

The Peano postulates, for instance, "define by postulates" the notions "zero," "natural number," and "successor of a number"; similarly, in Pieri's geometry, postulates are used as "definitions" of notions such as "point" and "movement." The idea is that the meaning of these terms are primitive and the postulates capture the arithmetic intuitions underlying them. Peano allowed "inductive definition," (recursive definition) in his conditional definitions of the addition sign "+" and his sign for multiplication. The former was as follows:

$$a \in \mathbb{N}_0 \,.\supset.\, a + 0 = a$$
$$a, b \in \mathbb{N}_0 :\supset: a + (b^+) \,.=.\, (a + b)^+,$$

where "a^+" is "successor of a." Recursive definitions violate the criteria of eliminability in a much more profound way than do conditional definitions. Such clauses are justified as "definitions" only if they can be transformed into proper nominal definitions. To do this requires a theory of classes and relations strong enough to prove a simple theorem of recursion:[6]

6. The notation "$f : w \times A \to A$" says that f is a function from the class $w \times A$ into A. The class $w \times A$ is the class of ordered pairs $\langle x, y\rangle$ such that $x \in w$ and $y \in A$. The class w here is the class of natural numbers 0, 1, 2, . . .

$$(f)(A)(a)[f\colon w \times A \rightarrow A .\&.\ a \in A :\supset:$$
$$(\exists^1 h)(h\colon w \rightarrow A .\&.\ h(0) = a .\&.$$
$$(n)(n \in w .\supset. h(n^+) = f(\langle n, h(n)\rangle)))]$$

(The superscripted "1" is for uniqueness.) Of course, once we have such a theory of classes and relations, we can prove the Peano postulates for any class w that has a distinguished element 0 and a mapping + from w onto itself for which the above simple theorem holds. Indeed, from within the theory of classes and relations, the theorem also yields the isomorphism of any two progressions satisfying the Peano postulates. On the other hand, in a first-order predicate logic with the Peano postulates one cannot give proper nominal definitions for operations such as "addition" (Robinson 1949). Without a fully developed theory of classes and relations, Peano cannot hope to regard his recursive clauses as "definitions" at all; they must be understood as introducing new axioms.[7] Moreover, such a system is not sufficient for the development of analysis.

Finally Peano follows Cantor in allowing "definition by abstraction." For instance, suggesting an alternative method to the Peano postulates, the 1901 *Formulaire* offers a definition of "sameness of cardinality" by abstraction. Proceeding by this alternative method, two primitives of the approach of postulates ("zero" and "successor") can be defined using only logical notions. But Peano explicitly states that no definition (i.e., no nominal definition) of "cardinal number" can be obtained using only the logical notions "cls", "∈", "⊃", "∩", "=" (*Formulaire* 1901, p. 39). Peano took both methods as reconciled to the primitive notion of "cardinal number." Definition by abstraction yields only a definition of "sameness of cardinality.'

Of the three forms of definition adopted by Peano—definition by postulates, definition by abstraction, and nominal definition—Russell accepts only the nominal, remarking that the first two "are only necessitated by Peano's refusal to regard relations as part of the fundamental apparatus of logic." (*PoM*, p. 112). (Russell omits definition by induction. But we have lately noticed that it is a legitimate form of "definition" when developed from within the logic of classes and relations.) The logic of relatives traces its history primarily to Peirce[8] and Schröder.[9] Both methods were algebraic, basing their development on the logic of Boole. Russell found their techniques "so cumbrous and difficult that the applications that ought to be made are practically not feasible" (*PoM*, p. 24). "Sur la logique des relations," published in Peano's *Revue de Mathématiques* in 1901, endeavored to express the logic of relations in Peano's notation, demonstrate the formal inadequacy of definition by abstraction,

7. Alessandro Padoa, however, did make a formal study of recursive definitions. See Padoa (1938).

8. See Charles Sanders Pierce (1870).

9. Friedrich Wilhelm Schröder (1890–1905).

show the isomorphism of any two progressions, and show in general that the logic of relations "permits us to give nominal definitions whenever definitions are possible" (*Sur*, p. 4).

1.6 Russell's demand for nominal definition was likely inherited from two of Peano's own disciples—Pieri and Burali-Forti (Rodriguez-Consuegra 1991, p. 132f). Indeed, at the Paris Congress Burali-Forti read a paper entitled "Sur les différentes méthodes logiques pour la définition du nombre réel." The paper claimed supremacy for nominal definition, and even set forth a proposed nominal definition of "cardinal number."[10] There is no doubt that Russell was influenced by Burali-Forti in this matter. Russell even possessed a marked copy of the article upon which Burali-Forti's presentation at the conference was based, viz., the 1899 "Les propriétés formales des opérations algébriques" (Rodriguez-Consuegra 1991, p. 134).

But unlike Burali-Forti, Russell sought a nominal definition by means of his "principle of abstraction." In *Principles* Russell explains. There is a principle "presupposed" in definitions by abstraction which is valid—even though definition by abstraction is not. This is the "principle of abstraction." The underlying idea of abstraction is that when we collect together terms under an equivalence relation, we do so by virtue of a property F common to those and only those terms. This common property is then said to be what gives rise to the equivalence relation[11] on the terms. The equivalence relation, then, can be seen as deriving from a function S such that for all the terms of our equivalence class, x, y, z, etc., $x\,S\,F$ & $y\,S\,F$ & $z\,S\,F$. That is, we have,

$$(x)(y)(x\,R\,y \equiv x\ (S/\overline{S})\ y).$$

By the definition of the relative product of a function S with its converse $\overline{S}$, we get:

$$(x, y)(x(S/\overline{S})y\ .{\equiv}.\ (\exists w)(x\,S\,w\,\&\,w\,\overline{S}\,y).$$

The w in question here is just our common property F. In the case of Peano's abstraction of "the number of a class," we start with a collection of terms that are similar. Our relation R is then 'similarity.' ("Sim" for short.)

Now Russell objects that there is an absolutely fatal formal defect in using abstraction to offer a "definition." Russell characterizes the principle underlying the use of abstraction as follows (*Sur*, p. 9):

10. The definition proceeded from a principle of primitive recursion, and it took certain notions such as addition as primitive.

11. An equivalence relation is one that has the following features: transitivity: $(x)(y)(z)(x\,R\,y\ \&\ y\,R\,z\,.{\supset}.\,x\,R\,z)$; reflexivity: $(x)(x\,R\,x)$; and symmetry: $(x)(y)(x\,R\,y\,.{\supset}.\,y\,R\,x)$.

(Principle of Abstraction)

$$*\,6.2 \vdash (R)(\text{trans}(R)\ \&\ \text{symm}(R)\ \&\ \text{reflex}(R)\ \&\ R \neq \Lambda :\supset: (\exists S)(\text{funct}(S)\ \&\ R = S/\breve{S})).$$[12]

The theorem says that for any equivalence relation R that is exemplified, there is some function S such that

$$(x)(y)(x\,R\,y \mathrel{.\equiv.} (\exists w)(x\,S\,w \mathrel{.\&.} y\,S\,w)).$$

The theorem does not, however, ensure that there is a unique function S for each given R; and there are, in fact, many such S. Thus the process of "definition by abstraction" is illicit; it does not provide uniqueness. Where R is 'similarity,' abstraction cannot produce a unique entity common to x, y, z, etc., which could be called "the" number of x and which is what gave rise to the equivalence relation 'similarity.'

An illustration will make the point clear. Two classes can be said to have the same cardinality (i.e., they are the same in number) when they are similar—that is, their members can be placed in a one-one correspondence. Accordingly, let R in the principle of abstraction be 'Similarity.' The principle ensures that there is a function S such that x is similar to y if and only if x has the relative product of S with its converse $\overline{S}$ to y. Taking the domain of R (written "Dom(R)" as the class of all classes,[13] the relation partitions classes into equivalence classes. All and only empty classes are in an equivalence class; all and only singleton classes are in the next equivalence class; all and only doubleton classes in the next; and so forth. Peano's definition by abstraction then yields a common property (entity) by virtue of which all the empty classes (there is only one) are similar. The entity indicated by abstraction (with respect to a given S) is just the n_0 such that for the empty class Λ, $\Lambda\,S\,n_0\ \&\ n_0\,\overline{S}\,\Lambda$. In the same way, abstraction (under the function S) indicates an entity z_1 for any two singleton classes—that is, where A and B are singletons, $A\,S\,n_1\ \&\ n_1\overline{S}\,A$. The same goes for each equivalence class. The series of n's, or in other words, the range of S (written "Rng(S)"), is therefore the cardinal numbers yielded by the process of abstraction under the function S.

These are not the only series yielded by the process. There are other functions P equally admissible. For P there will be a different series of entities. For the equivalence class of empty classes under the relation R (i.e., 'similarity'), we

12. Russell's identity sign here means material equivalence. That is, for dyadic relations R and S, he has:

$$R = S = \text{df}\ (x)(y)(x\,R\,y \equiv x\,S\,y).$$

Our notions also use:

$$R \neq \Lambda = \text{df}\ (\exists x)(\exists y)(x\,R\,y)$$

$$\text{funct}(R) = \text{df}\ (x)(y)(z)(x\,R\,y\ \&\ x\,R\,z \mathrel{.\supset.} y = z).$$

13. Such a class presents well-known difficulties, but for the discussion, Russell's theory of classes is naive.

get the entity m_0 such that for the empty class Λ, $\Lambda\ P m_0\ \&\ m_0\ \overline{P}\ \Lambda$. For the equivalence class of singleton classes, we get the m_1 such that for any singleton classes A, B, $A\ P\ m_1\ \&\ m_1\ \overline{P}\, B$, and so on. This forms the series of m's. This series is Rng(P). In fact, there are infinitely many such functions and so infinitely may series, each member of which is got by abstraction. Thus, no one notion of "number" is yielded at all; each of the series (the n series, the m series, etc.) have equal right to be called "the" cardinal numbers so far as abstraction goes. So far as the process of abstraction goes, a given class x in the equivalence class generated by the relation 'similarity' will have infinitely many numbers, one for each relation S.

Russell summarizes the result in *Principles*, writing that in employing the process of abstraction "we obtain a *class* of such properties, with no means of deciding how many terms this class contains" (*PoM*, p. 114). In his "Sur la logique des relations," he puts matters thusly (*Sur*, p. 11):

> If we wish to define cardinal number by abstraction, we can only define it as a class of classes, of which each has a one-one correspondence with the class 'cardinal number' and to which belongs every class that has such a correspondence.

That is, the best that definition of the notion "cardinal number" by abstraction can do is to form the class ς as follows:[14]

$$\varsigma = \text{df}\ \hat{S}(S \in Nc \rightarrow 1\ \&\ \text{sim} = S/\overline{S}).$$

This is the class of all functions S such that

$$(x)(y)(x \text{ Sim } y \ .\!\equiv\!.\ (\exists w)(x\, S\, w \ .\&.\ y\, S\, w)).$$

Then definition by abstraction would have to be content with maintaining that each range of S, for $S \in \varsigma$, are the "cardinal numbers" determined by S.

1.7 In "Sur la logique des relations," Russell reveals his definition of the "cardinal number of a class" in the process of proving the principle of abstraction as theorem (§1 *6.2) and in the proof of theorem (§2 *1.3). Using the now familiar notation $S(x)$ to designate the unique y such $x\, S\, y$, the theorem is proved by noting the existence of the function S which is such that,

$$\text{for all } x \text{ in Dom}(R),\ \ S(x) = \hat{z}(x\, R\, z).$$

In his section on cardinal numbers, Russell uses the proof of *6.2 to prove the theorem *1.3:

$$(\exists S)(S \in \text{Nc} \rightarrow 1\ .\&.\ \text{Sim} = S/\overline{S}).$$

He instantiates the equivalence relation R of *6.2 to the relation 'similarity' (i.e., $\approx$), and so the function used in proving *6.2 becomes:

$$\text{for all x in Dom}(\approx),\ \ S(x) = \hat{z}(x \approx z).$$

14. $Nc \rightarrow 1$ is the class of all functions.

Accordingly, Russell renders a nominal definition of "cardinal number of μ" writing (*Sur*, p. 10):

> Meanwhile we can always take the class $\breve{\rho}$, which appears in the demonstration of Prop *6.2, as the individual indicated by the definition by abstraction; thus for example the cardinal number of a class u will be the class of classes similar to u.

That is, the cardinal numbers are the series fixed by Russell's choice of S in ς.

A question now arises, however, as to the role the principle of abstraction was supposed to play. Russell wrote in *Principles* that "by means of the principle of abstraction, we can give, as we saw in Part II, a formal definition of cardinal numbers" (*PoM*, p. 305). Because of such passages, some interpret Russell as confused—thinking that his principle of abstraction provides the "solution" to the difficulties of definition by abstraction and at the same time admitting that definition by abstraction fails for want of uniqueness. As a result, the purpose of the principle of abstraction has been lost; and some dismiss it as failing to solve anything.[15] Worse, the "meanwhile . . ." passage (quoted above) is absent from the 1900 type edited copy of the original French version of "Sur la logique des relations." In fact, there is evidence that this passage was added after the appearance of Peano's 1901 *Formulaire*—a work which contains the suggestion that the notion of the cardinal number of a class μ could be defined as the class of classes similar to μ (Peano, 1901, §32). This being so, Rodriguez-Consuegra even contends that the codiscovery of Frege's famous definition belongs to Peano, not Russell!

This is just a misunderstanding. The principle of abstraction plays a central role for Russell's codiscovery of Frege's definition of "cardinal number." It is not incidental that Russell introduced his definition in the process of proving his principle of abstraction. Russell writes (*PoM*, p. 305):

> In order to fix upon one definite entity as the cardinal number of a given class, we decide to identify the number of a class with the whole class of classes similar to the given class. This class, taken as a single entity, has, as the proof of the principle of abstraction shows, all the properties required of a cardinal number.

Russell certainly knew that his function S such that

for all x in Dom(R), $S(x) = \hat{z}(x\,R\,z)$,

is not the only one that would serve to prove the principle of abstraction. Any function S in ς would also serve. But Russell's definition of "cardinal number" plays a central role in his demonstration that the choice of the range of *any* S in ς satisfies the Peano postulates. Other choices of S in ς cannot play this role. It is in this sense that the principle of abstraction provides the solution of the

15. See, for instance, Vuillemin (1972) and Rodriquez-Consuegra (1987).

problem of defining "cardinal number." In short, Russell provides grounds for claiming his choice of S in $\mathcal{S}$ is not on a par with other possible choices. It is, rather, the philosophically correct definition of "cardinal number," and "finite cardinal number" in particular.

Russell's justification is somewhat hidden in "Sur la logique des relations," because it is not stated in words, but is part of the proofs that formally demonstrate that cardinal arithmetic (for finite or infinite cardinals) proceeds for the range of every S in $\mathcal{S}$.

To see this, let us begin with Russell's theorem which says that the "cardinal numbers" given by S and those given by S' are in one-one correspondence (*Sur*, p. 11):

$$\cdot 52 \quad \vdash (S)(S')(S \in \mathcal{S} \,\&\, S' \in \mathcal{S} \,.\!\supset\!.\, \dot{\text{Rng}}(S) \approx \text{Rng}(S')).$$

He next deduces,

$$\cdot 54 \quad \vdash (S)(K)(S \in \mathcal{S} \,\&\, K \approx \text{Rng}(S) \,.\!\supset\!.\, (\exists S')(S' \in \mathcal{S} \,\&\, K = \text{Rng}(S')),$$

which ensures that for any $S \in \mathcal{S}$ and any class K similar to the cardinal numbers given by S, there is a function S' whose range is K and is the cardinal numbers given by S'. Russell's notation for the cardinal number 0 (relative to S in $\mathcal{S}$) is "0σ". (He uses "σ" for Domain, and "$\overline{\sigma}$" for Range.) At *2.4 he has the definition (*Sur*, p. 12):

$$0\sigma = \text{df}\, \imath\overline{\sigma}\Lambda.$$

Transformed into a more modern notation this is:

$$0\sigma = \text{df}\, \imath x\, x \in \text{Rng}(S) \,\&\, \Lambda S x.$$

Of course, for Russell's choice of S, this means that $0\sigma = \hat{z}(z \approx \Lambda)$ but for other choices of S there will be a different 0σ. In the same vein, Russell gives the definition at *2.7 (*Sur*, p. 12):

$$1\sigma = \text{df}\, \imath\overline{\sigma} \cap x \ni (u \in \text{Elm} \,.\!\supset_u\!.\, u\, S\, x).$$

Again, in modern notation this is just:

$$1\sigma = \text{df}\, \imath x\, x \in \text{Rng}(S) \,\&\, (u)(x \in \text{Elm} \,.\!\supset\!.\, u\, S\, x),$$

where Elm is the class of all singleton classes.[16] A conditional definition of *addition* for the cardinal numbers of any S in $\mathcal{S}$ is given at *4.1 as well (*Sur*, p. 13):

$$\begin{aligned} &S \in \mathcal{S} :\supset: m, n \in \text{Rng}(S) \,.\!\supset\!. \\ &\quad m + n = \imath x\, x \in \text{Rng}(S) \,\&\, (u, v)(u\, S\, m \,\&\, v\, S\, n \,\&\, u \cap v \\ &\quad = \Lambda \,.\!\supset\!.\, (u \cup v)\ S\, x\}. \end{aligned}$$

Russell's definition of "x plus one" ($x + 1\sigma$), relative to S in $\mathcal{S}$, is given at *4.4 as follows (*Sur*, p. 14):

16. The expression "Elm" is due to Alessandro Padoa. See Padoa (1886).

$$x \in \mathrm{Rng}(S) :\supset: x + 1\sigma =$$
$$\qquad \imath y\ y \in \mathrm{Rng}(s)\ \&\ (u, z)(u\,S\,x\ \&\ z - \in u\,.\supset.\,(u \cup \iota z)Sy\}.$$

Unlike the recursive ("inductive") definition of Peano, the former provides a nominal definition of the addition of cardinals, finite or infinite.

Russell's philosophical justification for his choice of S, and accordingly his definition of "cardinal number," are now revealed by the fact that his general definitions are appropriate to the Frege/Russell notion of cardinal numbers, and at the same time they apply to *any* S in ς. Let us show this for "$x + 1\sigma$". Observe first that, restricting the domain of 'similarity' to finite classes, the definition amounts to a Fregean definition of "successor":

$$x^{+} = \mathrm{df}\ \hat{n}((\exists z)(z \in n\,\&\,n - \iota z \in x)).$$

To verify this, choose Russell's favorite S in ς, that is, the function such that for all $A \in \mathrm{Dom}(\approx)$, $S(A) = \hat{z}(A \approx z)$. Instantiating the quantifiers of Russell's definition of "$x + 1\sigma$", we see that the x in $x + 1\sigma$ is just $\hat{n}(u \approx n)$, that is, the class of all n that are similar to u. Now the y in $\mathrm{Rng}(S)$ which is the successor of x is $\hat{n}(u \cup \iota z \approx n)$, that is, the class of all n similar to $u \cup \iota z$. This follows because $(x)(A, B)(A\,S\,x\,\&\,B\,S\,x\,.\supset.\,A \approx B)$. Hence, it follows that for any class n such that $n \in y$, there is some z such that $n - \iota z \approx u$. Accordingly, $n - \iota z \in x$.

Observe next that, again restricting the domain of 'similarity' to finite classes, the Zermelo numbers and the von Neumann numbers will also be given by choices of S in ς. Recall that for Zermelo, $0 = \mathrm{df}\,\Lambda$, $1 = \mathrm{df}\,\iota 0$, and $x^{+} = \mathrm{df}\,\hat{n}(n = \iota x)$. To generate this series, choose S as follows:

$$A \in \mathrm{Dom(Sim)},$$
$$S(A) = \Lambda, \quad \text{if } A = \Lambda$$
$$S(A) = \imath y(u, z)(z \in A\,\&\,A - \iota z\,S\,u\,.\supset.\,y = u^{+}), \quad \text{if } A \neq \Lambda.$$

Modifying this choice of S only by replacing Zermelo's u^{+} by von Neumann's $u^{+} = \mathrm{df}\ \hat{n}(n = u \cup \iota u)$, we get the series of von Neumann numbers.[17] Now Russell's general definitions of "0σ", "1σ", "$x + 1\sigma$", and "cardinal number" remain intact no matter which S in ς is chosen. At the same time, they parallel the notions given when the specific S chosen renders the Frege/Russell cardinal numbers. This was the centerpiece of Russell's codiscovery of Frege's conception of cardinal number; and it underlies Russell's belief in the correctness of that conception.

Peano was well aware that his postulates admit of infinitely many different *arithmetic* interpretations. He wrote (Peano, *Formulaire*, 1899, p. 30):

> Ces Pp, dont nous avons vu la nécessité, sont suffisantes pour déduire toutes les propriétés des nombres qu'on rencontrera dans la suite. Mais il y a une infinité de systémes qui satisfont à toutes les Pp. P. ex. elles sont

17. That the von Neumann numbers represent finite ordinals in Zermelo-Frankel set theory and not finite cardinals is not relevant to the issue here.

> toutes vérifiées si l'on remplace $\mathbb{N}_0$ et 0 par $\mathbb{N}_1$ et 1. Tous les systèmes qui satisfont aux Pp sont en correspondance réciproque avec les nombres.

$\mathbb{N}_1$ is the class of all whole numbers, which on Peano's construal is just the class of numbers excepting 0, that is, $\mathbb{N}_1 - \iota 0$. As Peano points out in the last sentence of the above quote, $\mathbb{N}_0$ and $\mathbb{N}_1$ are in one-one correspondence (*correspondence réciproque*). By a similar process of simply removing the first member of the series we clearly get infinitely many different series each satisfying the postulates and each in one-one correspondence. In the light of this, Peano is led to embrace a position that what matters to the natural numbers is the structure given by the Peano/Dedekind postulates. Peano explains (*ibid.*):

> Le nombre, $\mathbb{N}_0$, est ce qu'on obtient par abstraction de tous ces systémes; autrement dit, le nombre, $\mathbb{N}_0$, est le système qui a toutes les propriétés énoncées par les P primitives, et celles-là seulement.

The notion of the number $\mathbb{N}_0$ is thus obtained by abstraction from the structural characteristics common to all of the systems satisfying the Peano postulates—that is, for all "progressions."

Russell rejects Peano's view that all that is important to the finite cardinal numbers is the structure which ensures their arithmetic properties. Such an approach, he says, does not enable us to know whether there are any sets of terms verifying Peano's postulates (*PoM*, p. 127). We see that in "Sur la logique des relations" Russell took the Fregean conception to be primary because of its role in his demonstrating that arithmetic holds for all progressions. It holds for the range of any S in ς. Indeed, Russell shows that by means of the logic of relations, the very notion of a "progression" can be defined independently of the notion of "number." This goes well beyond Peano's recognition of alternative arithmetic models of the structure rendered by the Peano postulates. In Russell's view, the Fregean notion of "cardinal number" has a primacy revealed by its use in the demonstration that progressions satisfy the Peano postulates and are isomorphic in structure.

Logic as a Science

1.8 With the ontic commitments of arithmetic settled by the logical analysis of "natural number" the apparent ontological commitments of other branches of non-applied mathematics can be analyzed away by arithmetization. Now when properly understood, arithmetization seems unassailable. But it may well be questioned whether logic can provide the existence theorems needed for its analysis of "natural number." Is the theory of classes and relations-in-extension a proper part of logic?

Confronted with the logical paradoxes that arise with logic conceived of as having ontological commitments to propositions, classes, or propositional functions, the "modern" conception of "logic" takes it that insofar as the fundamental variables of logic should be entity/individual variables for whatever there is, logic should be ontologically uncommitted. It is mathematics that contains ontological commitments, say to sets, which are governed by non-logical intuitions. But we must not lose sight of the fact that Russell's conception of logic was quite different from the modern view. For logicism to be possible at all, logic must be a science and must itself have ontological commitments. So the question naturally arises: What was Russell's conception of logic?

On the modern conception, logic is the study of deductive systems. A given logic is a given uninterpreted deductive calculus, and there are different logics depending on the kind of calculus in question. The study of logic, under this conception, is the study of calculi in general (their semantic completeness, consistency, decidability, etc.). Since the study requires a fair amount of mathematics, this conception makes logic a branch of mathematics.

In Russell's conception, logic is not the study of deductive calculi and there are not different logics. The science of logic is a science of something in the world—in just the way arithmetic (on a realist commitment to an ontology of numbers) is a science about numbers. The subject matter of logic is ontological structure. At the time of *Principles*, Russell gets at this notion through an ontology of propositions. Structure is propositional structure. Russell's conception underwent modification with time, however. By the time *Principia Mathematica* appeared the ontology of propositions was abandoned. The subject matter of logic moves away from propositional structure to "structure" itself (or what Russell calls "logical form.") But the changes aside, Russell's conviction that logic is itself a science with content does not waver. As a science, logic is not to be identified with any given calculus proffered to represent it, or even with the study of calculi. According to *Principles*, logic is a synthetic a priori science that applies to every entity whatsoever. It is completely general and all-encompassing, and its subject matter is the structural forms of propositions (*PoM*, p. 457).

Russell (as Frege before him) spoke of himself as offering not merely a *calculus ratiocinator* in the manner of Boole, but a *characteristica lingua universalis* as Leibniz had conceived of it.[18] In an important paper, van Heijenoort interpreted Frege (and Russell) to be advocating a "syntactic approach" to logic according to which logic is a universal language with content in its own right; and logical truth and validity are to be construed in terms of deductive closure under the axioms. He contrasted this conception with the "semantic" approach of Tarski (originating in the algebraic tradition of Boole, Peirce, Schröder, and Löwenheim). Here logic is the study of uninterpreted deductive calculi—calculi which can express a content (e.g., mathematics) only when an interpretation over an appropriate domain (e.g., numbers) is provided. On the semantic view, logical truth and validity are construed in terms of invariant truth throughout all admissible interpretations. Van Heijenoort goes on to maintain that the Frege/Russell tradition excludes a distinction between the

18. See Frege (1969, p. 371), and see Russell (*MM*, p. 62).

logical system and the metasystem required to justify and study the system. Accordingly, he went so far as to say that the Frege/Russell approach actually obstructs metasystemic research (van Heijenoort 1987, p. 116).

In van Heijenoort's view, the Frege/Russell conception of logic excludes the very intelligibility of now rather commonplace distinctions between syntax and semantics, object-language and meta-language, schemata and axioms, axioms and transformation rules, and so on. His view has been influential. In a recent book, Hylton agrees. He writes (Hylton 1990, p. 203):

> If logic is to be unconditionally and unrestrictedly true in the sense that Russell requires it to be, then it must be universally applicable. This, in turn, implies that statements about logic must themselves fall within the scope of logic, so the notion of a meta-theoretical perspective falls away. If this were not so, if logic were thought of as set up within a more inclusive meta-language, then . . . it would appear that logic is not absolutely and unconditionally true. Logic, on this modern picture, is not unrestricted, for it is set up in a more inclusive language which must fall outside of its scope.

On the "logic as language" conception, logic must "contain its own meta-theory" in a way that makes such distinctions beyond the pale for Russell.

Now van Heijenoort is correct in so far as Russell held that logical truth is not invariant truth in every admissible interpretation. He rightly characterizes Russell's conception of logical validity as being syntactic (i.e., framed in terms of closure under the axioms of the calculus for logic). Logic, on Russell's early view, is a synthetic a priori science of the structures of propositions. But van Heijenoort goes too far. None of this means that Russell's conception of logic requires that a system for logic include its own meta-theory.

Frege made clear in his review of Peano that his *Begriffsschrift* is *both* a *lingua characteristica* and a *calculus ratiocinator*, while Boolean algebra amounts at most to the latter (Frege 1879, p. 7). The superiority over Boole's calculus that Frege awarded his *Begriffsschrift* does not lie in its being itself a "language" with content as opposed to an uninterpreted syntax with rules for the manipulation of the symbols. The distinction is not that Boolean calculi are uninterpreted and are only about something (say natural numbers) when coupled with a semantic interpretation (whose domain is the natural numbers), while on the Frege/Russell conception logic—as itself a language—has content in its own right. The notion of "content" that Frege and Russell have in mind is that of structure—and structures can be represented syntactically.

Adopting the notion of a function from mathematics, Frege constructed his *Begriffsschrift* for logic by inventing the notion of a "truth-function"— a function whose values are truth-values (the True or the False). Of course, "*fx*" and "*hxy*", etc., are terms since "*f*" and "*h*" are function expressions. Accordingly, to make a formula, Frege adds a turnstile "⊢". Sentences of ordinary language are then transcribed into the calculus via "definitions" of functions in analogy to the mathematical introduction of functions.

$$fx_1, \ldots, x_n = \left| \begin{array}{ll} \text{the True,} & \text{if } Ax_1, \ldots, x_n \\ \text{the False,} & \text{otherwise.} \end{array} \right.$$

The superiority of Frege's *Begriffsschrift* lies in its comprehension principle for functions. Comprehension introduces function terms which decompose complex sentential "contents" (logical structures) in different ways—so that now one structure and now another is singled out. Consider for instance,

> "Some object is left of George."

The "content" (structure) here permits decomposition in various ways. One way is represented by:

> "*ga*",

where we have "*a*" assigned to George and "*g*" assigned to the following truth function,

$$gx = \left| \begin{array}{ll} \text{the True,} & \text{if some object is left of } x \\ \text{the False,} & \text{otherwise.} \end{array} \right.$$

So "$\vdash ga$" says that some object is left of George. A different syntactic representation of a decomposition of this content takes it as the "mutual saturation" of two functions,

$$-\!\!\top\!\overset{a}{\smile}\!\top\!\!-\Phi(a) = \left| \begin{array}{ll} \text{the True,} & \text{if some object } a \text{ is such that } \Phi_a \text{ is the True.} \\ \text{the False,} & \text{otherwise.} \end{array} \right.$$

$$fz = \left| \begin{array}{ll} \text{the True,} & \text{if } z \text{ is left of George} \\ \text{the False,} & \text{otherwise.} \end{array} \right.$$

This structure would be represented syntactically by

> "$-\!\!\top\!\overset{x}{\smile}\!\top\!\!-fx$"

In this way, Frege can write

> "$\vdash\!\!\top\!\overset{x}{\smile}\!\top\!\!-fx$"

to say that some object is left of George. It is the availability of such decompositions that makes it possible to find logical structure in "contents" such as 'There are twelve apostles,' and makes it possible go on to uncover complex structure in such contents as '12 > 6,' contents that Kant's *Transcendental Aesthetic* had presumed to be simple. It is precisely the ability to express such structures that, in Frege's view, reveals the informativity of a calculus for logic in spite of its "analyticity." That is, the informativity of Frege's calculus for logic lies in its comprehension principle for functions. It was in virtue of this comprehension principle that Frege hoped to reveal that no non-logical and uniquely mathematical intuitions of number are needed to capture the notions and derivations of arithmetic.

Frege's calculus is a calculus for logic, which can express logical content (i.e., structure) in its syntactic forms. Logic itself, however, is neither a formal uninterpreted calculus nor a contentful language "containing its own meta-theory." Logic is a science. No mathematical realist would confuse the science of arithmetic with a language for arithmetic or an axiom system of formal number theory. Similarly, one must not confuse the science of logic with the formal system—the *lingua characteristica universalis*—which is proffered to capture it. Russell also took logic as a science of structure. Unlike Frege, he reifies structures as propositions and adopts predication (constituency in a proposition) rather than functionality as his starting point. But in any event, both Frege and Russell view logic as a science of structure in precisely the way that a mathematical realist takes arithmetic to be the science of number. Nothing in Frege's or Russell's conception of logic precludes the adoption of commonplace modern distinctions between a formal calculus and its meta-theory. These distinctions are untendentious; they do not distinguish the modern mathematical conception of logic from the Frege/Russell conception.

Indeed, it is quite clear that Frege and Russell understood the special status of inference rules—rules set out, as we would say, in the meta-language for the manipulation of the symbols of the object language. In Frege's case, consider the *Begriffsschrift* on the special status of *modus ponens* (Frege 1879, p. 28):

> We have already introduced a number of fundamental principles of thought in the first chapter in order to transform them into rules for the use of our signs. These rules and the laws whose transformations they are cannot be expressed in the ideography because they form its basis.

Inference rules cannot be expressed in the ideography of the *object language*, for they are meta-theoretic. They are rules for the manipulation of signs of that language. Accordingly, Frege is denying that the calculus contains its own meta-theory. Of course, a calculus *for logic* cannot be one whose inference rules for manipulating symbols are arbitrarily adopted. Frege takes a particular logical truth and patterns his inference rule after it. "From each of the judgments expressed in a formula in §13–22," he writes, "we could make a particular mode of inference" (Frege 1879, p. 17).

Russell was no less aware that some of the "principles" of his calculus for logic must be accorded special status. In *The Principles of Mathematics* (1903) he discusses Lewis Carroll's "What the Tortoise Said to Achilles," remarking that deduction will never begin unless we have a rule of detachment. Carroll's puzzle shows that "until we have some new principle, we shall only be led to an endless regress of more and more complicated implications" (*PoM*, p. 35). "We need," Russell continues, "the notion of *therefore*, which is quite different from the notion of *implies*" (*ibid.*). Indeed, he points out that the principle of detachment "is incapable of formal symbolic statement" (*PoM*, p. 16). Though infamously sloppy about use and mention, Russell did have a sense of what we call the difference between meta-language and object language, and endeavored to set forth axioms and inference rules of a formal calculus (in the modern sense) for logic.

The van Heijenoort view is mistaken. It is, nonetheless, prevalent in the literature on logicism. Let us, therefore, take on the arguments advanced in its favor individually. For starters, one commonly finds the claim that Russell's logicism must exclude a formal calculus because setting out any such calculus would have to use arithmetic notions. For instance, the wffs of the calculus will require a recursive characterization. The study of even predicate calculi, completeness, decidability, and the like, requires a fair amount of mathematics. Thus, the objection goes, Russell could not embrace a formal calculus for logic. If he did, then logic could not form an epistemological foundation for mathematics.

To be sure, Russell rejected the use of mathematical induction at the foundation of his system. This is often touted as evidence in favor of the van Heijenoort view. But this is easily explained. Russell holds that a calculus for logic is not the same thing as logic, and logic is not simply the study of deductive calculi. He need only admit, therefore, that in setting forth a calculus for logic we have to use *logic*. One cannot, without begging questions against logicism, suppose that the notions required for the expression of a calculus for logic are uniquely mathematical and non-logical. All the same, this is a double-edged knife. Russell was well aware that it would be equally question-begging in favor of logicism to employ what have been (traditionally) taken to be mathematical notions in setting forth a calculus for logic. The study of "logical calculi" (e.g., in proof theory) requires mathematics. But logic, in Russell's conception, is not proof theory. Since logic is not to be confused with any calculus, our *synthetic* a priori knowledge of its content is independent of any study of formal calculi and the (apparently) uniquely arithmetical notions needed for it.

What then of Frege and Russell's rejection of Hilbert-style independence proofs?[19] Does this not show that the very concept of a semantic interpretation of their systems was absurd to them? It does not. Russell, as Frege before him, rejected Hilbert-style independence proofs on the grounds that since proofs must rob the primitive symbols of their intended interpretations, the proofs cannot be germane when those primitives have their intended interpretations. Independence is a syntactic matter concerning derivability, and derivability within a calculus for logic must not be confused with 'implication.' It is important that the former track the latter. After all, one can adopt rules for a formal calculus that do not track logic at all. Nonetheless, results concerning derivability can be useful and Frege and Russell were simply over-zealous in their concern that changing the meaning of the logical particles of a purported formal calculus for logic severs the connection between the calculus and logic itself.

But, it might be objected, if Russell (and Frege) viewed a *lingua characteristica* as a calculus in the modern sense, won't logic fail to be "all-encompassing?" If logic is universal and "all-encompassing," so the argument goes, then it applies to all forms of reasoning, and accordingly meta-linguistic proofs and notions cannot have any special status. The whole meta-theoretical perspective

19. See Whitehead and Russell (*PM*, p. 115).

collapses for else there would be an essentially richer and more inclusive realm of reasoning (reasoning in the meta-language) which would have to fall outside the scope of logic.

The argument is wanting. To be sure, Russell's conception takes *logic* to be all-encompassing, for any reasoning, in virtue of being proper reasoning, *uses logic*. Logic encompasses all proper patterns of reasoning (and shows why others are improper). If a given formal calculus for logic *fully* captures logic—so that every relation is *expressible* and every function is *representable* in the ideography of an applied form of that calculus—then all proper reasoning patterns can be expressed within it. But suppose it is discovered[20] that for any given calculus for logic, there will be patterns of reasoning (and notions pertaining to the relationship between the expressions of that calculus and their referents) that must lie outside the calculus. What of it? If it happens that some properties such as 'designates-in-*s*' and 'true-in-*s*' cannot be expressed in the language of a proposed calculus *s* for logic,[21] the discovery in no way impugns the claim that logic is all-encompassing. It only shows that no calculus for logic can *fully* capture logic.

But again, one might object that if Russell offers an uninterpreted calculus in the modern sense then the variables of such a calculus (as with any calculus in the modern sense) will be interpretable over different domains. He won't be able to ensure that the whole universe must be the domain. Moreover, it will be possible to alter the interpretation of the logical particles themselves. So be it! Russell was well aware that once a science is represented by an axiomatized calculus there can be many interpretations. He quite clearly observes that arithmetic can be taken to be about all *progressions*, for all satisfy the Peano/Dedekind postulates under different interpretations of "zero," "number," and "successor" (*PoM*, p. 430). Russell did not try to avoid this by claiming that no formal calculus for arithmetic is possible. His response was that a logical (conceptual) analysis of the notion of "natural number" reveals the Frege/Russell cardinals. In the same way, a conceptual analysis of the notion of "continuity" and "limit" had (in the hands of Weierstrass, Dedekind, et al.) yielded that it is order (relation) not magnitude (quantity) that is basic. That there may be unintended interpretations of any formal system proffered to represent the structures revealed by the analysis of "number," is simply irrelevant.

What then of Russell's famous doctrine of the "unrestricted variable"? In the *Principles*, Russell holds that logic is all-encompassing and so applies to whatever has ontological status, irrespective of the sort of entity (concrete particular, universal, proposition, etc.) that it is. Russell maintains that unrestricted variables are required in any calculus for logic because any restricted variable would simply be in the antecedent clause of a conditional whose variables are

20. As we now know from Tarski and Gödel, this is indeed the case.

21. In fact, in 1906 Russell embraces the view that the "designates-in-*s*" cannot be a predicate of the language *s*, and he uses this to reply to the Richard paradox (*InS*, p. 209).

unrestricted, or if themselves restricted, would in turn require a new antecedent clause delimiting the restriction. Eventually, one arrives at an unrestricted variable.[22] Does this not show Russell trying to determine the domain of interpretation?

Hylton reads Russell this way. He argues that Russell's very argument for unrestricted variables precludes a distinction between object language and meta-language. It assumes from the onset that "the statement which establishes the universe of discourse is on the same level as the assertion which is made once the discourse is established. Thus the former can be taken as antecedent and the latter as consequent of a single conditional statement" (Hylton 1990, p. 202).

This is mistaken. To be sure, any modern quantificational calculus will employ meta-linguistic statements such as the schema of *universal instantiation*, and here we find "restrictions." One puts:

$$(\mu)A\mu \supset A[t|\mu],$$

where t is free for μ in the wff A. In the schema "A" is a meta-linguistic letter "restricted" to wffs of the object language and "t" a schematic letter "restricted" to terms of that language. But these are not at all the sort of "restrictions" that Russell finds objectionable. Put succinctly, the doctrine of the unrestricted variable simply demands that the object language of any calculus for logic should embrace one style of genuine variables—viz., "entity" or "individual" variables.[23] Russell's point is that even if there are distinct sorts of entities—universals, particulars, propositions, or what have you—a calculus for logic should not adopt many sorted variables such as "propositional" variables, "predicate variables," or "particular variables." If "p" is supposed to be a special propositional variable, then Russell argues, "p implies, for all p and q, q's implying p" really just means "x and y being propositions implies, for all x and y, x's implying y's implying x." Again, if "R" is supposed to be a special relation variable, then Russell argues, "For all R, Rab" just means "x's being a relation implies, for all x, a and b's belonging to x" (*PoM*, p. 87).

There are, to be sure, certain difficulties in Russell's approach. In *Principles*, Russell realizes that he will have to enlist primitive predicates such as "x is a concept (property or relation)" (*PoM*, p. 86). Moreover, he will need an infinite stock of n-place relations so that (e.g.), for any proposition '$R(a,b)$' there is another proposition '$R*(a,b,R)$' which is logically equivalent to it. Russell acknowledges this in his discussion of the "Bradley regress," finding the infinity undeniable but wholly innocuous so long as the predicational nature of concepts is accepted (*ibid.*, p. 100). Accordingly, Russell implicitly accepts "$x_1, \ldots, x_n$ exemplify y" as a primitive logical notion. But in any case, nothing in Russell's adoption of one style of variable shows that he did not (and could not) view his system for logic as an uninterpreted "calculus" in a modern sense.

22. See Russell (*PoM*, p. 7) and (*InS*, p. 205).

23. Quine (1963), pp. 31, 257, 268, has precisely the same view and on its basis rejects predicate variables. Surely one cannot claim that the view is committed to rejecting the meta-language/object language distinction.

What about the modern notion of the "semantic completeness" of a formal system? What explains Frege and Russell's archaic attitudes here? In Russell's *Principles*, we find (*PoM*, p. 15):

> We require certain indemonstrable propositions, which hitherto I have not succeeded in reducing to less than ten. Some indemonstrables there must be; and some propositions such as the syllogism, must be of the number, since no demonstration is possible without them. But concerning others, it may be doubted whether they are indemonstrable or merely undemonstrated; and it should be observed that the method of supposing an axiom false and deducing the consequences of this assumption, which has been found admirable in such cases as the axiom of parallels, is here not universally available. For all axioms are principles of deduction; and if they are true the consequence which appear to follow from the employment of an opposite principle will not really follow.

Frege has a similar view. In his *Begriffsschrift* he wrote (Frege 1879, p. 29):

> We arrive at a small number of laws in which, if we add those contained in the rules, the content of all the laws is included, albeit in an undeveloped state. . . . Since in view of the boundless multitude of laws that can be enunciated we cannot list them all, we cannot achieve completeness except by searching out those that *by their power*, contain all of them.

Many advocates of the van Heijenoort view seize on such passages. A recent instance is found in Hylton (1990, p. 202):

> This [semantic completeness] question relies upon the idea that we have, independently of the logical system, a criterion of what the system ought to be able to do, so it relies upon the essentially meta-linguistic notion of an interpretation, and of truth in all interpretations. These meta-linguistic ideas, however, are foreign to Russell's conception of logic; the question of the completeness of a system simply could not arise for him. Logic for him was not a system, or a formalism, which might or might not capture what we take to be the logically valid body of schemata.

On Russell's conception of logic, it is concluded, the very idea of semantic completeness cannot arise.

But the fact is that the question of semantic completeness *did* arise for Russell (and Frege). Of course, it did not arise as the question as to whether every wff which is true in every admissible interpretation of the non-logical words is a deductive consequence of the axioms (or schemata) and inference rules of the system. Russell had no formal Tarski-like semantics to characterize logical truth. Moreover, such a semantics would not be adequate to characterize all the logical truths of Russell's calculus for the logic of propositions. Russell's logical truths here include analogs of the "logically valid" wffs characterized by Tarski's formal semantics, but they include yet more. The logical words of a modern calculus are the logical connectives and apparatus of quantification theory. But Russell calculus aims to capture the truths of propositional structure. The logical words of modern predicate calculus won't suffice. For instance, Russell has:

(1) x's constituency in p, implies for all x and y and p, some q's being exactly like p except containing y at the positions x has in p.
(2) x's implying y, implies, for all x and y, x's implying x.
(3) x's constituency in y implies, for all x and y, y's constituency in x's implying x's identity with y.

One might say that these are "truths by virtue of the meaning of their logical words," but the logical words here ("constituency", for example) certainly differ from the modern sentential connectives and quantificational apparatus assumed by modern formal semantic definitions of "logical validity."

Another way to see this is to observe that on the modern conception of logic, a calculus with identity which has

$$(x)A \supset (\exists x)A,$$

as a theorem is an embarrassment. It would mean that logic has ontological commitments, for there are wffs A that make the antecedent true. This wants rectification; e.g., one might form a pure calculus with no singular terms (free variables or individual constants). Russell's calculus for the logic of propositions has no such embarrassment. It is existentially committed to the being of propositions. This is not to say (as Hylton would) that it demands a fixed interpretation over certain entities (propositions). It means that the formal system is committed to the presence of genuine singular terms (i.e., the nominalization of any wff of the calculus is just such a singular term). The existential commitments of Russell's calculus do not show that Russell's conception of logic is antithetical to semantics. They simply show that a straightforward Tarski-style formal semantics of logical validity (logical truth), as truth in every interpretation of the non-logical words, won't do.

Russell's own notion of logical validity (logical truth) was articulated syntactically (in terms of deductive closure under the axioms) as van Heijenoort rightly observed. In fact, Russell identified logical truth with fully general truth. More exactly, a formula of *Russell's calculus for the logic of propositions* will be a logical truth when it is *fully* general and true. (The language of a modern predicate calculus, or even a second-order calculus, won't suffice for this characterization.) Russell's notion of "necessity" is derived from this. Of course, in *Principles* we find (*PoM*, p. 454):

> There seems to be no true proposition of which there is any sense in saying it might have been false. . . . What is true, is true; what is false is false; and concerning fundamentals, there is nothing more to be said.

But in an undated manuscript[24] entitled "Necessity and Possibility," Russell allows talk of a proposition as being necessary or possible by appeal to the degree to which the structure of the proposition approximates a fully general

24. It was probably written in 1904.

and true proposition of like logical structure. This generates "degrees of necessity" culminating in logical necessity (logical truth). Consider the proposition

> 'Plato's exemplifying humanity
> implies Plato's exemplifying mortality.'

There are propositions like this in structure except containing different entities where this contains Plato. Recognizing that any such proposition is true, i.e.,

> *x*'s exemplifying humanity implies, for all *x*,
> *x*'s exemplifying mortality,

Russell allows that the proposition could be said (if only weakly) to be "necessary with respect to Plato." Logical necessity is the extension (abstraction) of this idea to its limit. If the structure of a proposition is such that every proposition with that structure is true—the structure in question being obtained by abstracting away all but the purely logical components—then we have a logical truth. For instance,

> '*x*'s exemplifying *y*, implies *x*'s exemplifying *z*'

results from such a full abstraction, and

> '*x*'s exemplifying *y*, implies, for all *x* and *y* and *z*,
> *x*'s exemplifying *z*'

is clearly not true. On the other hand, a complete abstraction of all but logical content from

> 'Plato's implying Plato, implies Plato's implying Plato'

yields

> '*x*'s implying *x*, implies *x*'s implying *x*'.

In virtue of the truth of (2) above, Russell regards this as a logical necessity—a truth in virtue of structure alone.

Russell would later modify this idea a bit, recongizing (after he abandoned the distinction between Being and Existence in *Principles*) that no proposition containing a contingent particular (such as Plato) could be regarded as a logical necessity. But it is quite clear that Russell is getting at necessity in terms of the idea of "truth in virtue of logical form." Logical form is captured in terms of the approximation of a fully general truth. In his manuscript, Russell introduces a new notion of "analyticity" in a sense explicitly designed to contrast Kant's. "Analytic propositions," he explains, "have the property that they are necessary with respect to all their constituents except such as what I call *logical constants*" (*N&P*, p. 16).

Russell's notion of logical necessity is similar, in one way, to the Tarski semantic notion of "invariant truth in all interpretations." But instead of interpretations of linguistic syntactic structure in different domains, Russell has the notion of propositional structure or "logical form" and the approximation of the original proposition to a fully general proposition which is true.

The full generality captures the idea of different interpretations in different domains. Russell's approach, however, reaches a limit with fully general propositions. For Russell, full generality and truth has to be sufficient for logical necessity (logical truth). In this respect, Russell's approach is quite different from the semantic. A denumerable sequence S of objects in the domain of the interpretation of a language satisfies a quantified formula $\ulcorner(x_1)Ax_i\urcorner$ of that language iff for every object a in the domain and every sequence S° exactly like S except containing a as its ith member, satisfies Ax_i. A sentence is "true" for the language relative to the interpretation when every sequence in the interpretation satisfies it. A sentence is logically true when it is true for every interpretation. This approach to variables bound and free is extendible to languages involving predicate variables. Fully closed wffs of the language pose no special problem for its definition of logical truth.

All the same, Russell's conception of logical necessity is a kindred spirit of the Tarski semantic conception of logical truth, if we consider both as capturing "logical truth" (logical validity) as "truth in virtue of form (structure)." Russell's syntactic approach is wider (in one sense) than Tarski's; and also weaker (in another) because it reaches a limit with full generality. But we can see that semantics in general is in no way antithetical to Russell's conception of logic. The appearance of discord lies in adopting the false dilemma that logic must either be a formal calculus or a language with semantic content in its own right (so that a meta-theory is impossible). Logic, on Russell's (and Frege's) conception, is neither.

CHAPTER TWO

The Logic of *The Principles of Mathematics*

2.0 In May of 1903, Russell's *The Principles of Mathematics* appeared. The book was the crescendo of Russell's studies on relations, setting forth a detailed explanation of how the logic of relations can encompass (non-applied) mathematics (including rational dynamics and geometry). This chapter focuses on the foundation of that explanation—viz., the characterization of logic in *Principles* as the science of propositions. Russell's formal calculus for logic is set forth and its improvements over Peano's work are noted. Next the ontology of propositions underwriting that logic is discussed. Here we shall find that the old maxim *quodlibet ens est unum* is the fundamental ontological doctrine of *Principles*.

The Calculus for the Logic Propositions

2.1 Russell's calculus for logic originates from his fateful encounter with Peano at a congress in Paris in 1900. Russell was quite impressed by the system for implication Peano had invented. Frege's more advanced quantification theory of the 1879 *Begriffsschrift* was available, but Russell did not devote himself to a careful reading until late in the writing of *Principles*. Peano's approach seemed congenial to a calculus for the logic of propositions, and Russell followed this lead. Peano introduced the expression, "$p \supset q$", writing (1889–1901, vol 2, p. 26):

> On pourrait indiquer la relation $p \supset q$ par le signe $q \mathrm{C} p$ qu'on lira "q est conséquence de p".

Peano reads the symbol as "q is a consequence of p," and p is called the "hypothesis." Unfortunately, he was in the habit of using context to determine the meaning of some of his signs. For instance, he used "=" to express a number of distinct relations; sometimes it expressed identity, other times equivalence, and still other times it was used in stipulative definitions. In the case of "$p \supset q$," he read it as "de la p déduit la q," a reading which he viewed as warranting detachment when the "hypothesis" is true.

Peano introduced an expression for universal quantification ("formal implication") as well. He had some innovations similar to Frege's *Begriffsschrift*, distinguishing free and bound variables and introducing a notation of quantification which made the expression of scope possible. He wrote (Peano, 1889–1901, vol. 2, p. iii):

> Soient p et q des propositions contenant des lettres variables $x \ldots z$. Nous écrirons
>
> $p \;.\!\supset_{x \ldots z}\!.\; q$
>
> pour indiquer la proposition "de p on déduit, quels que soient $x \ldots z$, la q."

The quantifier is here subscripted to the condition sign and thus the name "formal implication"—as if we have some new form of implication.

Following Peano, Russell adopts 'implication' as a primitive dyadic relation. No object language signs for a formal calculus are given in *Principles*, but presumably Russell intended to follow Peano's symbolisms as he had in "Sur la logique des relations," taking "$\supset$" as a primitive dyadic predicate constant for the relation of (material) implication. Similarly, "$\supset_{\mu_1, \ldots, \mu_n}$" would be adopted for formal implication.[1] Let us use "α", "β", and "δ", as meta-linguistic letters for terms, and "A", "B", and "C", as meta-linguistic letters for wffs. The wff $\ulcorner \alpha \supset \beta \urcorner$ is read α implies β, and the positions of α and β here are subject positions. Thus, "$\supset$" stands for a relation. The relata can be any two entities, propositions or otherwise. Propositions are mind-independent entities in Russell's view, and some are true and others false. Thus there is a proposition '$x \supset y$' no matter what entities x and y are; it may not, however, be a true proposition.

To make the syntax of Russell's dyadic predicate "$\supset$" salient, it is useful to introduce nominalizing braces "{", "}". By using the braces, a wff A is made into a term $\ulcorner \{A\} \urcorner$. Thus,

"$x \supset \{y \supset z\}$"

would be a formula. Russell did not employ such braces in *Principles*. Instead, he took subject position to be sufficient to mark the nominalizing transformation. Russell's approach is convenient. Braces can be dropped for subject

1. As we shall see, *Principles* offered an analysis of the constituents of propositions named by nominalizations of formal implications by employing a theory of denoting concepts.

position, and dots and brackets can be used for punctuation. Thus, one can write

"$x \,.\!\supset\!.\, y \supset z$".

We shall follow this. But readers should be wary not to confuse "$\supset$" with the modern conditional sign "$\rightarrow$" which is flanked by wffs, not terms. Unfortunately, the error is easy to fall into. Russell often uses the letters "p", "q", "r", "s", "t" and these appear as if they are special letters for propositions. They are not.[2] They are just variants of "x", "y", "z", etc., which are (now commonly used as) individual variables. In *Principles*, Russell uses the words "term," "unit," "one," "entity," and "logical subject," synonymously with the word "individual." The fundamental doctrine of the work is that whatever is (be it a proposition or otherwise), is an individual. The calculus for logic treats all entities alike as values of its wholly unrestricted variables. Thus, there are no special primitive propositional variables in the work. To remind the reader, I shall replace Russell's letters with individual variables "x", "y", "z", etc.

Now Russell decides that only propositions stand in true inferential relationships. No proposition '$x \supset y$' can be true unless both x and y are propositions. Accordingly, Russell notes that he need not introduce a primitive constant "x is a proposition." Rather he need only put (*PoM*, p. 15):

x is a proposition $=$ df $x \supset x$.

Russell never sets out his calculus for logic formally, and aspects of it are scattered throughout *Principles*. The first ten fundamental principles are as follows (the expression "xy" is used for conjunction) (*PoM*, p. 16):

Pp_1 $\quad x \supset y \,.\!\supset_{x,y}\, x \supset y$

Pp_2 $\quad x \supset y \,.\!\supset_{x,y}\, x \supset x$

Pp_3 $\quad x \supset y \,.\!\supset_{x,y}\, y \supset y$

Pp_4 A true hypothesis in an implication may be dropped, and the consequent asserted.

Simplification	Pp_5	$x \supset x :\supset_{x,y}: y \supset y \,.\!\supset\!.\, (xy \,.\!\supset\!.\, x)$
Syllogism	Pp_6	$x \supset y :\supset_{x,y,z}: y \supset z \,.\!\supset\!.\, x \supset z$
Composition	Pp_7	$x \supset y :\supset_{x,y,z}: x \supset z \,.\!\supset\!.\, (x \,.\!\supset\!.\, yz)$
Importation	Pp_8	$y \supset y :\supset_{x,y,z}: z \supset z \,.\!\supset\!.\, (x \,.\!\supset\!.\, y \supset z :\supset: xy \,.\!\supset\!.\, z)$
Exportation	Pp_9	$x \supset x :\supset_{x,y,z}: y \supset y \,.\!\supset\!.\, (xy \,.\!\supset\!.\, z :\supset: x \,.\!\supset\!.\, y \supset z)$
Reduction	Pp_{10}	$x \supset x :\supset_{x,y}: y \supset y \,.\!\supset\!.\, (x \supset y \,.\!\supset\!.\, x :\supset: x)$

2. In reconstructing the calculus of *Principles*, Church (1984) has special propositional variables, and employs the statement connective "$\rightarrow$". Then, noticing the oddity of $\ulcorner p \rightarrow q .{=}. r \urcorner$ he adds a special quadruple bar to express identity where propositions are concerned. Once we see that *Principles* has only individual variables and "$\supset$" as a dyadic predicate, there is no reason for the quadruple bar. We have "$\{x \supset y\} = z$" as a wff.

Special attention must be paid to Russell's Pp_4. Unlike the other principles, it is an inference rule similar to the modern *modus ponens*. Strictly speaking Peano had no formal calculus for logic, since he had no explicit inference rules. Frege informed him of this failing on several occasions to no avail. Reading "$\supset$" as "implies" in one context and "therefore" in another, Peano likely thought that detachment need not be explicitly stated as a rule. Russell, on the contrary, made a significant attempt at providing the missing rules.

Russell was aware of the special status of Pp_4 among the "principles." Concerning the principle, he wrote (*PoM*, p. 16):

> This is a principle incapable of formal symbolic statement, and illustrating the essential limitations of formalism—a point to which I shall return at a later stage.

The "essential limitation of formalism," as Russell put it, is precisely Russell's way of saying that an inference *rule* is meta-linguistic and not itself an axiom among others. Russell speaks at length on the point in his discussion of Carroll's paradox "What the Tortoise said to Achilles" (*PoM*, p. 35). He says that we need the notion of *therefore* which is quite distinct from the notion of "implies." The notion of *therefore* renders an inference license warranting detachment; the notion of "implies" does not.

The principle Pp_4 is important in another way as well. In stating his axioms, Russell does not index his sign "$\supset$" as we have above. For example, he states Pp_1 as follows (10, p. 16):

> If *p* implies *q*, then *p* implies *q*; in other words, whatever *p* and *q* may be, '*p* implies *q*' is a proposition.

But Russell is just following the mathematical convenience (commonplace as well in Peano's notation) of omitting indices. In a footnote to this he makes this clear (*ibid.*):

> Note that the implications denoted by *if* and *then*, in these axioms, are formal while those denoted by *implies* are material.

So Russell's axioms are formal implications. Accordingly, he needs a quantification theory for his calculus to work. For this reason, Russell allows Pp_4 to have an application to general formulae. He writes (*PoM*, p. 35):

> Another form in which the principle is constantly employed is the substitution of a constant, satisfying a hypothesis, in the consequent of a formal implication. If ϕx implies θx, for all values of x, and if a is a constant satisfying ϕx, we can assert θa, dropping the true hypothesis ϕa.

To accommodate both forms, Pp_4 can be written as follows:

> From $\alpha \supset_{\mu_0 \ldots \mu_n} \beta$ and $\alpha[t_0|\mu_0 \ldots t_n|\mu_n]$, infer $\beta[t_0|\mu_0 \ldots t_n|\mu_n]$, where t_i is free for μ_i in α and β, $0 \leq i \leq n$.

We let $\alpha[t|\mu]$ be t itself when the term α is the variable μ. The expression $\ulcorner\alpha \supset_{\mu_0 \ldots \mu_n} \beta\urcorner$ is for any implication, formal (when $n \neq 0$) or material (where $n = 0$).

Now in addition to the two forms of Pp_4, Russell has the rule of universal generalization. It is stated informally (*ibid.*, p. 7):

> So long as any term in our proposition can be turned into a variable, our proposition can be generalized; and so long as this is possible, it is the business of mathematics to do it.

The precise formulation of this rule is important. It would not do to characterize it as:

> If $\alpha \supset \beta$ is a thesis, so is $\alpha \supset_\mu \beta$,

where μ is an entity variable free in α or β (or equal to α or β). Such a formulation, together with Pp_4, would not suffice to generate quantification theory. Moreover, such a form would make universal generalization under the hypothesis of a conditional proof illegitimate. Russell's statement of the "other form" of Pp_4 suggests that he allows conditional proofs. Verification of this comes if we look at Peano.

Consider Peano's 1889 "The Principles of Arithmetic." As usual, Peano omits indices on his sign "$\supset$", writing (*1889*, p. 94):

> P4 $a, b, c \in \mathbb{N} \,.\!\supset :.\, a = b \,.\, b = c \,:\!\supset.\, a = c$

instead of

> P4 $a, b, c \in \mathbb{N} \,.\!\supset_{a,b,c} :.\, a = b \,.\, b = c \,:\!\supset.\, a = c.$

But examine the following "proof."

> Theorem 13 $a, b, c \in \mathbb{N} \,.\, a = b \,.\, b = c \,.\, c = d \,:\!\supset:\, a = d$
> proof: Hyp. P4 $:\!\supset:\, a, c, d \in \mathbb{N} \,.\, a = c \,.\, c = d \,.$ P4 $.\!\supset.$ Thes

The idea of the proof seems to be to assume the antecedent of the theorem to be shown. On the basis of the assumption, Peano's axiom P4 can be applied by means of Russell's rule Pp_4. This yields,

> $a, c, d \in \mathbb{N}$ & $a = c$ & $c = d$.

Calling again upon P4 and using Russell's Pp_4, one arrives at $a = d$. By an implicit rule of conditional proof, then, Peano has

> Hyp $\supset a = d$.

The final result is gotten by universal generalization. If this is correct, Peano (albeit tacitly) employs an inference rule legitimating deduction from a hypothesis.

Since Peano's calculus was Russell's paradigm and it is clear that Russell intended *Principles* to amend and correct Peano's proof techniques, it seems likely that he allowed derivations under hypotheses. When it came to the question of quantification rules, Peano did express concern. He wrote (1889–1901 vol, p. §18):

> The indices to the sign $\supset$ satisfy laws which have not yet been sufficiently studied. This theory, already abstruse in itself, becomes even more so unless the rules are accompanied by examples.

In a letter to Peano, Frege noted that he had already given the proper rules in his 1879 *Begriffsschrift* and, being that they are few in number and clear, knew of no reason why they should be said to be abstruse (Frege 1969, p. 11). But in allowing conditional proof, Peano and Russell faced issues that Frege did not; they must formulate the rule of universal generalization in a way that demarcates when generalization under the scope of a hypothesis is legitimate. This may well explain why Peano found the articulation of rules for quantification "abstruse." Unfortunately, formulations of universal generalization and conditional proof are hard to exact from Russell's writings of the period. In "The Theory of Implication" of 1905/6, Russell writes (*T of I*, p. 195):

> *7.11 What is true of any is true of all.

Alternatively, he puts:

> If ϕy is true however y is chosen, then $(x).\phi x$ is true.

This, of course, is misleading.[3] Consider:

1.	$(x)(Ax \supset Bx)$	Hyp.
2.	$(x)Ax$	Hyp.
3.	$Ax \supset Bx$	1, UI
4.	Ax	2, UI
5.	Bx	3, 4, MP
6.	$(x)Bx$	5, UG
7.	$2 \supset 6$	
8.	$1 \supset 7$	

Here line #5 may not be true at all. Russell seems to be aware of this. Perhaps then Russell's point was that a wff of the form ϕx (e.g., our line #5) can be said to be "true" in the sense that it would be arrived at in a similar demonstration no matter what variable is chosen instead of x. It seems appropriate, then, to adopt the following on Russell's behalf:

(Universal Generalization)

From $\alpha \supset_{v_0,\ldots,v_n} \beta$, infer $\alpha \supset_{\mu,v_0,\ldots,v_n} \beta$,

3. We should not be misled by the occurrence of "truth" in *7.11, as if this shows a conflation of theory and meta-theory. The rule UG is meta-linguistic, and as such it would be perfectly admissible to speak of wffs of a form as being true.

where μ is an individual variable which is free in $\alpha \supset_{v_0,\ldots,v_n} \beta$, and which does not occur free in any hypotheses within whose scope $\alpha \supset_{\mu,v_0,\ldots,v_n} \beta$ occurs. This completes the system.

It should be noted that the phrase "propositional calculus" is now commonly used to refer to a sentential calculus. Russell's use is nonstandard, for it refers to a system which has the expressive capacity of a calculus which permits bindable predicate variables in subject as well as predicate positions. At first blush, this seems surprising, for, as we saw, the fundamental concept behind Russell's calculus for logic is that only individual variables are adopted. But extended by the logical constant, say, "*C*" for the property of being a concept (property or relation), and the constants $\ulcorner P^n \urcorner$, $n \in \mathbb{N}$, for the n-adic relations of exemplification, the presence of only individual variables poses no untoward limitations. To transcribe "Every human is mortal," Russell can put:

$$P^2(x, \text{Humanity}) \supset_x P^2(x, \text{Mortality}),$$

which, read literally, says "*x*'s exemplifying Humanity implies, for all *x*, *x*'s exemplifying Mortality." Similarly, Russell can transcribe "Every entity has every property" with

$$C(x) \,.\!\supset_{x,y}\!.\; P^2(y, x)$$

which reads "*x*'s being a concept implies, for all *x* and *y*, *y*'s exemplifying *x*." So we can see that the system has an expressive capacity beyond that of a sentential logic.

Indeed, Russell intended yet more. He wanted a theory of classes as well. For this purpose he took Peano's "$\in$" (for membership) and "$\ni$" (for 'such that') as primitive signs so that $\ulcorner x \in z \ni (Az) \urcorner$ is a wff which reads "*x* is a member of the class of all *z* such that . . . *z* . . .". He adopts some primitive principles for classes (*PoM*, p. 20) as well, and in virtue of the theory of classes, he hopes to define the identity sign. He writes (*ibid*):

> Identity which occurs here may be defined as follows: *x* equals *y* if *y* belongs to every class to which *x* belongs, in other words *x* is a μ implies *y* is a μ for all values of μ.

Of course, this requires a comprehension principle for classes (or concepts). Russell had none to offer. He had discovered the paradoxes surrounding the unbridled assumption that every open wff comprehends an attribute and a class. Russell intended the theory of extension as a part of his calculus for logic. But in *Principles* his ideas on the paradoxes were in a state of flux.

Russell's Definitions

2.2 Russell put forth definitions of tilde, conjunction, and disjunction signs in *Principles*. There is a good discussion of the definitions in Byrd 1989. The following diverges somewhat from Byrd's renditions, however. What Byrd takes as errors I take as "slips" or infelicities of expression.

Russell begins with the conjunction sign (i.e., juxtaposition) as follows (*PoM*, p. 17):

> If *p* implies *p*, then if *q* implies *q*, then *pq* (the logical product of *p* and *q*) means that if *p* implies that *q* implies *r*, then *r* is true.

Unfortunately, there is a slip. Russell read "*r*" as if it stood only for propositions. Russell corrects this in an adjoining sentence of clarification, writing (*ibid*):

> In other words, if *p* and *q* are propositions, then their joint assertion is equivalent to saying that every proposition is true which is such that the first implies that the second implies it.

Accordingly Russell has:

$$x \supset x :\supset_{xy}: y \supset y .\supset. [xy \text{ is equivalent to } z \supset z .\supset_z. (x .\supset. y \supset z :\supset: z)].$$

We are still faced with difficulties. First, the definition is conditional; and second, what is the import of "means that" (or "is equivalent to" as Russell uses it above)?

Likely, the answer is simply that Russell felt his readers would find a conjunction sign between entity variables to be absurd. How, for example, can one assert 'Frege and Russell'? And obscure as it seems, would one be able to infer Frege from the conjunction?! Russell's *simplification* avoids this problem by restriction to propositions. There is no derived inference rule: from *xy*, infer *x*. A variable cannot occur isolated on a line of a demonstration. All the same, Russell may have felt he needed to assure readers that his conjunction *does* parallel the normal truth-conditions of "and." In his effort to assure readers, Russell adopted the conditional definition.

If this is right then we can be content to drop Russell's conditional definitions and his locution "is equivalent to." It would not suffice, however, to adopt the definition schema:

$$AB = \text{df}\, \mu \supset \mu .\supset_\mu. (A .\supset. B \supset \mu :\supset: \mu),$$

where *A* and *B* are any wffs of the language. This defines conjunction where nominalized wffs (for material or formal implications) are concerned. Russell's axioms (e.g., *simplification*) would not comport with definition, for they allow the expression of the conjunction sign (juxtaposition) with individual variables (Byrd 1989, p. 351). Our solution is to adopt the definition schema:

$$\alpha\beta = \text{df}\ \mu \supset \mu .\supset_\mu. (\alpha .\supset. \beta \supset \mu :\supset: \mu),$$

for any terms α, β.

Russell next defines disjunction. He introduces no special sign, but we shall adopt "$\vee$" on his behalf. He writes (*PoM*, p. 17):

> '*p* or *q*' is equivalent to '*p* implies *q*' implies *q*.'

At first blush, Russell seems to have offered a definition which is not conditional. His statement of clarification, however, suggests that he again slipped by reading "p" and "q" as if they stood for nominalized wffs (for propositions). He writes (*ibid*):

> It is easy to persuade ourselves of this equivalence since a false proposition implies every other.

We shall avoid the conditional definition putting:

$$\alpha \vee \beta = \text{df } \alpha \supset \beta . \supset . \beta .$$

The same point applies to the definition of the negation sign. Russell introduces no sign, but writes (*PoM*, 18):

> We proceed to the definition of negation; not-p is equivalent to the assertion that p implies all propositions; i.e., that 'r implies r' implies 'p implies r' whatever r may be.

Appearances to the contrary, Russell once again intends a conditional definition. We see this because the definition is accompanied by a footnote which reads (*ibid.*):

> The principle that false propositions imply all propositions solves Lewis Carroll's logical paradox in *Mind*, N.S. No. 11 (1894).

So again there is a slip, here with regard to the use of "p". Russell wants:

$$x \supset x . \supset_x . \text{ [not-}x\text{ is equivalent to } z \supset z . \supset_z . \ x \supset z].$$

Evidently, Russell wanted his readers to be sure to understand that his definition will yield the usual,

x implies not not x

where x is a proposition, even though it is not a thesis otherwise. All the same, we are dropping the conditional definitions, and so (adopting "$\sim$" on Russell's behalf) we have

$$\sim \alpha = \text{df } \mu \supset \mu . \supset_\mu . \ \alpha \supset \mu$$

as our definition schema.

No biconditional sign is introduced or defined, and we must be wary not to confuse this matter with the locution "is equivalent to" in Russell's statements of definition. Curiously, in discussing the "equivalence" of relations, Russell does suggest an awareness of Peano's approach—that is, of introducing a biconditional sign with the definiens the conjunction of two conditionals (*PoM*, p. 24). Peano used "=" to express a number of distinct relations, while Pieri used "≡" for definitions, reserving "=" for identity. Perhaps, Russell had not settled on a sign.

The existential quantifier remains. In a letter to Frege of 24 May 1903 we find Russell adopting the notation of $\ulcorner(\mu)A\mu\urcorner$ and defining Peano's notation which indexes the conditional sign. This provides a means of avoiding having to put $\ulcorner\sim Ax \supset_x Ax\urcorner$ when $\ulcorner(x)Ax\urcorner$ is wanted. Russell also puts (McGuinness (1980, p. 159):

$$(\exists x).\phi x = \text{df} \ \sim \{(x).\sim \phi x\}.$$

There are some hints of this in *Principles*. Peano had put:

$$\exists a = \text{df} \sim (a = \Lambda)$$

His symbol "∃" is not a variable binding operator. It flanks a class term "*a*" and is used to say that the class *a* is not empty. Now in *Principles*, Russell explains that a class *a* exists when "it" (i.e., when the class-concept of which *a* is the would-be extension) is "non-empty" (i.e., is exemplified). Haplessly, Russell uses "an *a*" for the class-concept and, running together the class concept expression with the class term (an error he chides Peano with), finds himself writing (*PoM*, p. 21):

> A class is said to exist when it has at least one term. A formal definition of this is: *a* is an existent class when and only when any proposition is true provided "*x* is an *a*" always implies it whatever value we may give to *x*.

Where "$\hat{x}(x \text{ is an } a)$" is a class term, what Russell intends is:

$$\exists \hat{x}(x \text{ is an } a) = \text{df} \sim (x) \sim (x \text{ is an } a).$$

It was not long before this would become

$$\exists ! \alpha = \text{df} \, (\exists x)(x \in \alpha)$$

which was put in *Principia* as *24.03. So the passage [4] from *Principles* certainly foreshadows Russell's later definition of the existential quantifier.

No rule for definitions is adopted explicitly. But Russell was well aware that since definitions are but notational conveniences, definiens and definiendum may replace one another in any context. He could easily read of this from most any volume of Peano's *Formulaire*.

In proving theorems within the system, the first thing to do is to prove propositional analogs of the principles. For instance, we get

$$\text{Simp}* \quad \vdash (x \supset y)(z \supset w) \,.\!\supset_{x,y,z,w}.\, (x \supset y).$$

This follows, since by Pp_1 both '$x \supset y$' and '$z \supset w$' are propositions. Once we have the analogs, we can manipulate propositions by means of them. It is easy to show that the analog system is complete with respect to the tautologies of modern propositional logic (Landini 1996). The basic principles of Russell's system—viz., (Simp.), (Syll), (Imp), (Exp), (Comp)—can be found in early volumes of Peano's *Formulaire*. Still, Russell has made a significant improvement over Peano in setting out a formal calculus for logic with explicit rules of inference.[5]

4. Byrd (1989) calls attention to this passage as well.

5. Note that if we replace Russell's "*xy*" in his axioms with "$\sim (x \supset \sim y)$", *reduction* is no longer independent. See Landini (1996).

The Theory of Implication

2.3 Not long after the publication of *Principles*, Russell became dissatisfied with calculus for logic set out in the work. The changes were published in the 1906 article "The Theory of Implication." It is worth discussing a few of the innovations here.

In "The Theory of Implication," *Principles's* tiresome antecedent clauses "$x \supset x$" are dropped for a new interpretation of the horseshoe. The proposition '$x \supset y$' is now to be regarded as true when x is not a proposition. All non-propositions are not-true; and a conditional is true whenever its antecedent is not-true. The rule of *modus ponens* is rendered as

> Anything implied by a true hypothesis is true.

The non-quantificational principles are now:

(Identity)	Pp_1	$p \supset p$
(Simplification)	Pp_2	$p . \supset . q \supset p$
(Syllogism)	Pp_3	$p \supset q : \supset : q \supset r . \supset . p \supset r$
(Commutation)	Pp_4	$p . \supset . q \supset r : \supset : q . \supset . p \supset r$
(Negation)	Pp_5	$\sim\sim p \supset p$
(Absurdity)	Pp_6	$p \supset \sim p . \supset . \sim p$
(Transposition)	Pp_7	$p \supset \sim q . \supset . q \supset \sim p.$

Russell then has the following definitions (*TofI.*, p. 176):

$$p . q = \text{df} \sim (p \supset \sim q)$$
$$p \vee q = \text{df} \sim p \supset q$$
$$p \equiv q = \text{df} p \supset q \;.\&.\; q \supset p.$$

The first principle is not independent.[6]

The influence of Frege's *Begriffsschrift* on Russell's discussion of quantification theory is clear. Unlike *Principles*, quantification theory is made separate by making it permissible to uniformly replace the letters "p", "q", etc., in principles and theorems with other terms. Russell labors at length to explain why such replacement preserves logical truth, grounding it on the meaning he assigns to his use of the turnstile. Church (1956) regards this as a rule of uniform substitution, but the text is unclear. Russell never discusses the application of such a rule for wffs involving quantifiers, and it is no simple matter to extend uniform replacement to such cases. Suffering from a hazy grasp of the distinction between schematic letters and genuine object-language variables, Russell may well have had schematic letters in mind all along. Consider the following (*TofI*, p. 163):

6. See Vuillemin (1968, p. 38). Russell should have known this. Likely he was not yet fully free of the system of *Principles* where Pp_1 (which is independent) ensures that every proposition implies itself.

> Any statement about a variable x will be expressed by $(C \between x)$ or by $(A \between x)$, $(B \between x)$, etc. Similarly any statement about two variables x and y will be expressed by $(C \between x \between y)$ or by $(A \between x \between y)$, $(B \between x \between y)$, etc., and for any number of variables.

Does this ontologize the variable or does Russell just mean to use "$(C \between x)$" schematically for any wff in which a variable such as "x" (a designated single letter) occurs free? We shall see that he intends a schematic use. For quantification theory Russell puts (*ibid*, p. 194):

$$(A \between x) \,.\supset_x.\, (B \between x) = \text{df}\, (x){:}(A \between x) \,.\supset.\, (B \between x),$$

and a similar definition for more than two variables. He then has:

Pp $\vdash{:}\,(x).(C \between x) \,.\supset.\, C(y)$
This proposition may be read: "What is true of all is true of any."
Pp $\vdash{:}\,(x).\ p \supset (C \between x) \,.\supset{:}\ p \supset (x).\ (C \between x)$
Pp What is true of any is true of all.

In the first, Russell relies on the word "any" to express (in conjunction with the meaning of the turnstile) that we replace every occurrence of the variable with a term so long as it is free for the variable. Similarly, in the second principle Russell has presumed that instances of the schema will be such that the bound variable "x" does not occur free in the wff p. Frege had been explicit on such matters and Russell should have followed his lead here.

Russell includes a revealing discussion of the third principle (i.e., the rule of *Universal instantiation*). He notes that it would not do to formulate the principle as

If $(C \between x)$ is a thesis, then so is $(x).(C \between x)$.

As we saw, this would prevent universal generalization under an assumption (*ibid.*, 195). All the same, since Russell never employs proof under an hypothesis in "The Theory of Implication." He may as well have intended the above formulation.

With quantification theory separate, the tilde is taken as a primitive logical particle along with "$\supset$". It is a monadic predicate standing for the property 'non-truth', which when flanked by a nominalized formula is the same as 'falsehood' (*ibid.*, 164). Russell notes that the separation of quantification theory has a certain appeal, but acknowledges that a system like *Principles* which does not separate them is still viable.[7] The resulting system, he says, would need the principles of quantification theory together with the following (*ibid.*, 200):

7. Byrd (1989) argues that the definition of tilde in *Principles* is defective. He says that negation can only be expressed if there are false propositions, and that a principle assuming such propositions would require tilde to express. Byrd overlooked the fact that the assumption of propositions is a logical assumption given with the grammatical rule that for any wff A, $\ulcorner\{A\}\urcorner$ is a singular term.

(Identity)	$p \supset p$
(Simplification)	$p . \supset . q \supset p$
(Commutation)	$p . \supset . q \supset r : \supset : q . \supset . p \supset r$
(Syllogism)	$p \supset q : \supset : q \supset r . \supset . p \supset r$
(Reduction)	$p \supset q . \supset . p : \supset : p$
	$\sim p = \text{df } p \supset (r).r$
	$p \vee q = \text{df } p \supset q . \supset . q$
	$p.q = \text{df } (r)(p . \supset . q \supset r : \supset : r)$.

This seems also to have been the plan of an early letter to Frege dated 12 December 1904.[8] The first principle is not independent, however. Moreover, there is a slip in the definition of the tilde. The expression "$(r).r$" is ill-formed because "r" is an individual variable, not a special propositional variable. Russell should put "$(\mu)(\mu \supset \mu . \supset . \mu)$".

Quodlibet Ens Est Unum

2.4 Russell takes logic to be the science of the inferential relations that obtain among *propositions*—purely logical, and mind-independent entities which subsist necessarily (as it were). The unrestricted variables of the formal implications of the calculus for pure logic, therefore, have to be grounded in an analysis of the nature of propositions.

Now the unrestricted variables of Russell's calculus for logic range over whatever has being. The notion is used synonymously with "individual," "term," "entity," "unit," and "logical subject." There is to be only one style of variable—the "individual" (or entity) variable. The following passage has become famous (*PoM*, p. 43):

> Whatever may be an object of thought, or may occur in a true or false proposition or can be counted as *one* is a *term*. This then, is the widest word in the philosophical vocabulary. I shall use as synonymous with it the words unit, individual, and entity. The first two emphasize the fact that every term is *one*, while the third is derived from the fact that every term has being, i.e., *is* in some sense.

The fundamental doctrine of *Principles* is *quodlibet ens est unum*, or "whatever is, is one" (*ibid.*, p. 132). Every term is a "logical subject" and as Russell says, it is the subject of the proposition that itself is one (*ibid.*, p. 44). Unfortunately, Russell cannot also say that since "being," "term," and the like, are given with the variables of logic they are not genuine properties. What prevents this is the fact that *Principles* analyzes the referential use of single letters as variables in formal implications by means of a theory of denoting. It is denoting that is fundamental, not the way a variable picks out its values. A denoting concept 'any *a*' denotes an entity *y* only insofar as the entity in question has the

8. See McGuinness (1988, p. 169).

property of being an *a*. So Russell's analysis of the pure unrestricted variables by means of the denoting concepts 'any term' and 'every term' requires that 'is a term' or 'Being' be a property.[9]

The notion of a logical subject is of central importance to the philosophy of *Principles*. The work separates beings into two exclusive sorts: "things" and "concepts." The former, writes Russell, "are *indicated* by proper names, the latter are indicated by all other words" (*ibid.*, p. 44). The difference, however, is not just in the words; it is in the ways in which concepts as opposed to things can occur in propositions. To distinguish these ways of occurrence, Russell introduces the expression "the term(s) of a proposition." He writes (*ibid.*, p. 44):

> I shall speak of the *terms* of a proposition as those terms, however numerous, which occur in a proposition and may be regarded as subjects about which the proposition is. It is a characteristic of the terms of a proposition that any one of them may be replaced by any other entity without our ceasing to have a proposition. Thus we shall say that 'Socrates is human' is a proposition having only one term; of the remaining components of the proposition, one is the verb, the other is a *predicate*. With the sense which *is* has in this proposition, we no longer have a proposition at all if we replace *human* by something other than a predicate. . . . Socrates is a thing because Socrates can never occur in a proposition otherwise than as a term in a proposition: Socrates is not capable of that curious twofold use which is involved in *human* and *humanity*.

To occur as "term of a proposition" (or for short, to "occur as term") is to be *both* a constituent of a proposition and what the proposition is about. In Russell's view, genuine proper names *indicate* their referents directly, so a proposition which is indicated by a sentence containing a proper name contains the entity indicated by the same as a "term of the proposition"—that is, it both contains the entity as its constituent and is about this constituent. Russell calls a term (occurring as a term of a proposition) the "subject" or "logical subject" of the proposition.

The notion of a logical subject is born from the notion of occurring as a term of a proposition. Thus, the fundamental doctrine "whatever is, is one" can be otherwise expressed as "Whatever has being is a logical subject," and even more revealingly as "Whatever has being can occur as a term of a proposition." As Russell explains, "every term . . . is a logical subject; it is, for example, the subject of the proposition that itself is one" (*ibid.*, p. 44). His point is that every being occurs "as a term" of some proposition. This is what it means

9. By 1905 the situation has been inverted. The use of a single letter as a variable and the notion of its "determination" (i.e., its assignment at this or that value) is primitive. Denoting concepts are abandoned and with them goes Russell's early hope of analyzing the use of single letters as variables. See Chapter 3.

to be a "logical subject."[10] The notion is tailor-made to fit the doctrine of the unrestricted variable.

2.5 Now a concept can be distinguished from a term which is not a concept by the fact that such a term (or "thing", as Russell calls it) can *only* occur as a term of a proposition. But concepts are themselves terms and every term can occur as a term of a proposition. This has important ramifications for Russell's early philosophy. It thrusts to the fore the importance of what Russell calls the "indefinable two-fold" nature of concepts. It is the capacity for a two-fold occurrence that makes the fundamental doctrine of *Principles* tenable. Concepts certainly have a predicable nature. In Russell's view, the unity of a proposition is due to the "indefinable way" in which a concept can occur "as a concept" in a proposition. The concept links the constituents of the proposition together and makes it 'one' complex entity. But concepts also have an *individual nature* in virtue of which they can occur as logical subjects and thereby be values of the "individual" variables of quantification. Thus, a concept is capable of *two* kinds of occurrences; it can occur "as concept" in a proposition and it can also occur as "term of a proposition." This is their "curious twofold use," as Russell puts it. For example, it is the very same concept 'human' that occurs as concept in 'Socrates is human' and occurs as a term of the proposition 'Humanity belongs to Socrates.'

Propositions also have the two-fold occurrence characteristic of concepts. In Russell's view, the very same proposition occurs as a concept when asserted, and occurs as a "propositional concept" (i.e., a term of a proposition) when it occurs unasserted, such as in a material conditional (*PoM*, p. 48). Denoting concepts must follow suit. Russell's writes (*ibid.*, 53):

> A concept *denotes* when, if it occurs in a proposition, the proposition is not about the concept, but about a term connected in a peculiar way with the concept.

To speak of a denoting concept itself, he continues, "we have to indicate the fact by italics or inverted commas" (*ibid.*). In the proposition '*Nothing* is not nothing,' the denoting concept '*Nothing*' occurs as logical subject and the proposition is about the denoting concept. In the proposition, 'Nothing is both round and square,' the denoting concept 'Nothing' occurs as concept. The two-fold nature of concepts is essential to Russell's doctrine that whatever is, is one.[11]

10. This is not to say that every term must have a proper name. To be sure, the only way we can name (or assert) a proposition in which a given entity *a* occurs "as a term of the proposition" is by means of a declarative sentence in which a name of *a* occurs as grammatical subject. But there may well be terms (including propositions) of which we shall never be acquainted; this has no bearing whatever on their being.

11. As we shall see in Chapter 3, problems arise here. They are the source of Russell's 1905 arguments against denoting concepts.

Denoting Concepts

2.6 We have seen that 'being' is a property. 'Existence' is also a genuine property. As Russell puts it, "it is the prerogative of some only amongst beings" (*PoM*, p. 449). Now we can truthfully deny existence, but with Being a genuine property, "'A is not' must always be false or meaningless, for if A were nothing it could not be said not to be" (*ibid.*). To properly understand this we must take note of Russell's use of the capital letter "A". It is used for a genuine proper name, and as such what it indicates occurs as a term of the proposition indicated by the nominalized sentence. It would be entirely inappropriate to put a definite description for "A". Russell did not embrace "objects" such as Meinong's "round–square."[12]

To be sure, not all beings exist. For instance, Russell writes: "Numbers, the Homeric gods, relations, chimeras, and four-dimensional spaces all have being, for if they were not entities of a kind we could make no propositions about them. Thus being is an attribute of everything and to mention anything is to show that it is" (*PoM*, p. 449). But "to mention" in this sense means "to use a genuine proper name for." It was because Russell regarded "Pegasus," "Apollo," and the like as genuine proper names, that he embraced non-existent possible beings. All the same, there is no compunction to admit "the round–square" as a genuine proper name. Russell can deny the being of such purported objects as "round–squares" by simply asserting that no being is both round and square. In fact, he denies the being of the null-class (*ibid.*, p. 68).

In the virtue of the theory of denoting concepts, *Principles* preserves a version of Brentano's *principle of intentionality* while at the same time denying that there are impossible or incomplete "objects of thought." Whatever is a genuine object of thought (or object of acquaintance) has Being and is therefore independent of the mental act. Mental acts, according to Brentano, are distinguished by their intending or "pointing toward" objects distinct from the act and its content. Denoting concepts can be constituents of propositions and in virtue of their occurrence a proposition can "point toward an object" which is not its constituent. A denoting concept has a capacity by means of which it purports to denote. Purporting to denote and actually denoting are quite different, however. Intensionality does not require an intensional "object" for every case of "pointing toward".

A denoting phrase which, on the occasion of its use, indicates a denoting concept is formed by combining an applicatival expression with a common noun or noun phrase. (Russell calls the concept expressed by a noun or noun phrase a "class-concept.") Thus, Russell has the denoting concepts 'all *a*,' 'every *a*,' 'any *a*,' 'a(n) *a*,' 'some *a*,' 'the *a*,' where *a* is a class-concept. The capacity for denoting, Russell tells us, is not a psychological or linguistic relation involved when we point, describe, or employ words as symbols. The fact that we are able to designate a thing which is not a concept is due, he says, "to

12. Cocchiarella (1980) was the first to realize this point, and happily it has won many supporters.

a logical relation between some concepts and some terms, in virtue of which such concepts inherently and logically denote such terms" (*PoM*, p. 53). Of course, it is only in its occurrence "as concept" that a denoting concept denotes, or even purports to denote, other terms. But how is it that a denoting concept purports to denote this object rather than that?

To have a property or stand in a relation is to occur as term (as the logical subject) of a proposition predicating the property or relation. The logical relationship of denoting establishes a connection between a denoting concept and its denotata (if any) by exploiting this fact. This is revealed in Russell's explanation of the conditions under which a denoting concept fails to denote. He writes (*ibid.*, p. 74):

> All denoting concepts, as we saw, are derived from class-concepts; and *a* is a class-concept when '*x* is an *a*' is a propositional function. The denoting concepts associated with *a* will not denote anything when and only when '*x* is an *a*' is false for all values of *x*.

There seems to be an oversight here due to the omission of uniqueness for denoting concepts such as 'the *a*'. Nonetheless, the point is clear. If the denoting concept 'a man' denotes Socrates, then Socrates satisfies the propositional function '*x* is-a man'; and to satisfy the propositional function, Socrates must occur as term of the proposition 'Socrates is human.'[13]

Russell's logical relationship of denoting is quite important. A denoting concept, say 'Every man,' denotes Socrates (among others) only insofar as Socrates occurs "as term" in the true proposition 'Socrates is-a man.'[14] That is, a proposition in which a denoting concept occurs as concept implies other propositions in which the denotation (or denotations) of that concept occur(s) as term. 'Every man' denotes Socrates in virtue of Socrates having the property of being human, and his having this property involves his occurrence as the term of a proposition predicating the property.

2.7 The logical relationship of denoting relies on an inferential relationship between propositions in which denoting concepts occur "as concept" and propositions in which what is denoted occur "as term." Russell, in fact, uses this to explain the differences between denoting concepts themselves—e.g., between 'all men' and 'every man' and 'any man.' A denoting concept, he says, denotes terms in such a way that they are "combined in a certain manner." Therefore, in cases where the noun concept has finite application, the differences in the denoting concepts can be revealed by looking at the propositions containing such "combinations of terms" with which propositions containing denoting

13. There is no circularity in Russell's definition. Russell uses "an" only because he used the vowel "a" to stand in for a common noun (*PoM*, p. 92). He distinguishes, for example, ". . . is-a man" which is just ". . . is human" from ". . . is a man," which involves the denoting phrase "a man." The property 'humanity' has a predicable nature and occurs "as concept" in the proposition 'Socrates is human.'

14. This point is further justified in our next section on propositional functions.

concepts are equivalent. This is just a version of the medieval theory of determinate versus confused *suppositio*.[15]

In medieval logic, the mode of reference of the phrase ⌈some *a*⌉ is said to be "determinate," that of ⌈ an *a*⌉ confused. In the former case, a sentential context in which the phrase ⌈some *a*⌉ occurs—that is, ⌈f(some *a*)⌉—will be true if some determinate interpretation of x in ⌈$f(x)$⌉ makes ⌈$f(x)$⌉ true (Geach 1962, p. 92). In the latter case, however, no proper name would do and this is the mark of confused *supposito*. William Ockham holds, Geach tells us, that in "I promise you a horse," the phrase "a horse" may be replaced *salva veritate* by a disjunctive list of names of (present and future) horses, but does not imply a disjunction of sentences each obtained by replacing "a horse" by a single name. This is the criterion of confused *suppositio* (*ibid.*, 92). Russell, of course, puts the matter speaking of propositions and terms (occurring as term) in them, not sentences and the occurrences of names. But his basic idea, as Geach points out, is akin to Ockham's. Thus, supposing Brown and Jones to be Miss Smith's suitors, Russell tells us that the proposition 'Some suitor has won Miss Smith's hand' is a "constant disjunction" for it is true iff either 'Brown has won Miss Smith's hand' is true or 'Jones has won Miss Smith's hand' is true. This is quite different from the proposition 'You must have met a suitor,' which Russell explains as a "variable disjunction" in that it implies 'You must have met Brown or Jones.'

The distinction between 'all *a*,' 'every *a*' is given in a similar fashion. A proposition in which 'all *a*' occurs will imply a proposition which contains a "numeric conjunction" of terms each of which is an *a*. Thus "all elephants are pachyderms" implies "Jumbo and Bimbo and . . . are pachyderms," where Jumbo and Bimbo, etc., are all the elephants. This is akin to *suppositio confusa*. A proposition in which 'every *a*' implies a "propositional conjunction." "Every elephant is a pachyderm" implies "Jumbo is a pachyderm and Bimbo is a pachyderm, and . . . ." This is *suppositio determinata*. Observe, however, that in distuinguishing determinate from confused *suppositio* it matters whether the phrase in question occurs as grammatical subject or as grammatical object (or part of "an . . ."). The phrases "some *a*" and "every *a*" will have determinate *suppositio* when they occur as grammatical subject, and may well have confused *suppositio* in other cases. For instance, in "Every contingent entity ceases to exist at some time," the phrase "some time" has confused *suppositio* and "every contingent entity" has determinate *suppositio*. It implies "n_1 ceases to exist at t_1 or t_2 or t_3 . . ., and n_2 ceases to exist at t_1 or t_2 or t_3 . . ., and n_3 ceases to exist at t_1 or t_2 or t_3, and . . . ." On the other hand, in "Some person can marry every sister of Miss Smith's," the phrase "some person" has determinate *suppositio* and "every sister of Miss Smith's" has confused *suppositio*. Where Mary and Jane are the sisters, it implies "n_1 can marry Mary and Jane; and n_2 can marry Mary and Jane."

15. P. T. Geach (1962, p. 90ff) has an excellent discussion of the matter.

This leaves the explanation of 'any *a*,' which denotes the combination of terms Russell calls a "variable conjunction." Thus, in "If you met any suitor of Miss Smith you met a very ardent lover," we can replace "any suitor of Miss Smith" by a variable disjunction to get the equivalent "If you met Brown or Jones, you met a very ardent lover." Russell observes that we can also replace the phrase by a propositional conjunction to get the equivalent "If you met Brown you met a very ardent lover; and if you met Jones you met a very ardent lover." But he notes that in some cases the replacement by a propositional conjunction will not suffice. Thus 'any *a*' is "half-way between conjunction and disjunction"; it is variable (as in a disjunction) but in some cases it equivalent to a propositional conjunction.

Geach criticizes Russell, challenging that use of "any *a*" is always equivalent with a propositional conjunction (op cit., 103). Geach is correct. Russell may be defended, however, on the grounds that he had scope in mind. He regards "every *a*" (occurring as grammatical subject) as implying a propositional conjunction. Also "every *a*" is quite different from "any *a*." The phrase "any *a*" typically takes largest scope, while "every *a*" takes a narrow scope; "Any value of ϕx implies *p*" differs from "every value of ϕx implies *p*." Moreover, I think that Russell was concerned to show that "any *a*" has disjunctive force as if a free variable.

The distinction between *suppositio confusa* and *suppositio determinata* was important to medieval logicians and used to expose fallacies which correspond to modern scope fallacies. Given the premise "Every contingent entity must cease to exist at some time" it would clearly be fallacious to ask "At which time?". The phrase "some time" has confused *suppositio*, so the question is out of place. Russell's explanation of the different denoting concepts in terms of the kinds of "combinations of terms" can be understood as serving this purpose as well. Still, it raises difficulties. What, after all, are "combinations of terms"? Is there a single logical subject 'Brown or Jones'? Russell's official answer should be "no." But at times he seems to demure in spite of the obvious threat to the fundamental doctrine of *Principles* (*PoM*, p. 55fn*). In the end, the most charitable interpretation is to take Russell's use of "combinations of terms" simply as a heuristic device to help in clarifying the kinds of denoting concepts.[16]

The theory of denoting concepts is a rich theory. As Russell knew, it shows that identity statements will be informative if a definite description flanks the identity sign (*PoM*, p. 64). The descriptive content brought in by use of a denoting concept underwrites informativity. Russell also introduced rules in order to handle the questions as to which among 'all *a*' 'any *a*,' 'some *a*,' 'every *a*,' and 'an *a*' is to have wider scope when more than one of them occur in the same proposition. The order, beginning from widest to narrowest, is to be: 'some *a*,' 'any *a*,' 'a(n) *a*,' and finally 'every *a*.' To take one of Russell's examples, in the proposition

16. For a dissenting opinion, see Paulo Dau (1986).

'Some moment does not follow any moment'

it is 'some moment' that has wider scope. The proposition implies that there is a first moment. On the other hand, in

'A moment precedes any moment'

'any moment' has wider scope than 'a moment' and requires that there is no first moment (*PoM*, p. 59).

More scope problems loom, however. What are to be the truth-conditions of a proposition containing a denoting concept *as concept* that does not denote anything? Such a proposition is not about anything, so it seems puzzling what makes the proposition true or false. The natural position to take is to lay down the law that any proposition in which a denoting concept not denoting anything occurs *as concept* is false. Troubles come, however, with the proposition

(1) 'The present king of France is not bald.'

Does the denoting concept 'the present king of France' occur *as concept*? If so, the proposition is false. Yet by the same reasoning,

(2) 'The present king of France is bald'

is also false! We have a violation of excluded middle.

A solution can be found if one can apply a scope distinction to the occurrence of denoting concepts. In (1) the denoting concept 'the present king of France' is to have a 'secondary scope.' Accordingly, the rule governing the truth-conditions for propositions containing denoting concepts that do not denote would be:

(Non-Denoting)
Any proposition in which a denoting concept that does not denote anything has a primary occurrence is false.

But how does the distinction of primary versus secondary occurrences apply to denoting concepts? A reasonable answer would be that a denoting concept has a secondary occurrence in a proposition q, if it occurs *as concept* in a proposition p which occurs *as term* in q. Applying Russell's definition of negation to proposition (1) we have a partial solution of our problem. The proposition (1) is:

$x \supset x :\supset_x: \{\text{The present king of France is bald}\} \supset x.$

Accordingly, since the proposition 'The present king of France is bald' occurs *as term* in this proposition, the denoting concept 'the present king of France' has a secondary occurrence in (1). There is now no conflict with (Non-Denoting) in regarding (1) as true.

There are yet more difficulties of scope, however. Consider the following proposition,

'All chimeras are animals.'

By (Non-Denoting) this would be false. Yet by the same reasoning, its contradictory,

'Some chimeras are not animals'

is also false. Russell agonized over this problem in *Principles* (*PoM*, p. 74). At one point he suggests that the truth-conditions of 'All chimeras are animals' might just be identified with those of the formal implication,

'x's being a chimera $\supset_x$ x's being mortal.'

This proposition is true since 'x is a chimera' is always false. On this view, (Non-Denoting) must be abandoned and Russell is left without an explanation of what it is about the proposition 'All chimeras are animals' that makes it true.

An alternative might be to keep (Non-Denoting) and the falsehood of 'All chimeras are animals,' claiming that its proper contradictory is

'$x \supset x :\supset_x: \{\text{All chimeras are animals}\} \supset x$',

or, avoiding formal implication, pick a necessarily false proposition f and put

Some a are not b = df All a are $b . \supset . f$.

The analog problem with

'No chimeras are animals,'

would have to be treated as well. This proposition would be false by (Non-Denoting) unless the scope of 'No chimera' is secondary. The problem could be avoided by dropping the denoting concept 'No a' putting:

No a are b = df $x \supset x :\supset_x: \{\text{Some } a \text{ are } b\} . \supset . x$.

Accordingly, 'some chimera' has a secondary occurrence. This plan preserves the modern square of opposition while at the same time allowing that there are denoting concepts that do not denote.

Of course the use of variables in the sentences expressing formal implications forms a language which is essentially richer than that of categoricals. Many relational notions are well out of reach without the device of variables, and Russell's denoting concepts cannot hope to express all the nuances of scope capturable by the apparatus of modern quantification theory. But Russell was under no illusions to the contrary. We must not lose sight of the fact that Russell never intended to capture formal implications categorically. The theory of denoting concepts was to form the basis of Russell's philosophical explanation of the constituents of the propositions which are formal implications. Denoting concepts are involved, but so also are a number of auxiliary notions. The degree of success he reached will be the topic of our next section.

The Analysis of the Variable

2.8 In Russell's view, mathematics is the logic of relations. The calculus for that logic involves sentences expressing formal implications and the formal notation of these sentences contain variables and not expressions for denoting concepts. Russell owes a philosophical analysis of the constituents of the propositions indicated by nominalization of formal implications. By providing such an analysis, Russell endeavored to introduce variables into the calculus for logic—a philosophical bridge (as it were) between Peano's formal implication and the old Aristotelian (and medieval) calculus of categoricals. The analysis rests on the theory of denoting concepts. Indeed, of the nine notions[17] Russell takes as the logical primitives of *Principles*, five are required for the analysis of formal implication: 'denoting,' 'class,' 'propositional function,' and the denoting concepts 'any term' and 'every term.'

In stark contrast, Frege abandoned categoricals from the start. His *Begriffsschfrift* allots no logical significance to the fact that "all men" is the grammatical subject of "All men are mortal." As we saw, Frege introduced the mathematical variable into his calculus for logic by taking the notion of a function as primitive and by modeling predication in terms of functionality. For instance, consider the following function names,

$$f\xi = \begin{bmatrix} \text{the True,} & \text{if } \xi \text{ is a man,} \\ \text{the False,} & \text{otherwise.} \end{bmatrix}$$

$$g\xi = \begin{bmatrix} \text{the True,} & \text{if } \xi \text{ is mortal,} \\ \text{the False,} & \text{otherwise.} \end{bmatrix}$$

$$\begin{array}{l} \top\zeta \\ \llcorner\xi \end{array} = \begin{bmatrix} \text{the False,} & \text{if } \xi \text{ is the True and } \zeta \text{ is any object other than the True.} \\ \text{the True,} & \text{otherwise.} \end{bmatrix}$$

$$\smile\!\!\!\!\!^{a}\;\Phi(a) = \begin{bmatrix} \text{the True,} & \text{if every object } a \text{ is such that } \Phi_a \text{ is the True.} \\ \text{the False,} & \text{otherwise.} \end{bmatrix}$$

Frege puts these function names together to form a complex name for a function whose value is the True if all men are mortal and is the False otherwise. Adding his assertion sign "$\vdash$" Frege writes

$$\text{“}\vdash\!\!\smile\!\!\!\!\!^{\mu}\!\!\begin{array}{l} \top g\mu\text{”} \\ \llcorner f\mu \end{array}$$

to assert that all men are mortal.

In his *Über Funktion und Gegenstand* (1892) Frege went on to give a philosophical explanation of the variable, claiming that the incompleteness of predicate expressions parallels the incompleteness of function names. This, he argues, is grounded in the fact that functions are essentially "unsaturated."[18]

17. The complete list is given by Russell on page 106 of *Principles*.
18. See Landini (1996c).

The term "variable," of course, carries the connotation of variation. But in Quine's words, "the connotation is best forgotten." It smacks of the wholly confused nineteenth century analyses which ontologized variables and populated the universe with "variable numbers," indefinite entities," and the like. Frege's approach is a marked step forward.

A Fregean function, however, takes its argument(s) to a value. There is no whole composed of function and argument any more than the number 3 is a whole composed of the numbers 2 and 1 and the function *plus*. Russell, on the contrary, could not model predication on mathematical functionality. To have a property or stand in a relation means to occur as a "term of a proposition" predicating the property or relation. Accordingly, Russell could not accept the notion that some entities are "unsaturated" so that variables for them cannot occupy subject positions. This would be tantamount to saying that some entities cannot "occur as a term" (cannot be "logical subjects"). Indeed, Russell thought "inextricable difficulties" envelop the view (*PoM*, p. 45). He thought its very statement to be self-refuting.

Russell takes the two kinds of occurrence in a proposition as primitive. On its basis he intends to capture functionality and the notion of the mathematical variable as used in sentences expressing formal implications. To do this *Principles* takes the notion of "variation" literally. Though appeals to "variable entities" are explicitly rejected, "variation" is to be understood by relying upon the notion of the substitution of one term for another term occurring "as term" (or as logical subject) in a proposition.

Russell begins by distinguishing the true or formal variable from the restricted variable. He writes that "*any term* is a concept denoting the true variable: if *u* be a class not containing all terms, *any u* denotes a restricted variable.[19] Thus, in the simple case of the expression "*x* is a man," Russell takes the expression to abbreviate "any term is a man." Nominalization of this expression indicates a proposition containing the denoting concept 'any term,' and this proposition will assert of some term (ambiguously denoted by 'any term') that it is a man. The use of a variable in a sentence is quite directly explained as indicating the denoting concept 'any term.'

Russell knew, however, that a similar tactic fails for the analysis of, say, the open formula "*x* is a man implies *x* is mortal." Russell observes that it would not do to maintain that the expression indicates 'any term is a man implies any term is mortal.' There is an individuality involved when the denoting concept 'any term' occurs. Each occurrence might denote a distinct term; and so the sameness expressed by the use of the letter "*x*" is lost. Thus, the explanation of how the variable "*x*" comes to refer when it is used in a formula like "*x* is a man implies *x* is mortal" is more complicated. Russell writes that "a variable is not *any term* simply, but any term as entering into a propositional function" (*PoM*, p. 94).

19. Russell continues, writing that "the terms included in the object denoted by the defining concept of a variable are called the *values* of the variable." He again speaks (I think heuristically) as if there is one complex object denoted—that is, a "variable disjunction."

The converse is true as well, though this has rarely been appreciated. The nature of a propositional function is itself caught up in the problem of the analysis of the variable. For instance, it would be mistaken to maintain that a propositional function is just that intensional entity which remains when a constituent of a proposition is extracted. Russell explicitly denies this in *Principles*. Beginning with a proposition in which a constituent occurs as a term, the extraction of this term leaves what Russell calls an "assertion" (*PoM*, p. 84). Thus, extracting Socrates from 'Socrates is a man' leaves the assertion '. . . is a man.' Russell admits that in such a simple case it might be possible to define the propositional function in terms of the assertion (*ibid.*, 39), but he points out that this fails in more difficult cases. For instance, extracting Socrates from the proposition 'Socrates is a man implies Socrates is mortal' leaves '. . . is a man implies . . . is mortal.' But this assertion will not do as a propositional function (*ibid.*, 85). The dots represent gaps, but there is nothing in the assertion which tells us how the gaps are to be filled—whether by the same entity or by different ones.[20]

This is a central problem for Russell. In the propositional function 'x is a man implies x is mortal' the variable marks the fact that the same term must go in each place. The function is quite different from 'x is a man implies y is mortal.' That is, something must remain to mark the way variability is to occur. Russell concludes that the notion of a propositional function cannot be defined by means of the idea of separating a proposition in which one or more entities occur as term into subject(s) and assertion.

Though the plan of separating propositions into subject and assertion fails, the idea of extracting a constituent of a proposition is connected both to Russell's explanation of the notion of the "variable" and to that of a "propositional function" as well. In an early chapter entitled "Implication and Formal Implication," Russell explains that 'x is a man implies x is a mortal' cannot involve a relation between propositional functions, but must be regarded as itself a single propositional function (*PoM*, p. 38). The reason is that "the x, though variable, must be the same on both sides of the implication, and this requires that we should not obtain our formal implication by first varying (say) Socrates in 'Socrates is a man' and then in 'Socrates is mortal', but that we should start from the whole proposition 'Socrates is a man implies Socrates is a mortal' and vary Socrates in this proposition as a whole" (*ibid.*).

The role of the analysis of the variable in Russell's reading of formal implication can now be made clearer. Russell writes (*PoM*, p. 38):

> Our formal implication asserts a class of implications, not a single implication at all. We do not, in a word, have one implication containing a variable, but rather a variable implication. We have a class of implications, no one of which contains a variable, and we assert that every mem-

20. Russell criticizes Frege on these grounds as well (*PoM*, p. 505). But the objection does not properly apply to Frege's functions.

> ber of this class is true. This is a first step towards the mathematical notion of a variable.

Consider, then, what appears to be Russell's preliminary formulation of the proposition indicated by "x is a man implies x is mortal, for all values of x" (*ibid.*, 92):

> Every term of the class of propositions, each asserting of some constant term that if it is a man then it is mortal, is true.

This will not quite do. In his characterization of the relevant class of propositions Russell uses anaphora and this is the linguistic representative of variables. Moreover, the expression "class of propositions" is best replaced by "class of terms." Perhaps a more accurate rendition would be:

> Every term of the class of terms (propositions) each of which is obtained by substituting any term for any term (say a) in 'a is a man implies a is mortal' is true.

This comes closer to the spirit of the thing. Of course, for two quantified variables, Russell has to repeat the substitution twice. He writes (*PoM*, p. 94):

> If the propositional function $\phi(x, y)$ is to be asserted for all values of x and y, we consider the assertion, for all values of y, of the propositional function $\phi(a, y)$, where a is a constant. This does not involve y, and may be represented by $\psi(a)$. We then vary a, and assert $\psi(x)$ for all values of x. This process is analogous to double integration; and it is necessary to prove formally that the other in which the variations are made makes no difference to the result.

Russell's plan is simply to repeat the process of substituting—a process, as he puts it, analogous to "double integration."

In short, the idea is to use the occurrence of a term (occurring as term) in a proposition to mark what is to be varied. We are to begin with a proposition, say, 'Socrates is a man implies Socrates is mortal' and we next consider the class of propositions resulting from substituting different terms for Socrates. The propositions in this class are said to have a "constancy of form" or "constancy of type" in virtue of being so obtained. In the chapter "The Variable," Russell writes (*PoM*, p. 89):

> When a term occurs as term in a proposition, that term may be replaced by any other while the remaining terms are unchanged. The class of propositions so obtained have what may be called constancy of form, and this constancy of form must be taken as a primitive idea. The notion of a class of propositions of a constant form is more fundamental than the general notion of a *class*, for the latter can be defined in terms of the former, but not the former in terms of the latter. Taking *any* term, a certain member of any class of propositions of constant form will contain that term.

In this way, variability is supposedly explained by means of the denoting concept 'any term' together with the notion of "substitution."

Observe that in Russell's view, a class is defined by (or determined by) a class-concept which is the common characteristic of members of the class. The class of men, for instance, is given by means of the class-concept 'man.') With more complicated classes, however, the common characteristic will be that all the terms of the class satisfy a given propositional function. Thus, for instance, we have the class of all x such that ϕx. Now Russell says that once ϕx is fixed, we can avoid appeal to the propositional function in determining the class by drawing on the notion of "substitution." That is, we can speak of the class of terms each of which is obtained by substituting any term for (say) a in 'ϕa'. For example, instead of "the class of all x such that x is a man implies x is mortal," Russell hopes to put "the class of terms each of which is obtained by substituting any term for any term (say a) in 'a is a man implies a is mortal'." But observe that the very expression,

> "the class of terms each of which is obtained by substituting any term for any term (say a) in 'a is a man implies a is mortal' "

is of the form "the class of all x such that ϕx." Because of this, Russell decides that the notion of a "class of propositions of the same form" must be taken as primitive. "Propositions," he says, "may have a certain constancy of form, expressed in the fact that they are instances of a given propositional function, without its being possible to analyze the propositions into a constant and a variable factor" (*PoM*, p. 85). Accordingly, he concludes, "it appears that propositional functions must be accepted as ultimate data" (*ibid.*, 88). That is to say, some notion of "class" (though not the general notion) and the class-concept "propositional function" must be taken as fundamental.

This is not to say, however, that an ontology of propositional functions is assumed in *Principles*. In the chapter "Propositional Functions," Russell denies this categorically. It is only the *general* notion of a "propositional function" that must be taken as primitive and undefinable (*ibid.*, 83). In the chapter "The Variable," he follows this up, writing that, "what is fundamental is not particular propositional functions, but the class-concept *propositional function* (*ibid.*, 92). It was because the analysis of formal implication relied on the notion of a "class of terms got by substituting . . . ," that is, the notion of a "class of propositions of the same form," that the class-concept 'propositional function' had to be taken as primitive. Propositional functions (as entities with both an individual and a predicable nature) are not assumed in *Principles*.

Russell's analysis of the variable is consonant with there being only one style of variable—the individual variable. The treatment demands that the entity to be varied occurs in a proposition "as term of the proposition," for it is only in such cases that the substitution of any other term results in a proposition. If the variable is to remain wholly unrestricted, any entity (whether a *thing* or a *concept*) must be admissible as a candidate for substitution. This is assured so long as the entity for which substitution takes place occurs as a term of the proposition. For instance, we can substitute any entity whatsoever for Socrates in the proposition 'Socrates is a man' without the result ceasing to be a proposition. Some substitutions will result in odd propositions, such as 'happiness is a man' and '$\{x$ implies $y\}$ is a man,' but Russell simply regards these as false.

The one style of variable does, however, pose a difficulty when it comes to defining a variable representing the variation of R in the proposition 'aRb'. Such variations play a central role in mathematics, and Russell mentions their role in the theory of progressions and series. It would not do to merely abandon them. Yet in the proposition 'aRb' the relation R occurs "as concept" (i.e., relating relation) and not as a "term of the proposition." If we substitute something which is not a relation there will be no proposition, and presumably no entity, 'aRb' at all.

Russell acknowledges that this is a difficult problem. His suggestion in *Principles* is to adopt a primitive logical notion 'being a relation.' He then allows variation of R in the proposition

$$R \text{ is a relation} \supset aRb.$$

Now, he says, the R can then be varied for it can take all values, relations or otherwise (*PoM*, p. 87). Using our individual variable "x", we see that Russell has put:

$$x \text{ is a relation} \;.\!\supset\!.\; axb.$$

The question naturally arises as to what sort of proposition 'axb' is when x is, say, Frege. Russell anticipated it, writing that "it is necessary to assign some meaning (other than that of a proposition) to aRb when R is not a relation" (*ibid.*). This, of course, won't do. In the 1905 manuscript "On Fundamentals," Russell tried another approach. He enjoins us to consider the proposition 'R holds between a and b.' This proposition is equivalent to (though not identical with) the proposition 'aRb'. Here the relation 'holds between' occurs as concept and R occurs as entity. Thus, any entity (be it a relation or not) may be substituted for it without difficulty. Similarly, to achieve what would be the variation of 'humanity' in the proposition 'Socrates is human,' we must employ the equivalent proposition 'Humanity belongs to Socrates.' In the latter, 'humanity' occurs as a term of the proposition. Unfortunately, this requires that 'belongs to,' 'holds between,' and the like, for as many places as needed, be new logical primitives—or, alternatively, a predicate constant for 'exemplifies' needs to be adopted as a primitive. (We suggested the latter in §2.1.) The problem would not be satisfactorily solved until late in 1905 when Russell invented his new substitutional theory—a theory which avoids predicate variables altogether and whose only predicate constants are analogs of the modern logical particles.

CHAPTER THREE

The New Theory of the Variable

3.0 *Principles* offered an analysis of the constituents of propositions which was to ground the use of unrestricted variables in the formal calculus for logic. The ground was an analysis of the use of variables by means of the notion of denotation and the substitution of one entity in a proposition (occurring as term of the proposition) for another. The fundamental notion was that of "occurring as term of a proposition" (or "logical subject").

Now *Principles* acknowledges that paradoxes plague logicism. The paradoxes of predication—for example the paradox of the attribute 'being an attribute not predicable of itself' (and similar analogs in polyadic cases)—involve the use of a formula in which a predicate variable "ϕ" occurs in both a predicate and a subject position. Russell calls such wffs "quadratic forms" (*PoM*, p. 104). In virtue of his analysis of the use of variables, however, Russell was able to suggest a possible philosophical escape from the paradox. The analysis of the use of predicate variables in terms of denoting and substitution may well show that one cannot automatically regard a wff $\ulcorner A\mu \urcorner$ as of the form "ϕx", with "x" the individual variable picked out by μ and "ϕ" a predicate variable assigned to the rest of the proposition indicated by the expression $A\mu$. Russell put it with characteristic sloppiness concerning his symbols for wffs, predicate variables, and attributes (*PoM*, p. 88):

> According to the theory of propositional functions here advocated, the ϕ in ϕx is not a separate and distinguishable entity: it lives in the propositions of the form ϕx, and cannot survive analysis.

Russell then goes on (*PoM*, p. 88):

> If ϕ were a distinguishable entity, there would be a proposition asserting ϕ of itself, which we may denote by $\phi(\phi)$; there would also be a proposition not-$\phi(\phi)$, denying $\phi(\phi)$. In this proposition we may regard ϕ as variable; we thus obtain a propositional function. The question arises: Can the assertion in this propositional function be asserted of itself. The assertion is non-assertibility of self; hence if it can be asserted of itself, it cannot, and if it cannot, it can. This contradiction is avoided by the recognition that the functional part of a propositional function is not an independent entity.

The analysis of variables avoids the paradox of predication which would otherwise ensue.

Sadly, Russell's articulation of the matter is encumbered by the fact that he jumbles together a number of distinct notions under the heading of "propositional functionality." There is the matter of the use of schematic letters as in the modern "$A\mu$". Some of Russell's uses of "$(C \lozenge x)$" and "ϕx" are intended schematically, so that "x" and "ϕ" are not object-language variables at all. But Russell was wholly undisciplined in his use of schematic letters as opposed to object-language variables. All the same, Russell is offering a diagnosis of the source of the paradox which is not unlike that which Quine would render some years later (Quine 1963, p. 258). Begin with the schematic

$$(x)(Ax \equiv Ax);$$

this is surely a truth of logic. Now, slighting the vital contrast between schematic letters and quantifiable variables, the following seems to follow straightaway by existential generalization,

$$(\exists\phi)(x)(\phi x \equiv Ax).$$

The new variables are terms of the formal language. And yet, if *whatever is, is one*, the entity variables "x" take all terms as admissible substituends in a *universal instantiation*, Russell's paradox follows.[1] Russell's diagnosis is akin to Quine's. But, unlike Quine, Russell is not prepared to abandon logicism. In *Principles*, Russell's hope was that an analysis of the conditions under which the introduction of predicate variables is legitimate—an analysis by means of denoting and the "substitution" of entities in propositions—will reveal purely logical principles which determine when it is safe (as Russell puts it) "to vary ϕ independently of x in ϕx."

In *Principles*, Russell's approach to the paradox of classes paralleled his approach to that of predication. In its primary sense, a "class' is what Russell called a "class-as-many." It is really just the many entities which are collectively denoted in virtue of each having a certain property (i.e., "class-

1. This is not to say that one cannot have a "standard second-order calculus." Frege's *Begriffsschrift* is just such a calculus. But in making predicate variables in subject positions ungrammatical, the calculus abandons the fundamental doctrine that both Russell and Quine share—viz., that whatever is, is one.

concept") or satisfying a propositional function. Indeed, it is because one can denote infinitely many entities satisfying a condition that Russell solves the problem of how knowledge of infinite classes should be possible. In Russell's words, denoting provides "the inmost secrets of our power to deal with infinity" (*PoM*, p. 73). The intensional viewpoint is primary here and yields Russell's "class-as-many."

The class-as-many is not *one* entity at all. It is just that many objects are collected together by a denoting concept. For every condition, there will be a class-as-many. This, however, does not suffice for the theory of extension. It does not even yield classes of classes, for here a class is to be *one* among another's members. Accordingly, *Principles* also allowed what is called the "class-as-one." And prior to the discovery of his paradox of the "class of all classes-as-one not members of themselves" Russell thought it axiomatic that to every class-as-many there is a class-as-one.

Now a class-as-one is a single entity; it is a single "whole" composed of the many and thus finds its existence in its having members. The paradox of classes is a paradox of classes-as-one, since a class-as-many its not itself "one" and cannot be "one" member of anything. The paradox shows that a class-as-many only sometimes has a corresponding class-as-one (*PoM*, p. 104). Since the paradox of classes-as-one parallels the paradox of predicates, Russell felt that his quest to find principles determining which quadratic forms are safe would ipso facto provide principles determining what classes-as-one there are. Hope of finding such principles led to Russell's "zig-zag" and "limitation of size" theories of classes-as-one.[2]

The common thread in Russell's thinking, then, is that the solution of the paradoxes lies in a proper philosophical analysis of variation. Russell hoped that the analysis in *Principles* (based on denoting and substitution) would show which quadratic forms are safe and show when predicate variables are legitimate. Russell's quest for principles governing quadratic forms and his examination of denoting, substitution, and variation, are linked in this way.

In fact, Russell's theory of types is also linked. We have seen that Russell did not fail to connect in *Principles* the analysis of the variable in terms of denoting and substitution and the problem of solving the paradoxes of classes-as-one (and of predication). The crude sketch of type regimentation in *Principles* was but a formal dodge which would require a philosophical theory to be made viable. Russell knew that type stratification is at odds with the fundamental doctrine that variables are wholly unrestricted. If classes are single entities, then nothing can prevent "$x \in x$" from being meaningful. Accordingly, the formal dodge (a theory of types) had to be paired with the theory of denoting. Denoting was to show how classes-as-one could always be avoided in favor of classes-as-many. This was the idea in *Principles* that "the fundamental doctrine upon which all rests is the doctrine that the subject of a proposition may be

2. Little is understood about Russell's work here. A good start may be found in Urquhart (1988).

plural, and that such plural logical subjects are what is meant by classes" (*PoM*, p. 517). Type stratified variables for classes-as-one could only be introduced if those variables were defined away by means of the theory of denoting concepts. In *Principles*, Russell worried whether his theory of denoting could bear the weight of proxying classes-as-one. Types were not adopted; the official approach was that there are classes-to-one posited by purely logical principles governing quadratic forms.

Russell's manuscripts of 1903 and 1904 examine the nature of the variable to solve the problem of finding purely logical principles as to which quadratic forms are "safe." By 1904 Russell all but abandoned the project of finding principles and focused intently upon the matter of denoting and variation. This chapter takes us Russell's later studies and follows their outcome to the new 1905 theory of the variable.

"On Fundamentals" Against Denoting Concepts

3.1 In his article "On Denoting," Russell set out a theory which refused to assign semantic significance to the fact that quantifier phrases such as "every *a*," "some *a*," "any *a*," and "the *a*," where "*a*" is a common noun or noun phrase, are grammatically subject expressions. The system of quantification theory in Frege's 1879 *Begriffsschrift* was some twenty-six years ahead of Russell in this insight. With his ontology of general propositions and the problem of their constituents, Russell was slow to catch on—detouring through a quagmire theory of denoting concepts in his efforts to analyze the use of variables in terms of the substitution of entities. But catch on he did in 1905. Galvanized by an entirely new perspective, he was able to extend Frege's insight to definite descriptions.

Just how did Russell come to this new perspective and what exactly was it? Alas, Russell's arguments are nothing if not obscure. His infamous "inextricable tangle" from "On Denoting" seems as intractable as the walls of Troy. There are helpful manuscripts of 1904 such as "Points About Denoting," "On the Meaning and Denotation of Phrases," and "On Meaning and Denotation." In fact, there is a manuscript entitled "On Fundamentals" which Russell dated June 1905, scrawling as if to posterity: "Pp 18ff contains the reasons for the new theory of denoting." But these manuscripts are as obscure as ever they could be.

Well protected by ramparts and barbed wire, the argument has remained an impenetrable fortress. Is an assault really worth the suffering? Church's assessment that there is nothing in it but use–mention confusions aside, early enthusiasm that the context of *Principles* would reveal Russell's intent has waned in recent years. Hoping for a return of focus to Frege, Blackburn and Code put matters starkly: "Authors have not failed to invent features of Russell's earlier view, so that his argument can be seen as relevant to them, but not to Frege" (Blackburn and Code 1978, p. 68).

Blackburn and Code are mistaken, however. The gate to the fortress was open all the time. We have only to ask: Are denoting concepts capable of the two-fold nature essential to concepts generally? According to *Principles* the answer has to be "yes." Its fundamental doctrine—whatever is, can occur as term of a proposition—requires it. Therein lies the secret. The answer "yes" spells doom for the theory of denoting concepts.

Surprisingly few have noticed that denoting concepts must be capable of the two-fold occurrence. Consider Blackburn and Code themselves (1978, p. 76):

> The PoM theory of denoting makes it impossible to directly name a sense or denoting concept. This is because (i) when a name is used in a sentence, the thing named is a constituent of the proposition expressed by the sentence, and (ii) if a denoting concept is a constituent of a proposition, then the proposition is about the object denoted and not about the denoting concept. If we suppose that he had come to see this by the time he wrote "On Denoting," we have an explanation for the fact that he is now insisting that senses be introduced by means of definite descriptions.

Now it may well be that Russell thought there are no genuine proper names for denoting concepts. But it would be the worst abductive reasoning to seek an explanation in the view that a denoting concept cannot occur as term of a proposition.[3] It is an unassailable fact that Russell speaks of denoting as a logical relationship based on the occurrence of a denoting concept in a proposition. Nonetheless, he does not say that whenever a denoting concept occurs in a proposition, the proposition is about the denotation of the concept (if any). He says that it denotes when "if it occurs in a proposition, the proposition is not *about* the concept, but about a term connected in a certain peculiar way with the concept" (*PoM*, p. 53). These words do not show that denoting concepts lack the two-fold nature.

In "On Fundamentals," Russell introduced a new terminology to replace the notions of "occurring as concept" and "occurring as term of a proposition." He now wishes to widen the discussion to include occurrences of terms in other sorts of "complexes."[4] Russell's use of the phrase "denoting complex" is not always consistent in the manuscripts, but it becomes clear that the expression was often used to mean "complex denoting concept." In such cases denoting complexes are simply denoting concepts that have other terms, e.g., persons, towns, or denoting concepts as constituents. In the new terminology the former notion of "occurring as concept" is replaced by the wider notion of "occurring

3. If, indeed, there are no proper names of denoting concepts it only follows that there are no nominalized statements expressing propositions in which denoting concepts occur as terms.

4. A fairly large literature has grown up concerning what Russell meant by "denoting complex." Russell's article "On the Existential Import of Propositions," published in *Mind* just two months before the appearance of "On Denoting," corroborates the present interpretation (*EI*, p. 100).

as meaning" in a proposition or complex denoting concept; the earlier notion of "occurring as term" is replaced by the notion of "occurring as entity" in a proposition or complex denoting concept (*OF*, p. 30).

Recalling our practice of using single quote marks to form genuine proper names of denoting concepts and propositions, let us illustrate the new terminology with examples from "On Denoting." Consider the denoting concept:

'the first line of Gray's *Elegy*'.

This denoting concept occurs as meaning in the proposition:

#1 The first line of Gray's *Elegy* states a proposition.

The denoting concept occurs as entity in the proposition

#2 'The first line of Gray's *Elegy*' does not state a proposition.

The case is similar when a denoting phrase occurs as grammatical subject of another phrase. For instance, the denoting concept 'the first line of Gray's *Elegy*' occurs "as meaning" in the complex denoting concept indicated by the phrase,

"the meaning of the first line of Gray's *Elegy*".

This complex denoting concept denotes the meaning of "The curfew tolls the knell of parting day." The same holds when we put a denoting phrase in the grammatical subject position of

"the denotation of . . .".

Follow Russell in putting the phrase "the denoting complex occurring [as entity] in the second of the above instances" in the dots. (The second instance is our #2 above.) The result speaks about the denotation of the denoting concept 'the first line of Gray's *Elegy*'—viz., "The curfew tolls the knell of parting day."

Russell concludes that to get what is wanted—i.e., to denote the denoting concept 'the first line of Gray's *Elegy*' we must employ

"The meaning of "the first line of Gray's *Elegy*" ".

or employ

"The denotation of "the denoting concept occurring [as entity] in the second of the above instances" ".

So far this is mundane. What is the trouble?

The trouble lies with entity occurrences of denoting concepts. Russell thinks himself to have demonstrated that since grammatical subject positions mark "meaning occurrences" there can be no entity occurrences! In "On Fundamentals," Russell puts the point another way. He introduces what he calls an "entity position." Properly speaking, an entity position is just the notion of being a grammatical subject of a larger phrase. Unfortunately, he conflates this notion with the ontological notion of an occurrence in a complex[5]. It is just this error which drives the following passage (*OF*, p. 30):

> The use of inverted commas may be explained as follows. When a concept has meaning and denotation, if we wish to say anything about the meaning, we must put it in an *entity position*; but if we put it itself in an entity position, we shall be really speaking about the denotation, not the meaning, for that is always the case when a denoting complex is put in an entity position. Thus, in order to speak about the meaning, we must substitute for the meaning something which *denotes* the meaning.

This is sloppy. A denoting concept does not occur in an entity position (grammatical subject position). It is the denoting phrase that is in entity position. Russell has not shown that entity occurrences are impossible. The linguistic notion of a grammatical subject position must not be conflated with the ontological notion of an entity occurrence.

The conflation seems also to be behind the following argument from "On Fundamentals." In discussing the possibility of explaining (defining) the use of inverted commas, Russell puts:

> "'any man'" indicates a denoting concept which denotes 'any man.'[6]

This purported explanation, he says, would be no explanation at all for it uses inverted commas in explaining them. True enough. But surely we can put:

> "'any man'" indicates a denoting concept which denotes the meaning of "any man."

Or better put:

> 'any man' = df the meaning of "any man."

This would be to follow Frege's own suggestion in "Über Sinn und Bedeutung". Frege wrote (1892, p. 59):

> In order to speak of the sense of an expression "A" one may simply use the phrase "the sense of the expression A."

Russell objects as follows (*OF*, p. 35):

> The meanings of denoting complexes can only be approached by means of complexes which denote those meanings. This is what complexes in inverted commas are. If we say '"any man" is a denoting complex,' "any man" stands for 'the meaning of the complex "any man,"' which is a denoting concept. But this is circular, for we use "any man" in

5. This conflation, in fact, is the source of many of the "dilemmas" which, Russell says, "force us to recognize a greater variety of modes of occurrence than we have yet admitted" (*OF*, p. 23). These other modes of occurrence are largely tangential to the main point and we have omitted their discussion here.

6. Here as before I use single quotes to form proper names for denoting concepts.

> explaining "any man." And the circle is unavoidable. For if we say "the meaning of any man," that will stand for the meaning of the denotation of any man, which is not what we want.

The passage is difficult because of the confusions of use and mention. Are Russell's quote marks for concepts or for phrases? Does he mean by a "complex" a phrase or a concept? On the face of it, the objection appears to be that on the above definition of inverted commas, the denoting concept indicated by the phrase "the meaning of "any man"" would contain the denoting concept 'any man' itself. This would indeed be circular; but this is wrong. It would contain a denoting concept indicated by ""any man""—that is, a denoting concept which denotes the expression "any man." There is no circle here.

The simple fact is that Russell has wrongly dismissed entity occurrences of denoting concepts by conflating an "entity occurrence" with an "entity position." Perhaps Russell reasoned that an entity occurrence can happen only when a denoting phrase is in an "entity position" (grammatical subject position). But this is a substantial slip. The ontological issue must be separated from the linguistic.

3.2 Happily, the slip does not vitiate the argument. Russell requires a theory of logical form that captures the difference in structure between meaning and entity occurrences of denoting concepts. It is far from clear how this could be accomplished. So Russell is justified in regarding entity occurrences as questionable, even if his argument for the result misses the mark.

Indeed, "On Fundamentals" shows that Russell was tracking the ontological issue. He is grappling with the problem of how to capture the structural difference in kinds of occurrences. Returning to a difficulty noted in *Principles* (*PoM*, p. 86), Russell wonders how to use a variable to mark an occurrence of an entity "as meaning" ("as concept"). This is a serious problem. Russell considers the proposition '*a* differs from *b*', and remarks that one cannot quantify over the concept 'differs from' since it occurs as meaning in the proposition. The values of the variables of quantification include all entities, whether concepts or things. And '*a* Socrates *b*' is not a proposition at all. For the time being, Russell decides that he shall have to vary the concept 'difference' in the proposition 'Difference holds between *a* and *b*,' for here the concept occurs as entity and 'Socrates holds between *a* and *b*' is simply a false proposition (*OF*, p. 30).

Sensitivity to the difference in structure led Russell to introduce the following "broad rule" governing meaning and entity occurrences (*OF*, p.23):

> When complexes occur as *meaning* their complexity is essential, and their constituents are constituents of any complex containing the said complexes; but when complexes occur as *entities*, their unity is what is essential, and they are not to be split into constituents. Hence generally when a complex *A* occurs in a complex *B*, if *A* occurs as *meaning*, its constituents are constituents of *B*, but if it occurs as *entity*, its constituents are not constituents of *B*.

When a complex denoting concept occurs as entity in a larger complex it is a unity, and is considered as "one" constituent of the larger. When it occurs as meaning its unity is lost. For instance, when the concept 'the author of *Waverley*' occurs as meaning it is the indication of the words "the," "author of," and "*Waverley*" that are among the constituents of the larger complex.[7] In this case, the denoting concept cannot be regarded as a single constituent of the larger.

Russell came to draw a crucially important consequence. The use of a single letter to represent a meaning occurrence of a complex is illegitimate. He writes (*OF*, p. 24):

> It is a fallacy to use a single letter to represent an occurrence of a complex as *meaning*, since a single letter will have all entities among its values; moreover, when a *complex* occurs as meaning, its structure is essential to its significance and a single letter, since it does not symbolize any structure, destroys the significance.

A single letter acts as a genuine proper name, and thus the designation of the letter would occur as entity and not as meaning. Since quantification (by means of formal implication) employs variables, it is impossible to quantify over meaning occurrences. To illustrate, consider the proposition

> 'The author of *Waverley* wrote *Ivanhoe*.'

The move to '$(\exists x)(x$ wrote *Ivanhoe*)' fails to capture the meaning occurrence of the denoting concept 'the author of *Waverley*.' Any entity satisfying the open formula "x wrote *Ivanhoe*" will have an entity occurrence in a proposition predicating authorship of *Ivanhoe*.

Generalizing, we see that whatever satisfies an open formula must occur as entity in a proposition. The point has a direct bearing on whether inverted commas can be explained as indicating denoting concepts which denote other denoting concepts. Recall from our Chapter 2 that *Principles* maintains that denoting is a logical relation grounded by the following law (*PoM*, p. 74):

> If a denoting concept associated with the class concept a denotes y, then y is an-a.

Here y occurs "as entity" in the proposition 'y is an-a'. Returning, then, to the purported definition,

> 'any man' = df the meaning of "any man,"

we get the following result. The denoting concept indicated by the phrase "the meaning of "any man,"" denotes 'any man' only if 'any man' uniquely satisfies the formula

> x is a meaning of "any man."

7. The view also occurs in "Points About Denoting" (*P*, p. 3).

But if so then the denoting concept 'any man' occurs as entity in the proposition ''any man' is a meaning of "any man"'. The moral is of inestimable importance:

> *Denoting concepts for other denoting concepts are impossible unless denoting concepts have entity occurrences.*

The logical relationship of denoting requires that a denoting concept denotes a given entity in virtue of the entity having the property or properties which form the class concept from which the denoting concept is formed. But an entity has a property in virtue of its occurring as entity in a proposition predicating the property. If denoting concepts are not capable of entity occurrences, then they cannot be denoted by other denoting concepts.

The logical relationship of denoting requires entity occurrences of denoting concepts. Assuming with Russell that entity occurrences are impossible, one would have to attempt a modification of the logical relationship of denoting of *Principles*. This is precisely what Russell did. Supposing there are no entity occurrences of denoting concepts, a denoting concept is, as it were, "that which occurs as meaning." Russell puts this in the following much maligned passage of "On Denoting" (*OD*, p. 112):

> This leads us to say that, when we distinguish meaning and denotation, we must be dealing with the meaning: the meaning has denotation and is a complex, and there is not something other than the meaning [i.e., something other than the meaning occurrence] which may be called the complex [logical subject] and be said to *have* both a meaning [occurrence] and a denotation.

Cumbrous as the wording is, the point is to just consider what happens if denoting concepts can *only* occur as meaning. What happens?

Denoting has to be a logical relationship. But how, one may ask, could one modify the logical relationship to accommodate the view that there are only meaning occurrences of denoting concepts? We cannot allow that one denoting concept, C, say

> 'the meaning of "all men,"'

can denote another D, say 'all men,' in virtue of D's occurrence "as meaning" in the proposition 'all men is a meaning of "all men."' The proposition is false; it is about all men, and not about the denoting concept 'all men.' So suppose C denotes D in virtue of the presence of a meaning occurrence of a third denoting concept E, denoting D. Surely this won't do either; it is either circular (if E is just C itself) or embarks one on a vicious regress of denoting concepts which are supposed to denote D. There is no hope in this. Russell needs both meaning and entity occurrences of denoting concepts, but he cannot find a theory of the structure of propositions which gets at the structural difference between them.

Interestingly, our present discussion shows that the interpretation of Chrystine Cassin (1970) was on the right track. Cassin is criticized by Blackburn and Code (among others) for claiming that it was part of the

Principles view that "only terms could be denoted" and holding that Russell came to realize that denoting concepts themselves must be denoted (Cassin 1970, p. 269). Blackburn and Code chide Cassin, charging that according to *Principles* whatever is, be it a thing or a concept, is a term (Blackburn and Code 1978, p. 68). Quite so. But they have forgotten that to be a term is to be able to occur "as term" (as "entity") in a proposition! Cassin should have said that whatever can be denoted must be capable of occurring "as term" (or, in the jargon of "On Fundamentals," "as entity"). It is no small irony that Blackburn and Code interpret Russell as having made no room for such an occurrence (i.e., in their argument against there being genuine names of denoting concepts) and yet failed to notice that this spells doom for the *Principles* theory.

In a way, our argument is similar to Cassin's "falling through" argument—similar, but with the added point that "falling through" or vicious regress occurs only if entity occurrences of denoting concepts have been abandoned so that the law of denoting has to get along with meaning occurrences alone. So long as there are entity occurrences of denoting concepts nothing "falls through." Moreover, the original *Principles* is committed to a denoting concept's being capable of the two-fold occurrence characteristic of concepts generally. A proposition containing a denoting concept "as entity" would be about the denoting concept itself, not its denotation. Aboutness does not "fall through" in such a case. The trouble is, rather, that Russell is at a complete loss when it comes to the structural difference between meaning and entity occurrences of denoting concepts.

We should pause here to observe that Hylton's recent revival of a version of Cassin's "falling through" argument is mistaken. In Hylton's view, *Principles* held a doctrine of "truth-value dependence" (Hylton 1990, p. 251):

> For a proposition containing a denoting concept to be about some other entity is for the truth-value of that proposition to be dependent upon the truth-value of the proposition obtained from it by replacing the denoting concept by the denoted entity.

Hylton is misled by the heuristic use of medieval doctrine of *suppositio* in *Principles*.[8] There is no "principle of truth-value dependence" in *Principles*. One cannot substitute, say, Scott for the denoting concept 'the author of *Waverley*' in the proposition 'The author of *Waverley* is tall' without a violation of the structural integrity of the proposition. Hylton's "principle of truth-value dependence" supposes that there is no difference in logical form. Yet it is the difference in logical form that is Russell's whole point. The proper relationship of dependence is given as the law of denoting in *Principles*—a law which, when applied to denoting concepts purporting to denote others, establishes a logical relationship between meaning and entity occurrences.

Without any way to capture the difference in logical form between meaning and entity occurrences, Russell is pressed to exclude entity occurrences of

8. See Chapter 2.

denoting concepts altogether. But then, as we saw, his troubles are worse (*OD*, p. 112):

> This only makes our difficulty in speaking of meanings more evident. For suppose *C* is our complex; then we are to say that *C is* the meaning of the complex.

Contrary to a growing literature, this passage does not introduce meanings of meanings.[9] The point is just that without entity occurrences, a meaning (i.e., a denoting concept) is only identifiable in "its" meaning occurrence. That is, it *is* its meaning occurrence. One cannot even properly speak of a denoting concept as itself an entity which *has* both a meaning occurrence and (sometimes) a denotation. There is the rub. Without entity occurrences, logic will be wholly incapable of distinguishing the denoting concept itself—occurring as meaning—from the entity (if any) that it denotes. (As we saw, it is always the denotation which is captured when we attempt with the variables of quantification to capture the denoting concept in its meaning occurrence.) Meaning and denotation collapse.

If denoting concepts cannot occur as entity, they cannot formally be logical subjects in the range of the variables of quantification. There must be no denoting concepts. The theory of denoting in *Principles* had been wrongly conceived. Russell came to just this conclusion. Within the framework of *Principles* the argument is not only valid, it is quite sound.

An Argument Against Frege?

3.3 Sound though Russell's argument is, it is wholly inapplicable against Frege's theory of *Sinn* and *Bedeutung*. To see this, we have only to remind ourselves of some features of Frege's philosophy of language.

Unlike Russell, Frege introduced variables into his calculus for logic by adopting mathematical functionality as a primitive. As we saw in Chapter 2, predication in natural language is transcribed into the ideography of Frege's *Begriffsschrift* via functionality. Frege's function symbol "$f\xi$" is not to be confused with a sentential context. One forms a *singular term* by putting a term in the position marked by "ξ". Thus "$f2$" is a term not a formula.

Now in *Über Sinn und Bedeutung* (1892) function symbols are said to have both a sense and a reference. The reference is a function, an "unsaturated" entity which takes objects as argument to truth-values. There is no complex whole composed of function and argument. So Frege entirely skirts Russell's worries about the way an entity occurs in a proposition ("as meaning" as opposed to "as entity"). Of course, senses of singular terms such as "2" can themselves be arguments to functions. But again, Frege has no propositions

9. See, for instance Turnau (1991) and Wahl (1993).

and so senses can be arguments to functions without the least concern about the nature of its "occurrence" in a complex (proposition).

To be sure, there is also the sense side. The function symbol "$f\xi$" has a sense and (like the function which is its reference) it is unsaturated. Moreover, there is a complex sense (*Gedanke*) of the term "$f2$", which has the senses of the parts as its constituents. Now Frege explains that it is the unsaturated nature of the sense of a function expression that accounts for the unity of the sense expressed by the whole. In *Über Begriff und Gegenstand* Frege explains (Geach and Black 1980, p. 54):

> Not all parts of a thought can be complete; at least one must be 'unsaturated', or predicative; otherwise they would not hold together.

Thus, the sense of "$f(\delta)$" and the sense of "2" are "parts" of the composite sense (*Gedanke*) expressed by "$f(2)$". There is now a complex whole with parts. Does Russell's argument apply here?

It does not. Russell's notion of "occurring as entity" has no analog whatsoever in Frege's theory. Though the sense of "2" is part of the whole *Gedanke*, it does not occur in the whole "as entity." Rather, it occurs in a way analogous to Russell's notion of an "occurrence as meaning" (or "as concept").

This point becomes clear in an exchange between Russell and Frege over a paradox of propositions arising in Russell's theory. The paradox is rendered in Appendix B (p. 527ff) of *Principles* and can be stated as follows. Every class of propositions m can be correlated with a proposition '$p \in m \supset_p p$'. Now consider the class w of all propositions not members of their respectively correlated class—i.e., consider

$$\hat{y}(\exists m)(y = \{p \in m \supset_p p\} .\&. \sim (y \in m)).$$

There is next a proposition '$p \in w \supset_p p$.' This proposition is a member of w if and only if it is not a member of w.

Russell first communicated this paradox to Frege in a letter of 29 September 1902, and it occupied the discussion of a good many letters thereafter. In a follow-up letter of 12 December of that year, Russell says that the paradox should apply *mutatis mutandis* to Frege's notion of a *Gedanke* since senses are themselves objects and since one can form classes of them (McGuinness 1980, p. 151). Frege would have nothing of it. He rightly insisted that it is not formulable within his sense-reference distinction (*ibid.*, p. 153):

> By "$p \in m \supset_p p$" you therefore designate a thought—the thought, namely, that *all thoughts belonging to the class m are true*, if I understand your signs correctly. But now a class cannot be a component part of a thought, though the sense of a class name can.

The formulation of the class w is illicit. In the expression,

$$\text{“}(\exists m)(y = \{p \in m \supset_p p\} \;\&\; \sim (y \in m)),\text{”}$$

the quantifier "$(\exists m)$" is to refer to a class in the conjunct "$\sim (q \in m)$" and to the sense of the class name "m" in the clause "$q = \{p \in m \supset_p p\}$." For in this clause the expression "$p \in m \supset_p p$" refers to a thought, and a class itself is never a part of a thought.

Russell allowed towns, people, classes, and all manner of physical objects, as occurring "as entity" in propositions. Frege's notion of a *Gedanke* does not accommodate this. In a letter to Jourdain he wrote (McGuinness 1980, p. 79):

> Now that part of the thought which corresponds to the name "Etna" cannot be Mount Etna itself; it cannot be the reference of this name. For each individual piece of frozen solidified lava which is part of Mount Etna would then also be part of the thought that Etna is higher than Vesuvius. But it seems to me absurd that pieces of lava, even pieces of which I had no knowledge, should be parts of my thought.

Frege needs only one manner of occurrence in a thought—occurrence as meaning. Senses of singular terms are objects which, when occurring as parts of thoughts, are used to refer. Since entities which are not senses cannot be employed by minds to refer, they are never parts of thoughts.

There is simply no analog of Russell's "occurrence as entity" (i.e., "as term of a proposition") in Frege's work. Frege's notion of being "argument to a function" is in no way like Russell's notion of "occurring as entity." Accordingly, Frege inherits none of the horrid problems of distinguishing "entity" and "meaning" occurrences of denoting concepts when he allows a sense to be an argument to a function. The crux of the difference boils down to the fact that Frege's theory models predication in terms of the primitive idea of functionality, while Russell analyzed predication by means of the primitive ideas of occurring "as meaning" versus "as entity" in a proposition.

The Variable as Primitive

3.4 With "On Denoting" Russell finally joined Frege in abandoning the subject–predicate form of categoricals and offering a transcription of denoting phrases into a symbolic notion using variables. But Russell did not follow Frege's idea of introducing variables into calculus for logic by adopting functionality as a primitive. In 1905, Russell brought variables into his calculus with open sentential forms—bracketing altogether the philosophical problem of what are to be the constituents of general propositions.

To see this, consider one of Russell's examples. In "On Denoting" we find (*OD*, p. 104):

> *C*(everything) means '$C(x)$' is always true.

It is clear that "$C(\ldots)$"is used schematically for a declarative sentential phrase which would become a sentence when the dots are filled with a proper name. Now Russell did not simply put "$(x).C(x)$" as the transcription. But he could have. The symbolism "$(x).\phi x$" occurred as early 1903. In a letter to Frege of 24 May, Russell wrote (McGuinness 1980, p. 159):

> By '$(x).\phi(x)$' I designate the thought that $\phi(x)$ is true for all values of x.

Moreover, in "The Theory of Implication" (composed in 1905) we find him introducing the notation as follows (*TofI*, p. 194):

> *Primitive Idea*: "$(x).(C\emptyset x)$" means "the truth of $(C\emptyset x)$ for all values of x."

The locution "always true" is but a variant of "truth for all values." In "On Denoting," Russell just writes out the semantic truth conditions for the expression so that the reader can grasp his meaning. Likely Russell's intent was to avoid a symbolism which he thought would be unfamiliar to readers of *Mind*.

It is no less certain that Russell had by and large what would be the *Principia* symbolism for contextually defining descriptions. In "On Denoting" Russell transcribes,

> C(the father of Charles II),

with the following,

> It is not always false of x that x begat Charles II and that Cx and that "if y begat Charles II, y is identical with x" is always true of y.

It is by no means improper to replace this archaic phrasing by a quasi-symbolic formulation based on *Principia's* contextual treatment of definite descriptions. Again in kindness to his readers, Russell was just writing out what amounts to his semantic interpretation of what would be the symbolic transcription. This is important, and since it has been contested[10] it is worth further discussion.

Peano had offered the basis for Russell's symbolism as early as 1900. For Peano, the symbol "℩" flanks a term to form a term. He conceives of "℩" as standing for a function which is the inverse of the singleton function—a function which takes a term to the class with it as its sole member. Together with some possible alternatives indicated by "Dfp", Peano had set out[11] the following conditional definition (*Formulaire*, 1900, p. 351):

$$
\begin{array}{llll}
\multicolumn{3}{l}{a \in \mathrm{cls}\,.\&.\,\exists a\,.\&.\, x \in a \,\&\, y \in a \;.\supset_{x,y}.\; y = x\,.:\supset:.} & \\
.0 & \multicolumn{2}{l}{z = \text{℩}a \;.\equiv.\; a = \iota z} & \mathrm{Df}\text{℩} \\
.01 & \multicolumn{2}{l}{b \in \mathrm{cls}:\supset:} & \\
 & \text{℩}a \in b & .\equiv.\, a = \iota x \,.\supset.\, x \in b & \mathrm{Dfp} \\
 & & .\equiv.\, \exists x \ni (a = \iota x \,.\&.\, x \in b) & \mathrm{Dfp} \\
 & & .\equiv.\, \exists a \cap b & \mathrm{Dfp} \\
 & & .\equiv.\, a \subseteq b. & \mathrm{Dfp}
\end{array}
$$

10. See Lambert (1990).
11. Peano defines as follows:
$\exists a = \mathrm{df}\; a \neq \Lambda$
$\iota a = \mathrm{df}\; x \ni (x = a)$.

Peano's notation $\ulcorner x \ni Ax \urcorner$ is for the class of all x such that Ax. Now Peano's pupil Padoa had come to a proof of:

$$\exists x \ni (a = \iota x .\&. x \in b) .\equiv. \exists x \ni (x \in a \,\&\, y \in a .\supset_y. y = x :\&: x \in b).$$

So if we do some manipulating, replacing the class term "a" with Russell's class term "$\hat{z}\phi z$", changing Peano's $\exists a$ to Russell's $(\exists x)\phi x$, and replacing "b" with "$\hat{z}\psi z$" so that "$x \in b$" can become "ψx", we can arrive at:

$$(\exists x)\phi x .\&. (x, y)(\phi x \,\&\, \phi y .\supset. y = x) .\supset.$$
$$\theta\text{℩}(\hat{z}\phi z) .\equiv. (\exists x)(\phi x \,\&\, (y)(\phi y .\supset. y = x) .\&. \psi x)$$

This is close to the *Principia* symbolism.[12]

This is not Russell's contextual definition, of course. Russell's "℩$x\phi x$" is not anything like Peano's "℩$(x \ni \phi x)$." The fact that Peano has a conditional definition is not incidental; it is essential to his view that "℩" stands for a function. On Russell's new view, $\ulcorner \psi$(℩$x\phi x)\urcorner$ is false when nothing satsifies ϕ or when more than one object satisfies it. This is the *consequence* of his view that "℩" does not express a function (as it had with Peano and Frege).[13] In Peano's view, $\ulcorner$℩$(x \ni \phi x)\urcorner$ is a singular term whose use must be counted as undefined (or even meaningless) when nothing satisfies ϕ or more than one entity does. Russell regards "℩" as an inseparable part of expressions of the form $\ulcorner \psi$(℩$x\phi x)\urcorner$.

There is no question but that Russell had come to see this in "On Fundamentals." In avoiding the intractable problems of distinguishing meaning and entity occurrences of purported denoting concepts indicated by definite descriptions, he renders a class-theoretic formulation of his famous contextual definition (*OF*, p. 38):

> On this view, we shall not introduce "℩'u" at all but put
>
> ϕ('℩'u) $.=: (\exists y) :\ y \in u : z \in u .\supset_z. y = z : \phi$'$y$ Df.
>
> This defines all propositions about ℩u which is all we need. But now ϕ(℩u) is a bad symbol; we shall have to substitute (say)
>
> (ϕ℩)'u
>
> On this view, "the author of *Waverley*" has no significance by itself, but propositions in which it occurs have significance. Thus, in regard to denoting phrases of this sort, the question of meaning and denotation ceases to exist.

The contemplated alteration of the notation was to reflect the new view that "℩" does not express a function and that "℩u" is not a singular term. On the

12. Thanks to Rodriguez-Consuegra for drawing my attention to Peano. Rodriguez-Consuegra goes too far, however, claiming that all the "essential ingredients" of Russell's new treatment of definite descriptions are to be found in Peano. Tichý (1988) makes the same claim for Frege.

13. Frege used the symbol "$\backslash\xi$" for a primitive function on classes.

next page, Russell goes on to give a formation free from the notion of classes (*ibid.*):

> Instead of first introducing $\iota`u$, as on the previous page, we shall of course begin with $(\iota\hat{x})(\phi\hat{x})$, defining $(f\iota)`\phi$ as follows:
>
> $(f\iota)`\phi \,.{=}{:}\, (\exists y){:}\phi`y{:}\phi`z \,.\supset_z.\, y = z{:}\, f`y$ Df.
>
> I am not at all sure that it is desirable to alter the notation. . . . Only we must remember not to regard them as constituents of the propositions in which they occur.

Russell used the expression "$\iota`\phi$" for "the term satisfying $\phi$" (*M&D*, 84). He contemplates putting "$(f\iota)`\phi$" instead of "$f`(\iota`\phi)$" to capture the new status of "ι". In the end, "$f(\iota x\phi x)$" is retained as the best notation, and we see that Russell regards the symbol "$\iota x\phi x$" as an incomplete symbol—an inseparable part of "$f(\iota x\phi x)$".

Russell goes on, revealing his command of the new contextual definition by applying the theory of descriptions. There is a discussion of "the class defined by ϕ" at page 39, and later a nice application to the Epimenides (liar) paradox (*OF*, p. 63):

> Consider "the statement I am now making is false."
>
> $\phi`p \,.{=}.$ I now state p.
>
> Then,
>
> $f`$ (the statement I am now making) $.{=}{:}$
>
> $(\exists p){:}\phi`p{:}\phi`q \,.\supset_q.\, p = q : f`q$
>
> Hence,
>
> "the statement I am now making is false" $.{=}.$
>
> $(\exists p){:}\phi`p{:}\ \phi`q \,.\supset_q.\, p = q : \sim p.$
>
> There would be no difficulty in this but for the fact that I am now making this statement, so that we have
>
> $\phi`\{(\exists p){:}\ \phi`p{:}\ \phi`q \,.\supset_q.\, p = q{:}\ \sim p\}.$
>
> It is further assumed that we have
>
> $(\exists p){:}\ \phi`p{:}\ \phi`q \,.\supset_q.\, p = q.$

Russell had certainly mastered the basics of what would become the formal mechanism for *Principia*'s treatment of definite descriptions.

3.5 Now it will be objected that if "$(C\emptyset x)$" is used schematically in "'On Denoting," and if Russell really had a full grasp (as in *Principia*) of the theory of incomplete symbols, then he is deprived of an account of the constituents of general propositions.

"On Denoting" abandons *all* denoting concepts. What then are the constituents of propositions whose linguistic expression employs bound variables? In the manuscript "On Fundamentals," Russell took this to be a problem of some moment. As a first stab, he suggests that the denoting concepts 'any term' and 'every term' might be retained—thus staving off a full collapse of denoting. Russell writes (*OF*, 41):

> The most convenient view might seem to be to take *everything* and *anything* as primitive ideas; putting:
> $(x).\phi`x\ .{=}.\ \phi`(\text{everything})$
> $\phi`x\quad .{=}.\ \phi`(\text{anything})$

Moments later he is less sanguine (*ibid.*):

> But it seems that on this view *everything* and *anything* are denoting concepts involving all the difficulties considered in 35–39 [i.e., the difficulties of meaning versus entity occurrences], on account of which we adopted the theory of 40 [i.e., contextual definition for definite descriptions]. . . . These objections, to all appearances, are as fatal here as they were in regard to *the*. Thus it is better to find some other theory.

With denoting concepts abandoned, the account of the variable in *Principles* is forfeit. But what then is the replacement? Hope of answering this question has led interpreters to insist that Russell's intent in writing "$C(x)$' is always true" as opposed to "$(x).C(x)$" must have been to reveal something about the constituents of general propositions. Some go so far as to say that the constituents are to be 'truth' and propositional functions.[14]

This is quite mistaken. Russell says that the variable is fundamental. It is not to be ontologized via an assumption of propositional functions qua attributes. To be sure, Russell remarks that he uses "$C(x)$" for a propositional function (*OD*, p. 104, fn2). But we must not forget that he often jumbled together distinct notions under the heading of "propositional functionality." The fact is that "$C(x)$" is just used schematically for a wff containing a variable—a single letter. Indeed, it is quite clear that "On Denoting" could not have been committed to an ontology of propositional functions. If it was, its contextual definitions would presuppose a solution of the paradox of prediction! Two months after the appearance of "On Denoting" we find Russell engaged in using the 1905 apparatus to generate a new "substitutional theory" which explicitly disavows any ontology of propositional functions.[15]

The significance of Russell's adoption of the variable as primitive has been wholly missed. It is obliterated by the imposition of a regulative principle governing all interpretations—a principle according to which the ontological

14. See for instance Kilmister (1984), p. 123, Tichý (1988), and most recently Hylton (1990), p. 296.

15. The December 1905 manuscript was entitled "On Substitution" and it will be taken up in detail in the next chapter.

question as to the constituents of general propositions *must* be solved—so that no interpretation can be correct that renders Russell without some approach (albeit rough) in mind.

Grattan-Guinness's discussion of the substitutional theory affords an example of the imposition of the regulative principle. The original substitutional theory, he says, eliminates free variables entirely and employs a forerunner of the modern "substitutional" account of quantification. He writes (Grattan-Guinness, p. 75):

> The central idea of this [1906 substitutional] theory was the abandonment not only of classes and relations but also of propositional functions and free variables. This left only propositions and their truth-values, and individuals named by constants. . . .
>
> Quantification is a good example of the new use of old notations. '$(x).\phi x$' means that 'ϕx is true for all values of x'; but 'ϕx' cannot refer now to a propositional function in the normal sense but to a proposition whose truth-value depends on the (quite possibly unknown) individual x. Russell interpreted quantification in a way which is somewhat similar to the modern substitutional interpretation of quantification, in speaking of constants rather than the individuals which the constants name, and in giving prominence to truth-values.

Russell's own account of the origins of his substitutional theory confounds this interpretation. In a letter to Jourdain of 15 March 1906 he wrote as follows:

> About June 1904, I tried hard to construct a substitutional theory more or less like my present theory. But I failed for want of the theory of denoting: also I did not distinguish between *substitution* of a constant for a constant and *determination* of a variable at this or that constant. . . . Then last autumn, as a consequence of the new theory of denoting, I found at last that substitution would work, and all went swimmingly. (Quoted from Grattan-Guinness 1977b, p. 79.)

Russell emphasizes the importance of separating the "determination" of a variable from the "substitution" of entities.

Curiously, Grattan-Guinness reads this letter as claiming that the post-"On Denoting" substitutional theory had to be "cleansed of the elements of determination" (Grattan-Guinness 1974, p. 391). This inverts the situation. Prior to 1905 Russell attempted (again and again) to explain "determination of the variable" (and indeed quantification) in terms denoting and substitution. This was the plan of *Principles* and we can find vestiges of it in manuscripts of 1904. "Variation" and "determination" of a variable is to be analyzed in terms of the substitution of one entity for another in a proposition. All such approaches halted with the abandonment of denoting concepts. Russell's letter to Jourdain shows that it was but for "want" of the distinction between "determination" and "substitution" (and the new theory of definite descriptions) that the early attempts at substitutional theories had failed.

There is no doubt that the substitutional theory begun in 1905 employed free variables in an entirely objectual sense. For the first time, the notion of "determination" of a variable is separated from the notion of "substitution" and *adopted* as primitive and unanalyzable. Unlike its predecessors, the 1905 substitutional theory adopts both determination and substitution. Put in modern terms, Russell's new position is that "substitution" is to be retained as an ontological notion (symbolically expressible in the object language) involving the constituents of propositions. The notion of "determination," on the other hand, is part of the interpretation of the formal language. Variables are just special single letters, and a variable's being "determined" or "assigned a value" occurs when the language containing it is given a semantic interpretation in the domain of a model.[16]

Once we see this, Russell's archaic locution "is always true" should be put aside as a misleading attempt at an informal semantics. Though his words betray confusions of use and mention and a haziness about the propriety of schematic letters, Russell intends objectual quantifiers and free variables in a fully modern sense. To be sure, he wants "x" in "$x \supset y$" to be an individual variable which can be bound. He cannot, therefore, adopt the modern reading of the expression,

"$(x)(x$ is a man $\rightarrow x$ is mortal)"

as "Every thing is such that if it is a man then it is mortal". Russell's conditional "$\supset$" is a dyadic predicate, read as "implies"; the modern "$\rightarrow$" is a statement connective, read as "if . . . then . . .". Nonetheless, Russell can put,

"$(x)(x$'s being a man implies x's being mortal),"

reading this as "Everything is such that it's being a man implies its being mortal." Similarly, "$(x)(x \supset x)$" is read "Everything is such that it implies itself."

3.6 What then of the problem of constituents of general propositions? Indeed, "On Denoting" would seem to make the matter worse for it proclaims that we must be acquainted with the constituents of every proposition we understand. What are we acquainted with in the case of general propositions? Moore put the difficulty sharply in his letter of 23 October 1905:

> You say '*all* the constituents of propositions we apprehend are entities with which we have immediate acquaintance.' Have we, then, immediate acquaintance with the variable?

16. Of course, if one deprives Russell of the distinction between the object language of a formal calculus and its interpretation then one cannot see this point. All too common is the view that Russell's conception of logic cannot embrace the distinction. This view has been thoroughly criticized in Chapter 1.

If the variable is simply a letter awaiting "determination" or assignment (in the meta-language), if it is not ontologized, then what answer is available to Russell?

This is an important and difficult problem. Nonetheless, the variable (and its "determination") is primitive and unanalyzable. One must not undermine this significant advancement by imposing upon Russell a solution of the problem of the constituents of general propositions. The bald fact is that Russell had no solution at all! And we can rely upon Russell's intellectual honesty for proof. Consider his response to Moore's criticism (Letter to Moore, 25 October 1905):

> I only profess to reduce the problem of denoting to the problem of the variable. This latter is horribly difficult, and there seem to be equally strong objections to all the views I have been able to think of.

Russell admits to having no solution. This is not surprising. A solution seems to require denoting concepts and an analysis of the variable in terms of them. As it happens, September of 1906 would find Russell thinking the problem was moot. As we shall see in Part III, once cornered by the paradoxes of general propositions, Russell felt ready to abandon general propositions entirely.

The Road to Substitution

3.7 Recall that Russell endeavored to solve the paradoxes besetting logicism by analyzing variation. The analysis was to yield an account of mathematical functionality and sort out the jumble of notions called "propositional functionality." The manuscripts Russell wrote shortly after *Principles* and in 1904 are tortuous. But it seems clear that his hope of analyzing variation via substitution led directly to the so-called substitutional theory of classes and relations he initiated in 1905. It's worth looking at this in overview.

Two notions appear in the manuscripts. In "Points About Denoting" and "On the Meaning and Denotation of Phrases" we find:

$$p\frac{y}{x}.$$

In a manuscript of 1905 entitled "On Fundamentals,"Russell introduces a similar notation "$(C)(\frac{z}{\text{var}})$". This seems to be a variant of "$p\frac{b}{a}$". Both are to indicate the result of substituting one entity for another in a complex (proposition or denoting concept).

There are scant remarks concerning the construction of a theory of classes via denoting concepts. But it is not difficult to imagine Russell substituting denoting concepts for one another to render a theory of classes. To proxy,

$$\hat{z}(z \text{ is a man}) \in \hat{a}(a \subseteq \hat{z}(z \text{ is a mortal})).$$

Russell might have contemplated putting something like:

$$\{(z)(z\text{'s being an animal } .\supset.\ z\text{'s being mortal})\}\left(\frac{\text{'a man'}}{\text{'an animal'}}\right).$$

But to go very far he shall need to use "$(C)\left(\frac{z}{\text{var}}\right)$". For instance to proxy the wff,

$$a \in \cup \hat{\alpha}(\alpha \subseteq \hat{z}(z \text{ is a mortal}))$$

Russell will need:

$$\{(\exists C)((z)(\ (C)\left(\tfrac{z}{\text{var}}\right)\ .\supset.\ z\text{'s being mortal})\ .\&.\ (C)\left(\tfrac{x}{\text{var}}\right))\}\left\{\tfrac{a}{x}\right\}.$$

Though there is little about classes, "On Fundamentals" attempts some formulations of the paradox of predication (*OF*, pp. 4, 55). Early on, we find the following:

$$R = \text{df} \sim (\hat{C})\left(\frac{\hat{C}}{\text{var}}\right)$$

Whence, from $(R)\left(\frac{R}{\text{var}}\right)$ we arrive at:

$$\sim (R)\left(\frac{R}{\text{var}}\right).$$

Continuing the substitutions leads to $\sim(\sim (R)(\frac{R}{\text{var}}))$, and so by double negation we are back again to $(R)(\frac{R}{\text{var}})$. Russell uses "var." to pick some entity to be varied in C and to mark that C itself is to be varied in R. The circumflex in the definition is used to mark that C is to be varied in R. We can avoid "var." With the following reformulation. Put:

$$R = \text{df} \sim (C)\left(\frac{C}{z}\right).$$

Now suppose that $(R)(\frac{(R)(\frac{z}{C})}{C})$. Performing the substitution, we get,

$$\sim \left((R)\left(\frac{z}{C}\right)\right)\left(\frac{(R)(\frac{z}{C})}{z}\right)$$

This renders $\sim (R)(\frac{(R)(\frac{z}{C})}{C})$. But continuing the substitutions yields,

$$\sim \left(\sim \left((R)\left(\frac{z}{C}\right)\right)\left(\frac{(R)(\frac{z}{C})}{z}\right)\right),$$

and then $\sim(\sim (R)(\frac{(R)(\frac{z}{C})}{C})))$. By double negation we arrive at $(R)(\frac{(R)(\frac{z}{C})}{C})$. In Russell's view, this was to be a version of the paradox of predication generated by the substitution of denoting complexes.

The 1904 manuscripts "Points About Denoting," "On Meaning and Denotation" as well as the 1905 "On Fundamentals," show Russell agonizing over the meaningfulness of his notation. In "Points About Denoting" he offers the following remark (*P*, p. 8):

> "$p\frac{y}{x}$" means "that which is denoted by the meaning which results from giving x the value y in p."

Let p be 'the godfather of any term.' Consider then the denoting concept indicated by the phrase "$p\frac{y}{x}$", viz.,

> 'that which is denoted by the meaning which results from giving any term the value any term in the godfather of any term.'

This is fraught with difficulties. The denoting concept 'any term' has three occurrences here. Each of the three occur in such a way as to denote and each might denote a different entity. Evidently, Russell thought of 'any term' as itself what is to be substituted for some term. Correction gives,

> 'that which is denoted by the meaning which results from substituting any term for 'any term' in the godfather of any term.'

But problems remain. We are enjoined to substitute some term, say Frege, for 'any term' in the godfather of say Russell—namely, J. S. Mill!

Russell was soon to detect these flaws. In "On Meaning and Denotation" he observes that his former reading of "$p\frac{y}{x}$" requires correction (*M*, p. 49):

> If we take any single letter p, a particular value of p may well be the value of some complex containing x for a particular value of x, but this cannot be the complex with x still variable. And this shows that p of $p\frac{y}{x}$ won't do, since it forgets that a variable x is not in any of its values.

Let us then modify the above as follows:

> 'that which is denoted by the meaning which results from substituting any term for 'any term' in 'the godfather of any term.''

We are now enjoined to substitute, say Frege, for the denoting concept 'any term' in the denoting concept 'the godfather of any term.' This is a vast improvement, but is still problematic. The denoting concept 'any term' occurs "as concept" in the 'godfather of any term,' but Frege is (as Russell puts it) a "thing" and things are capable of only one kind of occurrence—occurrence as term. Accordingly, the substitution has violated the structure of 'the godfather of any term.' As we saw, to occur as concept is structurally distinct from occurring "as term of a proposition" and this must apply to occurrences in denoting complexes as well.

In "Points About Denoting" and more saliently in "On Fundamentals," Russell exacts an additional point against his former rendering of "$p\frac{y}{x}$". He puts:

> $p =$ 'the present Prime Minister of England'
> $q =$ 'the nephew of the late Prime Minister of England'
> $x =$ England
> $y =$ France.

He then observes that

$$p = q \text{ .\&. } p\frac{y}{x} \neq q\frac{y}{x}.$$

This, he says, shows that his former reading will not do: "If p stands for the complex meaning, $p\frac{y}{x}$ stands for a complex meaning, not for what this meaning denotes" (*P*, p. 10). That is, if the substitution of France for England is to take place in the denoting concept *p*, then the denoting concept '$p\frac{y}{x}$' should denote the denoting concept 'the present Prime Minister of France' and not the present Prime Minister of France. How then to capture mathematical functionality in terms of substitution and denoting?

Consideration of such troubles eventually led Russell to insuperable difficulties with using the substitution of denoting concepts to analyze the use of variables. Troubles with "meaning" versus "entity" occurrences of denoting concepts led him to abandon denoting concepts entirely. The successor was the new theory of incomplete symbols, and its ally was an entirely new perspective on the use of variables that adopted both "determination" and "substitution." In the 22 December 1905 manuscript "On Substitution" we find the theory of definite descriptions set out as follows (*S*, p. 7):

$$\psi(\iota x\phi x) \;.=:\; (\exists x)\phi\text{‘}y \;.\equiv_y.\; y = x \;:\phi\text{‘}x$$
$$\text{Ex!}(\iota x\phi x) \;.:\; (\exists x).\; \phi\text{‘}y \;.\equiv_y.\; y = x$$

This is very close to the now familiar symbolic expression of *Principia Mathematica.*

The new theory makes the "determination" of a variable primitive and thereby enables Russell's to capture mathematical functionality. The functional,

$$f(x) = y$$

is captured by means of the contextual definition of

$$(\iota z)(xfz) = y.$$

Substitution of entities nowhere plays a role here. But substitution is retained as a solution of the problem of "propositional functionality." With the variable's determination as primitive the many distinct problems Russell places under category "propositional functionality" begin finally to be sorted out.[17] Russell glimpses the difference between a schematic use of the letter "ϕ" and its use as a predicate variable. To represent the notion of "substitution," Russell adopts the primitive formula "$p\frac{b}{a}!q$". A convenient reading is:

"*q* is exactly like *p* except containing *b* at every occurrence of *a* in *p*."

He then introduces the expression "$p\frac{b}{a}$", which is reminiscent of those found in "On Fundamentals" and in "On Meaning and Denotation." But Russell can now put (*S*, p. 4):

$$p\frac{b}{a} = \text{df } (\iota q)\left(p\frac{b}{a}!q\right).$$
$$\phi\left(p\frac{b}{a}\right) = \text{df } (\exists q)\left((r)\left(p\frac{b}{a}!r \;.\equiv.\; r = q\right) \;\&\; \phi q\right).$$

17. Russell explains this in a footnote of his November 1905 article "Sur la relation des mathématiques à la logistique," (*RML*, p. 261).

This avoids all the horrid problems formerly encountered. The expression "$p\frac{b}{a}$" does not indicate a denoting concept; it is an incomplete symbol contextually defined by the theory of definite descriptions. The contextual definition uses "ϕ" schematically. No predicate variables occur in substitution. All the same, an expression employing predicate variables such as,

$$(\exists\phi)(\exists x) \sim \phi x$$

can be proxied if it first undergoes type regimentation. For instance, substitutional grammar allows:

$$(\exists p)(\exists a)(\exists x) \sim (\imath q)\left(p\frac{x}{a}!q\right).$$

This proxies the statement that there is some attribute (of the type of attributes entities) that some entity does not have. There will be proxies for higher types. But, as we shall see, "$\phi(\phi)$", is inexpressible in the substitutional grammar. In this way, Russell hoped to solve the paradoxes of predication (and of classes). There are no types of entities; and the calculus adopts only entity ("individual") variables. Nonetheless, a type regimented language of predicate variables is proxied in the logical grammar of substitution in a rich enough way to capture mathematics. The substitutional theory was an immediate application of Russell's theory of definite descriptions and his adoption of "determination" as primitive and indefinable. It realizes the fundamental doctrines of *Principles*, adhering (as Russell put it) "with drastic pedantry" to the old maxim that "whatever is, is one" (*SCR*, p. 189).

PART II

Types as Logical Grammar

CHAPTER FOUR

The Logic of Substitution

4.0 1905 was a watershed year for Russell. By its end he had developed an intensional calculus for the logic of propositions adequate to proxy a type regimented theory which employs predicate variables in subject as well as predicate positions. The theory affords a proxy for a theory of classes and relations-in-extension. For this reason, Russell often referred to it as his "no-classes" theory.[1] He named it "the substitutional theory" because it allowed definite descriptions of propositions by employing the primitive notion of the substitution of an entity for another in a given proposition. The calculus is wholly type free and employs only one style of completely unrestricted variables—individual variables. It therefore adheres to the fundamental doctrines of *Principles*, and Russell had no small hope that it would salvage logicism from the grip of the paradoxes.

The present chapter sets forth the basic calculus of the substitutional theory. For the deductive system we could take the calculus of the 1903 *Principles*. But as Russell well knew, this calculus is overly tedious in having to preface theorems with clauses of the form "$x \supset x$". In "The Theory of Implication," we saw that Russell sought a remedy by reinterpreting "$\supset$" so that "$x \supset y$" is not automatically false when "x" or "y" do not name propositions. Indeed, Russell now maintains that

$$(x)(x \supset x),$$

1. Of course, there are many "no-classes" theories that Russell developed and abandoned in his quest to solve the paradoxes. One such was briefly attempted in May of 1903 and here Russell allowed predicate variables. See Landini (1992).

holds. "The Theory of Implication" is an attractive model from a historical standpoint, and probably Russell actually had it in mind for the calculus of substitution. For substitutional principles we must turn to unpublished manuscripts. Two manuscripts are fairly well known. The first is Russell's 24 November 1905 article "On Some Difficulties in the Theory of Transfinite Numbers and Order Types" (TN). Here the substitutional theory was suggested as one among three avenues toward solution of the paradoxes. The paper was read before the London Mathematical Society on 14 December 1905, and it was published in their *Proceedings* for March 1906. The second is "On the Substitutional Theory of Classes and Relations" (SCR). The manuscript was received by the London Mathematical Society on 24 May 1906 and read before the Society in May, but withdrawn from publication. The paper has been regarded as Russell's "principal" work on substitution for it explains in general how classes are to be proxied. But there is a third manuscript that is far more important for uncovering the basic system of substitution. This is the manuscript entitled "On Substitution" (S) and dated 22 December 1905. Here we find an axiomatization of the substitutional theory. In what follows, we rely heavily upon it.

Russell's Original Principles of Substitution

4.1 In his 1905 manuscript "On Substitution," Russell begins with the notation "$p\frac{x}{a}!q$" (or for notational convenience "$p/a;x!q$"). This primitive formula is read "q results from p by substituting x for a in all those places (if any) where a occurs in p" (*SCR*, p.168).

We are accustomed to think of substitution as something *we* do—something, therefore, which is easily confused with the replacement of linguistic symbols in other linguistic symbols. But Russell's "substitution" concerns entities not symbols. Substitution is ontological, not linguistic. Perhaps the point is more salient if we read "$p/a;b!q$" as "b occurs [*as entity*] in q wherever a occurs [*as entity*] in p." The fundamental notion is really the notion of occurring [*as entity*] at a position in an entity. This is none other than the notion Russell set out in *Principles* of "occurring as term of a proposition" (or occurring as "logical subject"). The distinction between occurring as term ("as entity") and occurring predicatively ("as concept," or "as meaning") was not abandoned with the demise of denoting concepts. I take Russell's relation '*in*' to be the relation 'occurs as entity in.' The universal 'Redness' occurs as a constituent of 'c is red,' but it occurs predicatively, not as entity.[2]

With "$p/a;b!q$" as a primitive formula, Russell defines as follows (*S*, p. 6):

$$a \; out \; p \; =\text{df}\; (x)\left(p\tfrac{x}{a}!p\right)$$
$$a \; in \; p \; =\text{df}\; \sim (a \; out \; p)$$
$$p \text{ ex } q \; =\text{df}\; (x) \sim (x \; in \; p \;.\&.\; x \; in \; q)$$

2. See Chapter 3.

$$*12.1 \qquad x = y \;.{=}{:}\; (p, q, r, a) : p\tfrac{x}{a}!q \,.\, p\tfrac{y}{a}!r \;.{\supset}.\; q \supset r \quad \text{Df.}$$

Russell then adopts the following principles for single substitutions (*S*, p. 7),

$$
\begin{array}{lll}
*12.2 & \vdash: & (\exists q).(p\tfrac{x}{a}!q) \qquad \text{Pp.} \\
*12.201 & \vdash: & p\tfrac{x}{a}!q \,.\, p\tfrac{x}{a}!r \;.{\supset}.\; q = r \quad \text{Pp.}
\end{array}
$$

Together these two ensure that there is always a unique proposition which is the outcome of a given substitution. Russell continues with:

$$
\begin{array}{lll}
*12.21 & \vdash: & p\tfrac{x}{p}!x \quad \text{Pp.} \\
*12.211 & \vdash: & p\tfrac{x}{x}!p \quad \text{Pp.} \\
*12.212 & \vdash: & a \neq \sim p \,.\, p\tfrac{x}{a}!q \;.{\supset}.\; (\sim p)\tfrac{x}{a}!(\sim q) \quad \text{Pp.} \\
*12.22 & \vdash: & \sim (y).\, a \textit{ in } (\psi!y).\, \supset .\, (\psi!a)\tfrac{x}{a}!(\psi x) \quad \text{Pp.} \\
*12.23 & \vdash: & (\exists q).\, p \text{ ex } q \quad \text{Pp.} \\
*12.24 & \vdash: & a \textit{ in } p \,.\, p \textit{ in } q \;.{\supset}.\; a \textit{ in } q \quad \text{Pp.} \\
*12.241 & \vdash: & a \textit{ in } p \,.\, p \textit{ in } a \;.{\supset}.\; a = p \quad \text{Pp.} \\
*12.25 & \vdash: & \sim (x, y).\, a \textit{ in } \phi!(x, y) \;.{\supset}. \\
 & & \quad \{(y).\phi!(a, y)\}\tfrac{a'}{a}!\{(y).\phi(a', y)\} \quad \text{Pp.} \\
*12.26 & \vdash: & \sim (c, d).\, a \textit{ in } \psi!(c, d) \text{: } p\tfrac{x}{a}!p' : q\tfrac{x}{a}q' \;{:}{\supset}. \\
 & & \qquad \{\psi!(p, q)\tfrac{x}{a}!\{\psi!(p', q')\} \quad \text{Pp.}
\end{array}
$$

Now it is clear that a singular term cannot occur in itself; it is not obvious, however, whether an entity can occur as a constituent of itself. The principle *12.21, together with *12.1, *12.201, and the definition of "*in*," yield that every entity occurs *in* itself in a trivial sense. A more substantive matter, however, concerns whether an entity can occur *in* itself in a non-trivial sense. That is, can one have $x \textit{ in } y \;.\&.\; y \textit{ in } x \;.\&.\; x \neq y$? Russell decides not, as we see from his *12.241. Axiom schema *12.24 ensures the transitivity of the relation '*in*'. As we shall see, these axioms have consequences for the propositional *liar* paradox.[3]

Russell also needs to be able to determine the nature of the outcome of a given substitution. The first axiom (schema) for this is *12.22,

$$\vdash: \sim (y).\, a \textit{ in } (\psi!y).\, \supset .\, (\psi!a)\tfrac{x}{a}!(\psi x) \quad \text{Pp.}$$

(The shriek in "$\psi!y$", etc., should be disregarded as it is here a non-functional vestige of earlier theories.) A few words of caution are in order about Russell's use of "ψx". Russell demurred over its use in the substitutional theory,[4] sometimes worrying that the theory must avoid the notation entirely. As we saw in Chapter 3, this is unfortunate. Russell's concerns were based on a confusion generated by his sloppiness with respect to the distinction between meta-language and object language. It is a confusion that we must not fall into. The issue is related to the difference between developing logic by axiom schemata and the alternative of using axioms and a rule of uniform replacement in any thesis. In *Principia*, for instance, the letters "ϕ" and "ψ" are officially predicate

3. See ($T_{4.20}$) below and also Chapter 9.
4. See especially Russell's 1906 manuscript "The Paradox of the Liar."

variables, but they are also used as schematic letters for wffs.[5] This is improper from a formal perspective, though perhaps innocuous. But it is intolerable in the substitutional theory. The language of substitution has no predicate variables at all. Substitutional logic *has* to be developed from axiom schemata. In the substitutional theory, Russell's "ψx" and "ϕx" must be viewed as schematic for wffs. (We shall use "Ax".)

In Russell's schema *12.22, it is vitally important not to miss the significance of the antecedent clause. It would not do to adopt the following as telling us the outcome of a substitution for a particular prototype:

$\{A\alpha|\mu\}/\alpha;\beta!\{A\beta|\mu\}$,
where α and β are terms free for μ in A.

The trouble is that A may contain free variables. Consider, the following

"$\{a = w\}/a;x!\{x = w\}$",

where "a", and "w" are variables. It might be that the entity assigned to the variable "a" is the same as the entity assigned to the variable "w", while the entity assigned to "x" is not identical to that assigned to the variable "w". For example, suppose "a" and "w" are both assigned to Russell, while "x" is assigned to Alys. Then "$\{a = w\}$" names the proposition 'Russell = Russell' and the outcome, i.e., the proposition just like 'Russell = Russell' except for containing Alys wherever it contains Russell, should be 'Alys = Alys'. But the proposition named by "$\{x = w\}$" is 'Alys = Russell'. Thus the above schema is wrong.

The antecedent clause of Russell's *12.22 was designed to rectify the problem. The same principle is stated in quasi-prose in the 1906 paper, "On the Substitutional Theory of Classes and Relations" (*SCR*, p. 168):

> If ϕx is any propositional function, the result of replacing b for a in ϕa will usually be ϕb, but will not be ϕb if a occurs in ϕa otherwise than as each x occurs in its ϕx. For in that case, this other occurrence of a also will be replaced by b, whereas ϕb, as a rule, will still contain this occurrence of a.

If we are to have $\phi a \frac{b}{a} ! \phi b$, for all b, then for some x, a must be out of ϕx (*SCR*, p. 172). To illustrate how this solves the problem of free variables, let us return to our above example. An instance of Russell's schema is

"$(\exists y)(a \text{ out } \{y = w\}) \;.\supset.\; \{a = w\}/a;x!\{x = w\}$".

If the entity assigned to "a" equals that assigned "w", then it is simply false that $(\exists y)(a$ out $\{y = w\})$.

Seeing what Russell is about, the following schema properly captures the idea of *12.22:

$(\exists\mu)(\alpha \text{ out } \{A\mu\}) \;.\supset.\; (\nu)(\{A\alpha|\mu\}/\alpha;\nu!\{A\nu|\mu\})$,

5. It may be that he intended a rule of replacement, but there are serious problems here. See Chapter 10.

where μ, ν are individual variables, α is a term free for μ in $A\mu$, and ν is free for μ in $A\mu$.

Every instance of this schema makes an assertion about the result of substituting entities (not wffs or variables). The antecedent clause and the qualifications serve to block erroneous constructions. Consider

$$\text{“}(\exists x)(a \textit{ out } \{(\exists y)(y \supset x)\}) \;.\!\supset\!.\; (z)(\{(\exists y)(y \supset a)\}/a;z!\{(\exists y)(y \supset z)\}).\text{”}$$

Supposing the antecedent is true, we can detach to get

$$\text{“}(z)(\{(\exists y)(y \supset a)\}/a;z!\{(\exists y)(y \supset z)\})\text{”}.$$

But the following is not obtainable by *universal instantiation*:

$$\text{“}\{(\exists y)(y \supset a)\}/a;\{y \supset y\}!\{(\exists y)(y \;.\!\supset\!.\; y \supset y)\}\text{”}.$$

Here "$\{y \supset y\}$" is not free for "z" in "$\{(\exists y)(y \supset z)\}$". Finally, observe that the following is not an instance of our schema,

$$\text{“}(\exists x)(\{y \supset y\} \textit{ out } \{(\exists y)(y \supset x)\}) \;.\!\supset\!.\; (z)(\{(\exists y)(y \;.\!\supset\!.\; y \supset y\}/\{y \supset y\};z!\{(\exists y)(y \supset z)\})\text{”}.$$

Here the term "$\{y \supset y\}$" is not free for "x" in "$\{(\exists y)(y \supset x)\}$".

Recall that Russell included two other principles governing substitution:

$$*12.25 \quad \vdash: \sim (x, y).\; a \textit{ in } \phi!(x, y) \;.\!\supset\!.\; \{(y).\phi!(a, y)\}\tfrac{a'}{a}!\{(y).\phi(a', y)\} \quad \text{Pp.}$$

$$*12.26 \quad \vdash: \sim (c, d).\; a \textit{ in } \psi!(c, d):\; p\tfrac{x}{a}!p' : q\tfrac{x}{a}q' :\supset.\; \{\psi!(p, q)\tfrac{x}{a}!\{\psi!(p', q')\} \quad \text{Pp.}$$

As we shall see, *12.25 is not needed. Russell does need a special axiom for the three-place case of, say, $\{p \,.\!\supset\!.\, q \supset r\}$. But the following is an instance of *12.26:

$$(\exists x, y)(a \textit{ out } \{x \supset y\}) \;.\&.\; p\tfrac{z}{a}!p^\circ \;.\&.\; \{q \supset r\}\tfrac{z}{a}!s^\circ \;:\supset:\; \{p \,.\!\supset\!.\, q \supset r\}\tfrac{z}{a}!\{p^\circ \supset s^\circ\}.$$

Now applying 12.26 again,

$$(\exists y)(a \textit{ out } \{x \supset y\}) \;.\&.\; q\tfrac{z}{a}!q^\circ \;.\&.\; r\tfrac{z}{a}!r^\circ :\supset:\; \{q \supset r\}\tfrac{z}{a}!\{q^\circ \supset r^\circ\}.$$

It will follow (once the laws of identity are established) that

$$\{p \,.\!\supset\!.\, q \supset r\}\tfrac{z}{a}!\{p^\circ \,.\!\supset\!.\, q^\circ \supset r^\circ\} \;.\&.\; p\tfrac{z}{a}!p^\circ \;.\&.\; q\tfrac{z}{a}!q^\circ \;.\&.\; r\tfrac{z}{a}!r^\circ.$$

But Russell has forgotten that in fact there are other sorts of propositions. In particular, there is '*p*/*a*;*b*!*q*'. Accordingly, he will need an axiom to cover the situation for up to four places.

Moreover, Russell has omitted an important clause that needs to be added to the antecedent of *12.26. There is evidence that he was aware of this in his comments on principle *21.212,

$\vdash: a \neq \sim p \,.\, p\frac{x}{a}!q \;.\!\supset\!.\; (\sim p)\frac{x}{a}!(\sim q).$

He wrote (*S*, p. 7):

> The case of $a = \sim p$ must be excluded, because in this case a is a constituent of $\sim p$ but not of p, and $p\frac{x}{\sim p}!p$, $(\sim p)\frac{x}{\sim p}!x$.

Russell has overlooked the fact that *12.26 requires a similar exclusion. Without it, one can prove *12.212 without "$a = \sim p$" as part of its antecedent clause.

Russell continues in "On Substitution" with some further principles he thinks might be unnecessary (*S*, p. 8):

> I do not think it is often necessary to show that one entity is not a constituent of another, and in certain cases the above Pp's enable us to show this, e.g., they give $(\sim p)$ out p. If it is necessary to prove it often, the following Pp's will probably be sufficient. I have not numbered them, as I do not think we need them.
>
> $\vdash: x \textit{ in } (\sim p) \;.\!\supset\!:\; x \textit{ in } p \;.\!\vee\!.\; x = \sim p \quad Pp$
> $\vdash: x \textit{ in } (p \supset q) \;.\!\supset\!:\; x \textit{ in } p \;.\!\vee\!.\; x \textit{ in } q \;.\!\vee\!.\; x = (p \supset q) \quad \mathrm{Pp}$
> $\vdash: x \textit{ in } (p\frac{y}{a}!q) \;.\!\supset\!:$
> $x \textit{ in } p \;.\!\vee\!.\; x \textit{ in } q \;.\!\vee\!.\; x \textit{ in } a \;.\!\vee\!.\; x \textit{ in } y \;.\!\vee\!.\; x = (p\frac{y}{a}!q) \quad \mathrm{Pp}$
> $\vdash: x \textit{ in } \{(y).\phi!y\} \;.\!\equiv\!.\; (y).\; x \textit{ in } \phi!y \quad \mathrm{Pp}$

This completes the main principles of Russell's basic substitutional system. Russell goes on in the December 1905 notes titled "On Substitution" to add some fundamental principles for "dual" substitutions—and presumably he envisioned principles for "triple" substitutions, etc. Shortly thereafter, however, in his "On the Substitutional Theory of Relations" which was composed early in 1906, he came to realize that these are not needed. Dual substitutions, triple substitutions, and the like, can be defined from the principles for single substitutions. We shall see this in Chapter 5.

The Basic Logic of Propositions

4.2 The expressions "A", "B", "C", are to be meta-linguistic expressions for wffs. We shall tag axiom schemata and theorem schemata with an "$*$" label. Greek symbols "α", "β", "π", "σ" and "δ" are used as meta-linguistic variables for any terms, individual variables or otherwise. We use "μ", "ν", and "τ", and any of the above with one or more superscripts "°" or "$*$" as meta-linguistic expressions for individual variables. The *individual* variables for the object language of substitution are the lower-case letters of the English alphabet, and any such letter followed by one or more superscripts "°" or "$*$". The intended interpretation of the variables, of course, is to be all entities whatsoever. All entities have an individual nature, and are properly called "individuals." So we must not be misled into thinking our use of "p", "q", "r" are special "propositional variables." These are just individual variables too. The substitutional theory shall never need to restrict quantification to propositions. All its definitions, axioms, and theorems are formulated with respect to individual variables.

The primitive symbols of the language of substitution are the dyadic predicate expression "⊃" for 'implication,' the individual variables, the signs "°", "*", ";", "!", "/", and the brackets "(",")", "[","]" and braces "{" and "}". To characterize the terms and wffs of the system, we must begin with the following: The *basic* terms are just the individual variables. The *atomic* wffs are just $\alpha \supset \beta$ and $\pi/\alpha;\beta!\delta$, where $\alpha, \beta, \pi, \delta$ are basic terms. The wffs of the language of substitution are then just those of the smallest set K containing all atomic wffs and such that $\{A\} \supset \{B\}$, $\{A\} \supset \alpha$, $\alpha \supset \{A\}$, $(\mu)C\mu$, and

$$\begin{array}{llll}
\pi/\alpha;\beta!\{A\} & \pi/\alpha;\{B\}!\{A\} & \pi/\{D\};\{B\}!\{A\} & \{A\}/\{B\};\{D\}!\{E\} \\
\pi/\alpha;\{A\}!\delta & \pi/\{B\};\beta!\{A\} & \{D\}/\alpha;\{B\}!\{A\} & \\
\pi/\{A\};\beta!\delta & \{B\}/\alpha;\beta!\{A\} & \{D\}/\{B\};\beta!\{A\} & \\
\{A\}/\alpha;\beta!\delta & \{B\}/\{A\};\beta!\delta & \{D\}/\{B\};\{A\}!\delta & \\
\{B\}/\alpha;\{A\}!\delta & \pi/\{A\};\{B\}!\delta & &
\end{array}$$

are in K whenever A, B, C, D, and E are in K, μ is an individual variable free in C, and $\alpha, \beta, \pi, \delta$ are basic terms. The terms are as follows: (i) all individual variables are terms; (ii) if A is a wff, then $\{A\}$ is a term. There are no other terms.

The nature of the substitutional logic requires that we be able to nominalize any wff to make a term. Such terms shall name propositions, and we want the constituents of such propositions to be as intelligible as possible. Accordingly, in forming wffs $\ulcorner(\mu)A\urcorner$ we require that there be at least one free occurrence of μ in A. In this way we avoid vacuous quantifiers. (All wffs come with outer brackets, but for convenience we often omit them as above.) Where A is a wff and $\alpha_1, \ldots, \alpha_n$ are terms, $A\alpha_1, \ldots, \alpha_n$ wff containing the terms $\alpha_1, \ldots, \alpha_n$ free. If A is a wff and α is a term and ν is an individual variable, $A\alpha|\nu$ is the wff exactly like $A\nu$ except containing the term α at every free occurrence of the variable ν in $A\nu$. An occurrence of a variable μ is *bound* in a wff iff[6] either it immediately follows a quantifier in the wff, or it is within the scope of a quantifier formed by using the variable μ. Otherwise the occurrences is *free* in the wff. (Note that a variable is free in a wff if any occurrence of it is free in the wff; it is bound in a wff if any occurrence of it is bound in the wff.)

If A is a wff and α and β are terms, then $A(\beta|\alpha, \alpha)$ is the wff exactly like A except containing β for zero, one, or more free occurrences of α in A. The term α in which any free variables occur is free in a wff A just when no variable in α is bound in A. If A is a wff and α and β are terms, then $A\alpha|\beta$ is the wff exactly like $A\beta$ except containing α wherever the term β occurs free in $A\beta$. A term α is *free for a variable μ in a wff $A\mu$* if: (1) if α is a variable, then α occurs free in $A\alpha|\mu$ wherever μ occurs free in $A\mu$; and (2) if α is a term in which any variables

6. This is a convenient abbreviation for "if and only if."

occur, then no individual variable occurring free in α is bound in $A\alpha|\mu$. If α is a closed term, then α is free for μ in any wff $A\mu$. (A term is *closed* when it contains no free occurrences of any variable; a term that is not closed is *open*. Similarly a wff is closed when it contains no free occurrence of any variable; if it is not closed it is open.) A term α containing free variables is free for a term β in $A\beta$ when no individual variable occurring free in α is bound in $A\alpha|\beta$. Finally, a term α is said to occur free in a wff A just when α occurs in A and is either closed or if not closed is such that no variable free in α is bound in A.

Braces are used to transform wffs into singular terms—that is names of propositions. For example, "$\{x \supset x\}$" and "$\{(x)(x \supset x)\}$" are terms; they name propositions. It is useful however to adopt the following conventions. First, outer braces are frequently dropped and dots shall be used for punctuation. (Dots are used symmetrically, the greater number indicating the main connective.) Since the position of "x" in the wff "$x \supset y$" is a subject position on the model of "Rxy" of predicate logic, it is convenient to informally drop braces in such occurrences. The subject position is sufficient to indicate that the nominalizing transformation has taken place. Hence we write "$x .\supset. y \supset x$" instead of "$x \supset \{y \supset x\}$".

The propositional deductive system for our language is adopted from Russell's "The Theory of Implication."

$*(\mathrm{A}_1)$ $\alpha .\supset. \beta \supset \alpha$
$*(\mathrm{A}_2)$ $\alpha .\supset. \beta \supset \delta :\supset: \beta .\supset. \alpha \supset \delta$
$*(\mathrm{A}_3)$ $\alpha \supset \beta :\supset: \beta \supset \delta .\supset. \alpha \supset \delta$
$*(\mathrm{A}_4)$ $\alpha \supset \beta .\supset. \alpha :\supset: \alpha$
$*(\mathrm{A}_5)$ $(\mu)A\mu \supset A\alpha|\mu$,
where α is free for μ in A
$*(\mathrm{A}_6)$ $(\mu)(\alpha \supset A\mu) .\supset. \alpha \supset (\mu)A\mu$,
where μ does not occur free in α and μ is not α.

The inference rules are:

Modus ponens (mp) $A, \{A\} \supset \{B\} \vdash B$.
Universal generalization (ug) $A \vdash (\mu)A$,
where μ is an individual variable free in A.
Replacement rule for defined signs Definiendum may replace and be replaced by definiens wherever it occurs.

For definitions we have:

df(=)
$\alpha = \beta =$ df $(p, a, q, r)(p/a; \alpha!q \ \& \ p/a; \beta!r \ .\supset. \ q \supset r)$
df($\neq$) $\alpha \neq \beta =$ df $\sim (\alpha = \beta)$
df($\exists$) $(\exists\mu)A =$ df $\sim (\mu) \sim A$
df(A) $(\mu_1, \ldots, \mu_n)A =$ df $(\mu_1), \ldots, (\mu_n)A$
df(E) $(\exists\mu_1, \ldots, \mu_n)A =$ df $(\exists\mu_1), \ldots, (\exists\mu_n)A$
df($\sim$) $\sim \alpha =$ df $(\mu)(\alpha \supset \mu)$

df(∨) $\alpha \vee \beta = \text{df} \sim \alpha \supset \beta$
df(&) $\alpha \,\&\, \beta \;= \text{df} \sim (\alpha \supset \sim \beta)$
dt(≡) $\alpha \equiv \beta \;= \text{df}\ \alpha \supset \beta \,.\&.\, \beta \supset \alpha.$

This completes the axiomatic basis for the basic logic of propositions.

The special rule for definitions is wanted because the logic of propositions is an intensional logic. Logical equivalents will, of course, be replaceable in any truth-functional context, but not in intensional contexts. For instance, we want to allow

$$\vdash \{p \vee q\} = \{\sim p \supset q\},$$

on the basis of our df(∨). Our rule achieves this, and we shall annotate by merely citing the appropriate definition.

In proofs we shall employ derived rules. It is worth emphasizing that unlike standard formulations of propositional logic, our letters "p", "q", "r", etc., are not schematic letters for wffs. Our letters are simply variants for individual variables "x", "y", "z", etc. It is important, therefore, to observe that (say) *simplification* is not to be: From $\alpha \,\&\, \beta$, infer α. The meta-linguistic "α" stands for any term. Since "p" is a term (it is an individual variable) it cannot be asserted or stand isolated on a line of a proof. Our *modus ponens* is designed to avoid this problem, and the derived rules must do the same. We have (for instance):

(simp) $\{A\} \,\&\, \alpha \vdash A$
(add) $A \vdash \{A\} \vee \alpha$
(mt) $\alpha \supset \beta, \sim \beta \vdash \sim \alpha$
(ds) $\alpha \vee \{B\}, \sim \alpha \vdash B$
(conj) $A, B \vdash \{A\} \,\&\, \{B\}$
(syll) $\alpha \supset \beta,\ \beta \supset \delta \vdash \alpha \supset \delta$
(ui) $(\mu)A\mu \ \vdash A\alpha|\mu$,
where α is free for μ in $A\mu$.
(eg) $A\alpha|\mu \vdash (\exists\mu)A\mu$,
where α is free for μ in A.

The following Deduction Theorem is provable by induction in the meta-language:

Deduction Theorem
If $A_1, \ldots, A_{n-1}, A_n \vdash B$ and no variable occurring free in A_n is bound by application of (ug) in any step of the demonstration, then $A_1, \ldots, A_{n-1} \vdash \{A_n\} \supset \{B\}$, and in this demonstration exactly those variables are bound by application of (ug) that were bound by application of (ug) in the original demonstration.

This validates an object-language proof technique of conditional proof:

Conditional Proof (p.)
$i. \rightarrow$ A
$n.$ ____ B
$n + 1.\ i \supset n$ cp.

No occurrence of (ug) on a variable free in A may occur within the scope of the assumption introduced at line i. If, in a conditional proof, we arrive at a contradiction, we shall simply write "F", close with a conditional, and deny the antecedent.

$i.\ \rightarrow\quad A$
$n.\ \underline{\quad\quad}\ F$
$n+1.\ i \supset F$ cp.
$n+2.\ \sim A\ n+1$, reductio.

We shall follow Copi[7] in subsuming existential instantiation under conditional proof:

$n.\ (\exists\nu)A\nu|\mu$
$m.\ \rightarrow\ A\mu$

$o.\ \underline{\quad\quad}\ B$
$o+1.\ A\mu \supset B\ m \sim o$, cp.
$o+2.\ (\mu)(A\mu \supset B)\ o+1$, ug.
$o+3.\ (\exists\mu)A\mu \supset B\ o+2$, theorem.
$o+4.\ B\ n, o+3$, mp.

Here we shall want the variable μ not to be free in B. The process is, however, notationally cumbrous. We therefore abbreviate with:

$n.\ (\exists\nu)A\nu$
$m.\ A\mu\ n$, EI.

$o.\ B$ (*m*ei).

The reader can then easily write out the full proof. In the course of proofs in subsequent sections, we will simply annotate with "log." when the derivation is obvious.

Substitutional Principles

4.3 Let us now set out Russell's system. We begin with some axiom schemata governing constituency (as entity). Curiously, in the 1905 "On the Substitutional Theory of Classes and Relations," Russell renamed "ex" and "ind" given in "On Substitution." The expression "*out*" is replaced by "ex", and the old "ex" is replaced by "ind". The old definition of "ind" is dropped entirely. The old locutions are preferable. Define as follows:

df(*out*) α *out* $\beta =$ df $(\mu)(\beta/\alpha;\mu!\beta)$
df(*in*) α *in* $\beta =$ df $\sim(\alpha$ out $\beta)$

7. Copi (1979).

$\text{df}(ex)\quad \alpha\ ex\ \beta = \text{df}\quad \sim(\exists\mu)(\mu\ in\ \alpha\ .\&.\ \mu\ in\ \beta)$
$\text{df}(ind)\quad \alpha\ ind\ \beta = \text{df}\quad \alpha\ out\ \beta\ .\&.\ \beta\ out\ \alpha.$

The following axiom schemata is obvious,

$*(\text{A}_7)\quad \alpha\ in\ \{A\alpha|\nu\},$

where α is free for ν in A. Note that the axioms which are instances of this schema are about entities occurring in propositions. Schema $*(\text{A}_7)$ ensures that the entity a is *in* the proposition $\{a \supset a\}$. Next we have:

$*(\text{A}_8)\quad \alpha\ in\ \{A\beta_1, \ldots, \beta_n\} :\supset:$
$\alpha = \{A\beta_1, \ldots, \beta_n\}\ .\vee.\ \alpha\ in\ \beta_1\ .\vee, \ldots, \vee.\ \alpha\ in\ \beta_n,$
where A is any wff all of whose distinct free terms are $\beta_1, \ldots, \beta_n$.

Next we have the antisymmetry and transitivity of the *in* relation:

$(\text{A}_9)\quad (x, y)(x\ in\ y\ .\&.\ y\ in\ x :\supset: x = y)$
$(\text{A}_{10})\quad (x, y, z)(x \text{ in } y\ .\&.\ y \text{ in } z :\supset: x \text{ in } z).$

Now the definition of "*in*" also ensures that if an entity a is *in* p then some entity substituted for a alters p. But surely, we know more. If a is in p and the outcome is the same in substituting x for a and in substituting y for a, then x equals y. Thus we add:

$(\text{A}_{11})\quad (p, a)(q)(x, y)(p/a;x!q\ .\&.\ p/a;y!q\ .\&.\ a\ in\ p :\supset: x = y).$

It follows that if a is in p, then every entity other than a will alter p when substituted for a in p. We also have:

$(\text{A}_{12})\quad (p, a)(z)(q)(p/a;z!q\ .\&.\ a\ in\ p\ .\&.\ a \neq p :\supset: z \neq q\ .\&.\ z\ in\ q).$

If a is *in* p in an interesting way (i.e., not equal to p), then z is *in* q when q results from substituting z for a in p.

The next axioms ensure that there will always be an outcome of any substitution:

$(\text{A}_{13})\quad (x, y)(x/x;y!y).$
$(\text{A}_{14})\quad (x, y)(x/y;y!x)$
$(\text{A}_{15})\quad (p, a)(x)(\exists q)(p/a;x!q\ .\&.\ (r)(p/a;x!r\ .\supset.\ r = q)).$

Axiom (A_{15}) fuses Russell's $*12.2$ and $*12.201$. For convenience we shall call it by the name "(p)". Also there is:

$(\text{A}_{16})\quad (p)(\exists q)(q \text{ ex } p).$

For every entity there is some entity excluded from it. We call this axiom "(q)" for convenience.

Axiom (q) is of central importance because of the support it provides for the following axiom schemata. The first reconstructs Russell's $*12.22$. We shall call it "$*(r)$" for convenience:

$*(A_{17})$

$$(\exists\mu)(\alpha \textit{ out } \{A\mu|\nu\}) \;.\!\supset\!.\; (\mu)(\{A\alpha|\nu\}/\alpha;\mu!\{A\mu|\nu),$$

where α and μ are free for ν in A.

It will not be necessary to adopt counterparts for both of Russell's $*12.25$, $*12.26$. Rather we simply adopt the following axiom schema which covers all the needed cases. We call it "$*(s)$" for convenience.

$*(A_{18})$

$$(\exists\mu_1, \ldots, \mu_n)(\alpha \textit{ out } \{A\mu|\nu_1, \ldots, \mu|\nu_n\}) \;:\&:$$
$$\alpha \neq \{A\sigma_1|\nu_1, \ldots, \sigma_n|\nu_n\} \;.\&.\; \alpha \neq \beta_1 \;.\&., \ldots, .\&.\; \alpha \neq \beta_m \;.:\supset:.$$
$$(\mu)(\exists\tau_1, \ldots, \tau_n)(\{A\sigma_1|\nu_1, \ldots, \sigma_n|\nu_n\}/\alpha;\mu!\{A\tau_1|\nu_1, \ldots, \tau_n|\nu_n\}$$
$$:\&: \sigma_1/\alpha;\mu!\tau_1 \;.\&., \ldots, \;.\&.\; \sigma_n/\alpha;\mu!\tau_n),$$

where each μ_i, σ_i, τ_i, $1 \geq i \geq n$, $n \in \mathbb{N}\text{–}\{0\}$, is free for its respective ν_i in the wff $A\nu_1, \ldots, \nu_n$, and $\beta_1, \ldots, \beta_m$, $m \in \mathbb{N}$, are all the distinct singular terms occurring free in the wff $A\sigma_1|\nu_1, \ldots, \sigma_n|\nu_n$ excepting $\sigma_1, \ldots, \sigma_n$.

The qualification on the singular terms occurring in the wff is needed to rule out misapplications such as:

$$(\exists x, y, z)(\{b \supset c\} \textit{ out } \{x \supset y \;.\!\supset\!.\; z\}) \;.\!\supset\!.$$
$$(z)(\exists b^\circ, c^\circ)(\{b \supset c \;.\!\supset\!.\; c\}/\{b \supset c\};z!\{b^\circ \supset c^\circ \;.\!\supset\!.\; c^\circ\} \;\&$$
$$b/\{b \supset c\};z!b^\circ \;\&\; c/\{b \supset c\};z!c^\circ).$$

Here we can see that the outcome of the substitution is not what these misapplications say it should be. Observe that schema $(s*)$ does not yield $*(r)$. To derive $*(r)$ we would need the laws of identity,[8] but in the present system we shall see that these laws are proved using $*(r)$.

We are almost finished with the presentation of axioms and axiom schemata. Only the matter of general propositions remains. We have:

$*(A_{19})$ $\{(\mu)A\mu\} = \{(\nu)A\nu|\mu\}$, where ν is free for μ in A

$*(A_{20})$ $\{(\mu)A\mu\} = \{(\mu)B\mu\} \;.\!\supset\!.\; (\mu)(\{A\mu\} = \{B\mu\})$

$*(A_{21})$ $\{(\mu)A\mu\} \neq \{\alpha \supset \beta\}$

$*(A_{22})$ $\alpha \textit{ in } \{(\mu)A\mu\} .\&.\, \alpha \neq \{(\mu)A\mu\} :\supset: (\mu)(\alpha \textit{ in } \{A\mu\})$, where μ is not α and is not free in α

$*(A_{23})$ $\supset \neq \{\alpha \supset \beta\} .\&.\, \supset \neq \{(\mu)A\mu\}$

Unfortunately, identity conditions for general propositions are made difficult by the fact that, with denoting concepts abandoned, it is wholly unclear what are to be the constituents of general propositions. One thing is clear. Whatever ontological counterparts quantifier phrases have in general propositions, such "constituents" do not occur "as entity" in them. Accordingly, the above schemata seem natural.

8. Thanks to Russell Wahl for a helpful discussion on this matter.

Identity

4.4 With the matter of the principles of substitution settled, we turn to identity. We plan to show that Russell's definition of the identity symbol is adequate to identity. We begin with some useful theorems.

From (A_{13}), df(=), df(*out*) and df(*in*), it follows that every entity occurs in itself.

$(T_{4.0}) \quad \vdash (x)(x \; in \; x).$

An immediate consequence of (A_9) and df(*in*) is:

$(T_{4.1}) \quad \vdash (x, y)(x \neq y \; :\supset: \; x \; in \; y \; .\supset. \; y \text{ out } x).$

From (A_{12}) we get two important theorems:

$(T_{4.2})a \quad \vdash (p, a)(z)(q)(p/a;z!q \; :\supset: \; a \; in \; p \; .\supset. \; z \; in \; q).$

If *a* is *in p* and we substitute *z* for it, then surely *z* is *in* the outcome. Note, however, that it is false that if *z in q* then *a in p*.

$(T_{4.2})b \quad \vdash (p, a, b)(p/a;b!b \; :\supset: \; a \text{ out } p \; .\vee. \; a = p).$

If *a* is *in p* in a significant way (i.e., it is *in* and not equal to), then *b* should also be *in* the outcome in a significant way. The following theorems are immediately forthcoming from (A_{15}).

$(T_{4.3}) \quad \vdash (p, a)(x)(q, r)(p/a;x!q \; \& \; p/a;x!r \; .\supset. \; q = r).$
$(T_{4.4}) \quad \vdash (p, a)(z, q)(p/a;z!q \; :\supset: \; a \text{ out } p \; .\supset. \; p = q).$

These theorems shall be useful in proving identities. The next theorems are immediate from the definitions df(*in*), df(*ex*), and df(ind).

$(T_{4.5})a \quad \vdash (p, q)(q \text{ ex } p \; .\supset. \; p \text{ ex } q)$
$(T_{4.5})b \quad \vdash (p, q)(q \text{ ex } p \; :\supset: \; q \text{ ind } p \; .\&. \; q \neq p)$
$(T_{4.5})c \quad \vdash (p, q)(q \text{ ex } p \; .\supset. \; (z)(z \; in \; p \; .\supset. \; q \text{ ex } z)).$

Perhaps most importantly, $*(r)$ yields the following theorem schema.

$*(T_{4.6}) \quad \vdash \; (\exists\mu^\circ, \nu^\circ)(\nu^\circ \; in \; \mu^\circ \; .\&. \; (\mu)(\mu^\circ/\nu^\circ;\mu!\{A\})),$
where μ° and ν° do not occur free in *A* and μ is free in *A*.

Proof:
1. $\sigma \text{ ex } \{A\nu\}$ $\quad (q)$, EI.
2. $\sigma \text{ out } \{A\nu\}$ $\quad$ 1, $(T_{4.5})$b, df(ind).
3. $(\exists\nu)(\sigma \text{ out } \{A\nu\})$ $\quad$ 2, eg.
4. $(\exists\nu)(\sigma \text{ out } \{A\nu\}) \; .\supset. \; (\mu)(\{A\sigma|\nu\}/\sigma;\mu!\{A\mu|\nu\})$ $\quad *(r)$.
5. $(\mu)(\{A\sigma|\nu\}/\sigma;\mu!\{A\mu|\nu\})$ $\quad$ 3, 4, mp.
6. $\sigma \; in \; \{A\sigma|\nu\}$ $\quad *(A_7)$.
7. $\sigma \; in \; \{A\sigma|\nu\} \; .\&. \; (\mu)(\{A\sigma|\mu\}/\sigma;\mu!\{A\mu|\nu\})$ $\quad$ 5, 6, conj.
8. $(\exists\mu^\circ, \nu^\circ)(\nu^\circ \; in \; \mu^\circ \; .\&. \; (\mu)(\mu^\circ/\nu^\circ;\mu!\{A\mu|\nu\}))$ $\quad$ 7, eg, (1ei).

Hereafter, we shall call this "$*(CP_{sub})^1$" for convenience.

Theorem schema $*(\mathrm{CP}_{\mathrm{sub}})^1$ is central. From it we can prove that identity is an equivalence relation (reflexive, symmetric, and transitive).

$(\mathrm{T}_{4.7})$ $\vdash (x, y)(x = y\ .\supset.\ x \equiv y)$.

Proof:

1. $\rightarrow\ x = y$.
2. $(p, a, q, r)(p/a;x!q\ \&\ p/a;y!r\ .\supset.\ q \supset r)$ 1, df(=).
3. $(\exists p, a)(a$ in $p\ .\&.\ (z)(p/a;z!\{x \equiv z\})$ $*(\mathrm{CP}_{\mathrm{sub}})^1$.
4. $(z)(p/a;z!\{x \equiv z\})$ 3, EI, simp.
5. $p/a;x!\{x \equiv x\}$ 4, ui.
6. $p/a;y!\{x \equiv y\}$ 4, ui.
7. $p/a;x!\{x \equiv x\}\ \&\ p/a;y!\{x \equiv y\}\ :\supset:\ x \equiv x\ .\supset.\ x \equiv y$ 2, ui.
8. 5 & 6 conj.
9. $x \equiv x\ .\supset.\ x \equiv y$ 7, 8, mp.
10. $x \equiv x$ taut.
11. __ $x \equiv y$ 9, 10, mp, (4ei).
12. $1 \supset 11$ cp.
13. $(x, y)12$ ug.

From $(\mathrm{T}_{4.7})$ we prove the reflexivity and symmetry of identity.

$(\mathrm{T}_{4.8})$a $\vdash (z)(z = z)$.

Proof:

1. $\rightarrow\ p/a;z!q\ \&\ p/a;z!r$
2. $q = r$ 1, $(\mathrm{T}_{4.3})$, ui.
3. $q \equiv r$ 2, $(\mathrm{T}_{4.7})$, ui, mp.
4. ___ $q \supset r$ 3, df(≡), log.
5. $1 \supset 4$ cp.
6. $(p, a, q, r)5$ ug.
7. $z = z$ 6,df(=).
8. $(z)(z = z)$ 7, ug.

$(\mathrm{T}_{4.8})$b $\vdash (x, y)(x = y\ .\supset.\ y = x)$.

Proof:

1. $\rightarrow\ x = y$
2. $(p, a, q, r)(p/a;x!q\ \&\ p/a;y!r\ .\supset.\ q \supset r)$ 1, df(=).
3. $(\exists p, a)(a\ in\ p\ .\&.\ (z)(p/a;\{z = x\})$ $*(\mathrm{CP}_{\mathrm{sub}})^1$.
4. $(z)(p/a;z!\{z = x\})$ 3, EI, simp.
5. $p/a;x!\{x = x\}$ 4, ui.
6. $p/a;y!\{y = x\}$ 4, ui.
7. $p/a;x!\{x = x\}\ \&\ p/a;y!\{y = x\}\ :\supset:\ x = x\ .\supset.\ y = x$ 2, ui.
8. 5 & 6 conj.
9. $x = x\ .\supset.\ y = x$ 7, 8, mp.
10. $x = x$ $(\mathrm{T}_{4.8})$a, ui.
11. __ $y = x$ 9, 10, mp. (4ei)
12. $1 \supset 11$ cp.

13. $(x, y)12$ ug.

By a similar procedure we get transitivity,

$$(\mathrm{T}_{4.8})\mathrm{c} \quad \vdash (x, y, z)(x = y .\&. \ y = z :\supset: x = z).$$

The proof is left to the reader.

Next we prove the following theorem schema governing the replacement of identicals:

$$*(\mathrm{T}_{4.8})\mathrm{d} \quad \vdash \alpha = \beta .\supset. \ A(\alpha, \alpha) \supset A(\beta|\alpha, \alpha),$$

where β is free for free occurrences of α in A.

Proof:

1. $\rightarrow \alpha = \beta$
2. $\rightarrow A(\alpha, \alpha)$
3. $(\exists \tau, \nu)(\nu \ in \ \tau .\&. \ (\mu)(\tau/\nu;\mu!\{A(\mu|\alpha, \alpha)\})$ $*(\mathrm{CP_{sub}})^1$.
4. $\nu \ in \ \tau .\&. \ (\mu)(\tau/\nu;\mu!\{A(\mu|\alpha, \alpha)\})$ 3, EI.
5. $(\mu)(\tau/\nu;\mu!\{A(\mu|\alpha, \alpha)\})$ 4, simp.
6. $\tau/\nu;\alpha!\{A(\alpha|\alpha, \alpha)\}$ 5, ui.
7. $\tau/\nu;\beta!\{A(\beta|\alpha, \alpha)\}$ 5, ui.
8. $\tau/\nu;\alpha!\{A(\alpha|\alpha, \alpha)\} \ \& \ \tau/\nu;\beta!\{A(\beta|\alpha, \alpha)\} .\supset.$
   $A(\alpha, \alpha) \supset A(\beta|\alpha, \alpha)$ 1, df(=), ui.
9. $A(\alpha, \alpha) \supset A(\beta|\alpha, \alpha)$ 6, 7, 8, conj, mp.
10. $A(\beta|\alpha, \alpha)$ 2, 9, mp. (4ei).
11. $2 \supset 10$ cp.
12. $1 \supset 11$ cp.

Thus if $\alpha = \beta$, we can replace some (and not necessarily all) occurrences of α in A with β.

It is worth noting that in later manuscripts Russell had a new definition of the identity symbol, a definition he apparently thought more elegant in that it does not involve quantifiers. He put (*SCR*, p. 169):

$$x = y = \mathrm{df} \ x\frac{y}{x}!x.$$

(Here "x" and "y" are used schematically for any terms.) It follows most immediately from (A_{13}) that $(x)(x = x)$; and it is easy to prove $(x)(y)(x = y .\supset. \ y = x)$. But $*(\mathrm{T}_{4.8})$d is not forthcoming and would have to be assumed as an axiom schema.[9] The following fairs no better:

$$\alpha = \beta = \mathrm{df} \ (\mu)(\mu/a;\beta!\mu).$$

9. If we put:

$$\alpha = \beta = \mathrm{df} \ \alpha/\alpha;\beta!\alpha$$
$$\alpha = \beta .\supset. \ A[\alpha, \alpha] \supset A[\beta|\alpha, \alpha]$$

then $*(\mathrm{A}_{17})$ would be provable from $*(\mathrm{A}_{18})$.

We cannot validate $*(T_{4.8})$d.

In any event, we see that Russell's original definition of the identity symbol is adequate. The following are all derived rules:

$\alpha = \beta,\ A\alpha \vdash A\beta,$
where $A\beta$ results from replacing one or more free occurrences of α by β and β is free for α in A.
$\sim\{A\alpha\},\ A\beta \vdash \sim\{\alpha = \beta\}$
$\alpha = \beta \vdash \beta = \alpha$

We shall indiscriminately name these with "$*(=)$", and use them together rather than proceeding through $*(CP_{sub})^1$ and df(=).

Proofs of Propositional Identities

4.5 Identical propositions have identical constituents occurring (as entity) in the same positions in the propositions. With the notion of substitution as fundamental, Russell hopes to generate proofs of the identity of propositions. This is no easy matter, however, even in the case of propositions which are not general. Church was quite unaware of Russell's substitutional theory when he set forth an intensional logic of propositions. He adopts analogs of the following as axiom schemata for the identity of propositions (Church, 1984, p. 514):

(i) $(p, q)(\{\sim p\} = \{\sim q\}\ .\supset.\ p = q)$
(ii) $(p, r, s)(\{\sim p\} \neq \{r \supset s\})$
(iii) $(p, a, r, c)(\{p \supset a\} = \{r \supset c\}\ :\supset:\ p = r\ .\&.\ a = c)$.

Russell's original substitutional theory affords proofs of these as we shall see. Moreover, it affords a means of proving:

(v) $(p, a, b, q)(\{p/a;b!q\} = \{r/c;b!q\}\ :\supset:\ p = r\ .\&.\ a = c)$.

Both (iii) and (v) shall be important for the propositional paradoxes discussed in Part III.

The logic of substitution is an intensional logic of propositions and as such material equivalence is not sufficient for identity. To illustrate, we prove the following theorem:

$(T_{4.9})\quad \vdash (\exists p, q)(p \equiv q\ .\&.\ p \neq q)$.

Proof:
1. $\rightarrow \{\sim\sim p\} = p$
2. $\{\sim p\}\ in\ \{\sim\sim p\}$ $*(A_7)$.
3. $\{\sim p\}\ in\ p$ 1, 2, $*(=)$.
4. $p\ in\ \{\sim p\}$ $*(A_7)$.
5. $\{\sim p\} = p$ 3, 4, (A_9).
6. $\sim p \equiv p$ 5, $(T_{4.7})$
7. $\underline{\quad}\ F$ 6, log.
8. $1 \supset F$ cp.

9. $\{\sim\sim p\} \neq p$ 8, reductio.
10. $\sim\sim p \equiv p$.&. $\{\sim\sim p\} \neq p$ 9, log.
11. $(\exists p, q)(p \equiv q$.&. $p \neq q)$ 10, eg.

This is surely expected. There are a host of entities which have no constituents occurring *in* them besides (trivially) themselves. Let us abbreviate putting,

df@ $@ = \mathrm{df}\ (x)(x \supset x)$.
df(f) $f = \mathrm{df}(x)(x \supset x\ .\!\supset\!.\ x)$.

From, ($T_{4.7}$), we have:

($T_{4.10}$)a $\vdash @ \neq f$.

Moreover, from $*(A_{22})$ we have:

($T_{4.10}$)b $\vdash (z)(z \neq @\ .\!\supset\!.\ z\ \mathit{out}\ @)$.

Proof:
1. $\rightarrow z \neq @$
2. $\quad \rightarrow z\ \mathit{in}\ \{(x)(x \supset x)\}$
3. $\quad\quad (x)(z\ \mathit{in}\ \{x \supset x\})$ 1, 2, $*(A_{22})$, mp.
4. $\quad\quad (\exists m)(m\ \mathrm{ex}\ z)$ (q), ui.
5. $\quad\quad m\ \mathrm{ex}\ z$ 4, EI.
6. $\quad\quad z\ \mathit{in}\ \{m \supset m\}$ 3, ui.
7. $\quad\quad z\ \mathit{in}\ \{m \supset m\}\ :\!\supset\!:\ z = \{m \supset m\}\ .\!\vee\!.\ z = m\ .\!\vee\!.\ z\ \mathit{in}\ m$ $*(A_8)$.
8. $\quad\quad m\ \mathit{out}\ z$.&. m ind z .&. $m \neq z$ 5, ($T_{4.5}$)b, c.
9. $\quad\quad \sim (z\ \mathit{in}\ \{m \supset m\})$ 7, 8, log.
10. $\quad\quad \_\ F$ 6, 9, conj, (4ei).
11. $\quad 2 \supset F$ cp.
12. $\_\ z\ \mathit{out}\ @$ 11, reductio, df(@), df(in).
13. $1 \supset 12$ cp.
14. $(z)13$ ug.

($T_{4.10}$)c $\vdash (z)(z \neq f\ .\!\supset\!.\ z\ \mathit{out}\ f)$.

This is proved by a similar argument.

Next let us prove some theorems which are important for our later use of $*(s)$ in proving identities.

($T_{4.11}$)a $\vdash (p, a)(p \neq \{p \supset a\})$.

Proof:
1. $\rightarrow p = \{p \supset a\}$
2. $\quad (\exists l, m)(m\ \mathit{in}\ l$.&. $(z)(l/m;z!\{z \supset a\})$ $*(\mathrm{CP}_{\mathrm{sub}})$[1].
3. $\quad m\ \mathit{in}\ l$.&. $(z)(l/m;z!\{z \supset a\})$ 2, EI.
4. $\quad l/m;p!\{p \supset a\}$ 3, simp, ui.
5. $\quad l/m;p!p$ 1, 4, $*(=)$.
6. $\quad m\ \mathit{out}\ l\ .\!\vee\!.\ m = l$ 5, ($T_{4.2}$)b.
7. $\quad m = l$ 3, 6, simp, ds.
8. $\quad (\exists c)(c\ \mathit{ex}\ a)$ (q).

9. $c\ ex\ a$ 8, EI.
10. $l/m;c!\{c \supset a\}$ 3, ui.
11. $l/m;c!c$ 7, $*(=)$, (A_{13}), ui.
12. $c = \{c \supset a\}$ 10, 11, $(T_{4.3})$.
13. $a\ in\ c$ 12, $*(A_7)$, $*(=)$.
14. $\underline{\quad}$ F 9, 13, $(T_{4.5})$b, (9ei), (3ei).
15. $1 \supset F$ cp.
16. $p \neq \{p \supset a\}$ 15, reductio.
17. $(p, a)16$

By similar proofs we get:

$(T_{4.11})$b $\vdash (p, a)(p \neq \{a \supset p\})$
$(T_{4.11})$c $\vdash (p, a, b, q)(p \neq \{p/a;b!q\})$
$(T_{4.11})$d $\vdash (p, a, b, q)(a \neq \{p/a;b!q\})$
$(T_{4.11})$e $\vdash (p, a, b, q)(q \neq \{p/a;b!q\})$.

With theorem $(T_{4.11})$b, we can go on to:

$@^\circ = \text{df } (x)(x\ .\!\supset\!.\ x \supset x)$
$f^\circ = \text{df } (x)(x \supset x\ :\!\supset\!:\ x \supset x\ .\!\supset\!.\ x)$.

These entities, like @ and *f*, have no entities *in* them other than themselves. Moreover, they are all distinct. Even logically equivalent propositions, such as @ and $@^\circ$, may nonetheless not be identical.

$(T_{4.12})$a $\vdash (p, a, r)(\{p \supset a\} = \{r \supset a\}\ :\!\supset\!:\ p = r)$.

Proof:
1. $\rightarrow \{p \supset a\} = \{r \supset a\}$
2. $(\exists l, m)(m\ in\ l\ .\&.\ (z)(l/m;z!\{z \supset a\})$ $* (CP_{sub})^1$.
3. $m\ in\ l\ .\&.\ (z)(l/m;z!\{z \supset a\})$ 2, EI.
4. $l/m;p!\{p \supset a\}$ 3, simp, ui.
5. $l/m;r!\{r \supset a\}$ 3, simp, ui.
6. $l/m;p!\{r \supset a\}$ 1, 4, $*(=)$.
7. $m\ in\ l\ .\&.\ l/m;p!\{r \supset a\}\ .\&.\ l/m;r!\{r \supset a\}$ 3, 5, 6, simp, conj.
8. $\underline{\quad} p = r$ 7, (A_{11}), ui, mp, (3ei).
9. $1 \supset 8$ cp.
10. $(p, a, r)9$, ug.

By a similar proof we arrive at

$(T_{4.12})$b $\vdash (p, a, c)(\{p \supset a\} = \{p \supset c\}\ .\!\supset\!.\ a = c)$.

Let us now prove the following useful theorems which will be lemmas for our proofs of Church's (ii) and (iii) and for our (v).

$(T_{4.13})$a $\vdash (p)(\exists y)(p\ out\ \{\sim y\})$.

Proof:
1. $(\exists m)(m \text{ ex } \{\sim p\})$ (q), ui.
2. $m \text{ ex } \{\sim p\}$ 1, EI.

3. m ex p 2, $(T_{4.5})$a, $*(A_7)$.
4. p *in* $\{\sim m\}$ $:\supset:$ $p = \{\sim m\}$ $.\vee.$ p *in* m $*(A_8)$.
5. p *out* $\{\sim m\}$ 3, 4, $(T_{4.5})$abc, mt.
6. $(\exists y)(p$ *out* $\{\sim y\})$ 5, eg, (2ei).
7. (p)6 ug.

$(T_{4.13})$b $\vdash$ $(p)(\exists y, z)(p$ *out* $\{y \supset z\})$.
$(T_{4.13})$c $\vdash$ $(p)(\exists x, y, z, w)(p$ *out* $\{x/y;z!w\})$.

These are proved by similar arguments.

Now to proofs of the identities that Church must take as fundamental principles. We have Church's (i):

$(T_{4.14})$ $\vdash$ $(p, q)(\{\sim p\} = \{\sim q\}$ $.\supset.$ $p = q)$.

1. $\rightarrow$ $\{\sim p\} = \{\sim q\}$
2. $(\exists l, m)(m$ *in* l .&. $(z)(l/m;z!\{\sim z\}))$ $*(CP_{sub})$[1].
3. m *in* l .&. $(z)(l/m;z!\{\sim z\})$ 2, EI.
4. $l/m;p!\{\sim p\}$ 3, simp, ui.
5. $l/m;q!\{\sim q\}$ 3, simp, ui.
6. $l/m;p!\{\sim q\}$ 1, 4, $*(=)$.
7. $l/m;p!\{\sim q\}$.&. $l/m;q!\{\sim q\}$.&. m *in* l $:\supset:$ $p = q$ (A_{11}), ui.
8. $p = q$ 3, 5, 6, mp, (3ei).
9. $1 \supset 8$ cp.
10. (p, q)9 ug.

As we can see, (A_{11}) comes in handy.

Church's (ii) is immediate from (A_{21}) given df($\sim$). But we include an alternate proof which would obtain had tilde been adopted as a primitive monadic predicate.

$(T_{4.15})$ $\vdash$ $(p)(\{\sim p\} \neq \{r \supset s\})$.

Proof:

1. $\rightarrow$ $\{\sim p\} = \{r \supset s\}$
2. $(z)(\exists r^\circ, s^\circ)(\{r \supset s\}/p;z!\{r^\circ \supset s^\circ\}$.&. $r/p;z!r^\circ$.&. $s/p;z!s^\circ)$
 1, $*(T_{4.7})$, $(T_{4.13})$b, $*(s)$.
3. $\{r \supset s\}/p;@!\{r^\circ \supset s^\circ\}$.&. $r/p;@!r^\circ$.&. $s/p;@!s^\circ$ 2, ui, EI.
4. $(z)(\{\sim p\}/p;z!\{\sim z\})$, $(T_{4.13})$a, $*(r)$.
5. $\{\sim p\}/p;@!\{\sim @\}$ 4, ui.
6. $\{r \supset s\}/p;@!\{\sim @\}$ 1, 5, $*(=)$.
7. $\{\sim @\} = \{r^\circ \supset s^\circ\}$ 3, 6, $(T_{4.3})$.
8. r° *in* $\{\sim @\}$ 7, (A_7), $*(=)$.
9. s° *in* $\{\sim @\}$ 7, (A_7), $*(=)$.
10. $r^\circ = \{\sim @\}$ $.\vee.$ r° *in* $@$ 8, $*(A_8)$.
11. $s^\circ = \{\sim @\}$ $.\vee.$ s° *in* $@$ 9, $*(A_8)$.
12. $r^\circ \neq \{\sim @\}$.&. $s^\circ \neq \{\sim @\}$ 7, $*(T_{4.11})$ ab.
13. r° *in* $@$ 10, 12, log.
14. s° *in* $@$ 11, 12, log.

15. r° *in* @ $.\supset.$ $r^\circ = @$ $(T_{4.10})$b.
16. s° *in* @ $.\supset.$ $s^\circ = @$ $(T_{4.10})$b.
17. $r^\circ = @$.&. $s^\circ = @$ 13, 14, 15, 16, log.
18. $\{\sim @\} = \{@ \supset @\}$ 7, 17, $*(=)$.
19. $\sim @ .\equiv. @ \supset @$ 18, $(T_{4.7})$.
20. ___ F 19, df(@), log, (3ei).
21. $1 \supset F$ cp.
22. $\{\sim p\} \neq \{r \supset s\}$ 21, reductio, df($\neq$)
23. $(p, r, s)22$ ug.

To get a proof of Church's (iii) we first prove two lemmas:

$(T_{4.16})$a $\vdash (p, r, c)(\{p \supset p\} = \{r \supset c\} \ .\supset.\ p = r)$.

Proof:

1. $\rightarrow \{p \supset p\} = \{r \supset c\}$
2. $\rightarrow p \neq r$
3. $p \neq c$ 1, 2, $(T_{4.12})$a.
6. $\rightarrow p = \{r \supset c\}$
7. $\{p \supset p\} = p$ 1, 6, $*(=)$.
8. ___ F $(T_{4.11})$a.
9. $6 \supset F$ cp.
10. $p \neq \{r \supset c\}$ 9, reductio.
11. $\{r \supset c\}/p;f!\{r^\circ \supset c^\circ\}$.&. $r/p;f!r^\circ$.&. $c/p;f!c^\circ$
    10, $(T_{4.13})$b, $*(s)$, ui, EI.
12. $\{p \supset p\}/p;f!\{f \supset f\}$ $(T_{4.11})$a, $(T_{4.13})$b, $*(s)$, ui, EI.
13. $\{f \supset f\} = \{r^\circ \supset c^\circ\}$ 1, 11, 12, $*(=)$, $(T_{4.3})$.
14. $r^\circ \neq \{f \supset f\}$ 13, $(T_{4.11})$a.
15. r° *in* f 13, 14, $*(A_7)$, $*(A_8)$, $*(=)$.
16. $r^\circ = f$ 15, $(T_{4.10})$c.
17. $c^\circ = f$ 13, 16, $(T_{4.12})$b.
18. $r/p;f!f$ 11, 16, $*(=)$.
19. p *out* r 2, 18, $(T_{4.2})$b.
20. $r = f$ 18, 19, $(T_{4.4})$.
21. $c = c^\circ$.&. $c = f$ 3, 11, 17, $(T_{4,2})$b, $*(=)$.
22. $r = c$ 20, 21, $*(=)$.
23. $\{p \supset p\} = \{r \supset r\}$ 1, 22, $*(=)$.
24. $r = \{p \supset p\}$ .v. r *in* p 23, $*(A_7)$, $*(A_8)$, $*(=)$.
25. $p = \{r \supset r\}$ .v. p *in* r 24, $*(A_7)$, $*(A_8)$, $*(=)$.
26. $r \neq \{p \supset p\}$ 23, $(T_{4.11})$a.
27. $p \neq \{r \supset r\}$ 23, $(T_{4.11})$a.
28. r *in* p .&. p *in* r 24, 25, 26, 27, log.
29. $p = r$ 28, (A_9).
30. ___ F 2, 29, conj, (12ei), (11ei).
31. $2 \supset F$ cp.
32. ___ $p = r$ 31, reductio.
33. $1 \supset 32$ cp.

34. $(p, r, c)33$ ug.

One more lemma is needed. For this we again turn to the technique using $*(s)$.

$(T_{4.16})b \quad \vdash (p, a)(\{p \supset a\} = \{a \supset p\} \;.\supset.\; p = a).$

Proof:

1. $\rightarrow \{p \supset a\} = \{a \supset p\}$
2. $\rightarrow p \neq a$
3. $\rightarrow p \; in \; a$
4. $a \; out \; p$ 2, 3, df(*in*).
5. $(z)(\exists p^\circ, a^\circ)(\{p \supset a\}/a;z!\{p^\circ \supset a^\circ\} \;.\&.\; p/a;z!p^\circ \;.\&.\; a/a;z!a^\circ)$ 1, $(T_{4.13})$b, $(T_{4.11})$a, $*(s)$.
6. $(z)(\exists p^\circ, a^\circ)(\{a \supset p\}/a;z!\{a^\circ \supset p^\circ\} \;.\&.\; p/a;z!p^\circ \;.\&.\; a/a;z!a^\circ)$ $(T_{4.13})$b, $(T_{4.11})$a, $*(s)$.
7. $(z)(\{p \supset z\} = \{z \supset p\})$ 4, 5, 6, ui, EI, (A_{13}), $(T_{4.3})$, $(T_{4.4})$, $*(=)$, ug.
8. $\{p \supset @\} = \{@ \supset p\}$ 7, ui.
9. $p \supset @ \;.\equiv.\; @ \supset p$ 8, $(T_{4.7})$.
10. $\{p \supset f\} = \{f \supset p\}$ 7, ui.
11. $p \supset f \;.\equiv.\; f \supset p$ 10, $(T_{4.7})$.
12. $\sim p \;.\supset.\; \sim 9$ 9, df(@), log.
13. $p \;.\supset.\; \sim 11$ 11, df(f), log.
14. $\underline{\quad} F$ 9, 11, 12, 13, log, (7ei).
15. $3 \supset F$ cp.
16. $p \; out \; a$ 15, reductio, df(*in*).
17. $(z)(\exists p^\circ, a^\circ)(\{p \supset a\}/p;z!\{p^\circ \supset a^\circ\} \;.\&.\; p/p;z!p^\circ \;.\&.\; a/p;z!a^\circ)$ $(T_{4.13})$b, $(T_{4.11})$a, $*(s)$.
18. $(z)(\exists p^\circ, a^\circ)(\{a \supset p\}/p;z!\{a^\circ \supset p^\circ\} \;.\&.\; p/p{:}z!p^\circ \;.\&.\; a/p;z!a^\circ)$ 1, $(T_{4.13})$b, $(T_{4.11})$a, $*(s)$.
19. $(z)(\{z \supset a\} = \{a \supset z\})$ 16, 17, 18, ui, EI, (A_{13}), $(T_{4.3})$, $(T_{4.4})$, $*(=)$, ug.
20. $\{@ \supset a\} = \{a \supset @\}$ 19, ui.
21. $@ \supset a \;.\equiv.\; a \supset @$ 20, $(T_{4.7})$.
22. $\{f \supset a\} = \{a \supset f\}$ 19, ui.
23. $f \supset a \;.\equiv.\; a \supset f$ 22, $(T_{4.7})$.
24. $\sim a \;.\supset.\; \sim 21$ 21, df(@), log.
25. $a \;.\supset.\; \sim 23$ 23, df(f), log.
26. $\underline{\quad} F$ 21, 23, 24, 25, log, (19ei).
27. $2 \supset F$ cp.
28. $\underline{\quad} p = a$ 27, reductio, df($\neq$)
29. $1 \supset 28$ cp.
30. $(p, a)29$ ug.

From these lemmas we arrive at:

$(T_{4.17})a \quad \vdash (p, a, r, c)(\; \{p \supset a\} = \{r \supset c\} \;.\supset.\; p = r).$

Proof:

1. $\rightarrow \{p \supset a\} = \{r \supset c\}$
2. $\rightarrow p \neq r$
3. $a \neq c$ 1, 2, $(T_{4.12})$a.
4. $r \neq c$ 1, 2, $(T_{4.16})$a.
5. $p \neq a$ 1, 2, $(T_{4.16})$a.
6. $\rightarrow p \; out \; a$
7. $(z)(\exists p^{\circ}, a^{\circ})(\{p \supset a\}/p;z!\{p^{\circ} \supset a^{\circ}\} \;.\&.\; p/p;z!p^{\circ} \;.\&.\; a/p;z!a^{\circ})$ $(T_{4.13})$b, $(T_{4.11})$a, $*(s)$.
8. $(z)(\exists r^{\circ}, c^{\circ})(\{r \supset c\}/p;z!\{r^{\circ} \supset c^{\circ}\} \;.\&.\; r/p;z!r^{\circ} \;.\&.\; c/p;z!c^{\circ})$ 1, $(T_{4.13})$b, $(T_{4.11})$a, $*(s)$, $*(=)$.
9. $\{a \supset a\} = \{r^{\circ} \supset c^{\circ}\} \;.\&.\; r/p;a!r^{\circ} \;.\&.\; c/p;a!c^{\circ}$ 7, 8, ui, EI, (A_{13}), $(T_{4.3})$, $(T_{4.4})$, $*(=)$.
10. $a = r^{\circ}$ 9, $(T_{4.16})$a.
11. $r/p;a!a$ 9, 10, $*(=)$.
12. $p \; out \; r$ 2, 11, $(T_{4.2})$b.
13. $r = a$ 11, 12, $(T_{4.4})$.
14. $\{a \supset a\} = \{a \supset c^{\circ}\}$ 9, 10, $*(=)$.
15. $a = c^{\circ}$ 14, $(T_{4.12})$b
16. $c/p;a!a$ 9, 15, $*(=)$.
17. $p = c \;.\vee.\; p \; out \; c$ 16, $(T_{4.2})$b.
18. $\rightarrow p = c$
19. $\{p \supset a\} = \{a \supset p\}$ 1, 13, 18, $*(=)$.
20. F 5, 19, $(T_{4.16})$b.
21. $18 \supset F$ cp.
22. $p \neq c$ 21, reductio.
23. $p \; out \; c$ 17, 22, ds.
24. $c = a$ 16, 23, $(T_{4.4})$.
25. $r = c$ 13, 24, $*(=)$.
26. F 4, 25, conj, (9ei).
27. $6 \supset F$ cp.
28. $p \; in \; a$ 27, reductio, df(*in*).
29. $a \; out \; p$ 5, 28, $(T_{4.1})$.
30. $(z)(\exists p^{\circ}, a^{\circ})(\{p \supset a\}/a;z!\{p^{\circ} \supset a^{\circ}\} \;.\&.\; p/a;z!p^{\circ} \;.\&.\; a/a;z!a^{\circ})$ $(T_{4.13})$b, $(T_{4.11})$b, $*(s)$.
31. $(z)(\exists r^{\circ}, c^{\circ})(\{r \supset c\}/a;z!\{r^{\circ} \supset c^{\circ}\} \;.\&.\; r/a;z!r^{\circ} \;.\&.\; c/a;z!c^{\circ})$ 1, $(T_{4.13})$b, $(T_{4.11})$b, $*(s)$.
32. $\{p \supset p\} = \{r^{\circ} \supset c^{\circ}\} \;.\&.\; r/a;p!r^{\circ} \;.\&.\; c/a;p!c^{\circ}$ 30, 31, ui, EI, (A_{13}), $(T_{4.3})$, $(T_{4.4})$, $*(=)$.
33. $p = r^{\circ}$ 32, $(T_{4.16})$a.
34. $r/a;p!p$ 32, 33, $*(=)$.
35. $a = r$ 2, 34, $(T_{4.2})$b.
36. $p = c^{\circ}$ 32, 33, $(T_{4.12})$b.
37. $c/a;p!p$ 32, 36, $*(=)$.
38. $a = c \;.\vee.\; a \; out \; c$ 37, $(T_{4.2})$b.
39. $a \; out \; c$ 3, 38, ds.

40. $\quad\quad c = p$ 37, 39, $(T_{4.4})$.
41. $\quad\quad \{p \supset a\} = \{a \supset p\}$ 1, 35, 40, $*(=)$
42. $\quad\underline{\quad} F$ 5, 41, $(T_{4.16})$b, 32(ei).
43. $\quad 2 \supset F$ cp.
44. $\underline{\quad} p = r$ 43, reductio, df($\neq$).
45. $1 \supset 44$ cp.
46. $(p, a, r, c)45$ ug.

Finally we have Church's (iii):

$(T_{4.17})$b $\quad \vdash (p, a, r, c)(\{p \supset a\} = \{r \supset c\} \;{:}{\supset}{:}\; p = r \;.\&.\; a = c)$.

The proof is immediate from $(T_{4.17})$a and $(T_{4.12})$b.

This completes Church's (i)–(iii). We also need a proof of (iv), viz.,

$(b, q)(p, a)(r, c)(\{p/a;b!q\} = \{r/c;b!q\} \;{:}{\supset}{:}\; p = r \;.\&.\; a = c)$.

The proof here is somewhat difficult. But once we have the lemmas, we need only follow the lines of our proof of Church's (iii). We have:

$(T_{4.18})$a $\quad \vdash (b, q)(p, a, r)(\{p/a;b!q\} = \{r/a;b!q\} \;{:}{\supset}{:}\; p = r)$.

Proof:

1. $\rightarrow \{p/a;b!q\} = \{r/a;b!q\}$
2. $\quad (\exists l, m)(\, m \textit{ in } l \;.\&.\; (z)(l/m;z!\{z/a;b!q\})$ $*(CP_{sub})^1$.
3. $\quad m \textit{ in } l \;.\&.\; (z)(l/m;z!\{z/a;b!q\})$ 2, EI.
4. $\quad l/m;p!\{p/a;b!q\}$ 3, simp, ui.
5. $\quad l/m;r!\{r/a;b!q\}$ 3, ui.
6. $\quad l/m;p!\{r/a;b!q\}$ 1, 4, $*(=)$.
7. $\quad m \textit{ in } l \;.\&.\; l/m;p!\{r/a;b!q\} \;.\&.\; l/m;r!\{r/a;b!q\}$ 3, 5, 6, simp, conj.
8. $\underline{\quad} p = r$ 7, (A_{11}), ui, mp, (3ei).
9. $1 \supset 8$ cp.
10. $(b, q)(p, a, r)9$ ug.

The proof parallels that of $(T_{4.12})$a. By a similar proof we arrive at:

$(T_{4.18})$b $\quad \vdash (b, q)(p, a, c)(\{p/a;b!q\} = \{p/c;b!q\} \;.{\supset}.\; a = c)$.

Unfortunately, we cannot also parallel our proof of $(T_{4.16})$a. Instead, we put:

$(T_{4.18})$c $\quad \vdash (b, q)(p, r)(\{p/p;b!q\} = \{r/r;b!q\} \;.{\supset}.\; p = r)$.

1. $\rightarrow \{p/p;b!q\} = \{r/r;b!q\}$
2. $\quad \rightarrow p \neq r$
3. $\quad\quad (\exists l, m)(m \textit{ in } l \;.\&.\; (z)(l/m;z!\{z/z;b!q\}))$ $*(CP_{sub})^1$.
4. $\quad\quad m \textit{ in } l \;.\&.\; (z)(l/m;z!\{z/z;b!q\})$ 3, ui.
5. $\quad\quad l/m;p!\{p/p;b!q\}$ 4, simp, ui.
6. $\quad\quad l/m;r!\{r/r;b!q\}$ 4, simp, ui.
7. $\quad\quad l/m;r!\{p/p;b!q\}$ 1, 6, $*(=)$.
8. $\quad\quad p = r$ 4, 5, 7, (A_{11}).
9. $\quad\underline{\quad} F$ 2, 9, conj.
10. $\quad 2 \supset F$ cp.

11. $\underline{\quad}\ p = r$ 10, reductio, df($\neq$).
12. $\overline{1} \supset 11$ cp.
13. $(b, q)(p, r)12$ ug.

$(\mathrm{T}_{4.18})$d $\vdash (b, q)(p, r, c)(\{p/p;b!q\} = \{r/c;b!q\}\ .\supset.\ p = r)$.

1. $\rightarrow \{p/p;b!q\} = \{r/c;b!q\}$
2. $\rightarrow p \neq r$
3. $p \neq c$ 1, 2, $(\mathrm{T}_{4.18})$a.
4. $\rightarrow r\ out\ p$
5. $(z)(\exists r^\circ, c^\circ, b^\circ, q^\circ)(\{r/c;b!q\}/r;z!\{r^\circ/c^\circ;b^\circ!q^\circ\}$ $.\&.\ r/r;z!r^\circ\ .\&.\ c/r;z!c^\circ\ .\&.\ b/r;z!b^\circ\ .\&.\ q/r;z!q^\circ)$ $(\mathrm{T}_{4.13})$c, $(\mathrm{T}_{4.11})$c, $*(s)$
6. $(z)(\exists p^\circ, b^\circ, q^\circ)(\{p/p;b!q\}/r;z!\{p^\circ/p^\circ;b^\circ!q^\circ\}$ $.\&.\ p/r;z!p^\circ\ .\&.\ b/r;z!b^\circ\ .\&.\ q/r;z!q^\circ)$ 1, $(\mathrm{T}_{4.13})$c, $(\mathrm{T}_{4.11})$c, $*(s)$
7. $\{p/p;b^\circ!q^\circ\} = \{p/c^\circ;b^\circ!q^\circ\}\ .\&.\ c/r;p!c^\circ\ .\&.\ b/r;p!b^\circ\ .\&.\ q/r;p!q^\circ$ 4, 5, 6, ui, EI, (A_{13}), $(\mathrm{T}_{4.3})$, $*(=)$.
8. $p = c^\circ$ 7, $(\mathrm{T}_{4.18})$b.
9. $c/r;p!p$ 7, 8, $*(=)$.
10. $r = c\ .\vee.\ r\ out\ c$ 9, $(\mathrm{T}_{4.2})$b.
11. $\rightarrow r\ out\ c$
12. $c = p$ 9, 11, $(\mathrm{T}_{4.4})$.
13. $\underline{\quad}\ F$ 3, 12, conj.
14. $\overline{11} \supset F$ cp.
15. $\sim (r\ out\ c)$ 14, reductio.
16. $r = c$ 10, 15, ds.
17. $\{p/p;b!q\} = \{r/r;b!q\}$ 1, 16. $*(=)$.
18. $\underline{\quad}\ F$ 1, 17, $(\mathrm{T}_{4.18})$c, (7ei).
19. $\overline{4} \supset F$ cp.
20. $r\ in\ p$ 19, reductio, df(*in*).
21. $p\ out\ r$ 2, 20, $(\mathrm{T}_{4.1})$.
22. $(z)(\exists r^\circ, c^\circ, b^\circ, q^\circ)(\{r/c;b!q\}/p;z!\{r^\circ/c^\circ;b^\circ!q^\circ\}$ $.\&.\ r/p;z!r^\circ\ .\&.\ c/p;z!c^\circ\ .\&.\ b/p;z!b^\circ\ .\&.\ q/p;z!q^\circ)$ 1, $(\mathrm{T}_{4.13})$c, $(\mathrm{T}_{4.11})$c, $*(s)$
23. $(z)(\exists p^\circ, b^\circ, q^\circ)(\{p/p;b!q\}/p;z!\{p^\circ/p^\circ;b^\circ!q^\circ\}$ $.\&.\ p/p;z!p^\circ\ .\&.\ b/p;z!b^\circ\ .\&.\ q/p;z!q^\circ)$ $(\mathrm{T}_{4.13})$c, $(\mathrm{T}_{4.11})$c, $*(s)$
24. $\{r/r;b^\circ!q^\circ\} = \{r/c^\circ;b^\circ!q^\circ\}\ .\&.\ c/p;r!c^\circ\ .\&.\ b/p;r!b^\circ\ .\&.\ q/p;r!q^\circ$ 21, 22, 23, ui, EI, (A_{13}), $(\mathrm{T}_{4.3})$, $*(=)$.
25. $r = c^\circ$ 24, $(\mathrm{T}_{4.18})$b.
26. $c/p;r!r$ 24, 25, $*(=)$.
27. $p\ out\ c$ 3, 26, $(\mathrm{T}_{4.2})$b.
28. $c = r$ 26, 27, $(\mathrm{T}_{4.4})$.
29. $\{p/p;b!q\} = \{r/r;b!q\}$ 1, 28, $*(=)$.
30. $p = r$ 29, $(\mathrm{T}_{4.18})$c.
31. $\underline{\quad}\ F$ 2, 30, conj, (24ei).

32. $2 \supset F$ cp.
33. __ $p = r$ 32, reductio.
34. $1 \supset 33$ cp.
35. $(b, q)(p, r, c)34$ ug.

Two more lemmas are needed.

$(T_{4.18})$e $\vdash (p, a, q)(m)(m \text{ ex } p \text{ .\&. } p/a;m!q \text{ :}\supset\text{: } q = m \text{ .}\vee\text{. } q \textit{ out } m)$.

1. $\rightarrow m \text{ ex } p \text{ .\&. } p/a;m!q$
2. $\rightarrow a = p$
3. $p/p;m!q$ 1, 2, $*(=)$.
4. $p/p;m!m$ (A_{13}).
5. __ $q = m$ 3, 4, $(T_{4.3})$.
6. $a = p \text{ .}\supset\text{. } q = m$ 2–5, cp.
7. $\rightarrow a \neq p$
8. $\rightarrow a \textit{ in } p$
9. $m \textit{ in } q \text{ .\&. } m \neq q$ 1, 7, 8, (A_{12}).
10. __ $q \textit{ out } m$ 9, (A_9).
11. $a \textit{ in } p \text{ .}\supset\text{. } q \textit{ out } m$
12. $\rightarrow a \textit{ out } p$
13. $p = q$ 1, 12, $(T_{4.4})$.
14. $p \textit{ out } m$ 1, $(T_{4.5})$b, df(ind).
15. __ $q \textit{ out } m$ 13, 14, $*(=)$.
16. $a \textit{ out } p \text{ .}\supset\text{. } q \textit{ out } m$ 12–15, cp.
17. __ $q \textit{ out } m$ 11, 16, log.
18. $a \neq p \text{ .}\supset\text{. } q \textit{ out } m$ 7–17, cp.
19. __ $q = m \text{ .}\vee\text{. } q \textit{ out } m$ 6, 18, log.
20. $1 \supset 19$ cp.
21. $(p, a, q)20$ ug.

$(T_{4.18})$f $\vdash (b)(p, a)(\{p/a;b!a\} = \{a/p;b!a\} \text{ .}\supset\text{. } p = a)$.

1. $\rightarrow \{p/a;b!a\} = \{a/p;b!a\}$
2. $\rightarrow p \neq a$
3. $(\exists m)(m \text{ ex } \{p \supset a\})$ (q).
4. $m \text{ ex } a \text{ .\&. } m \text{ ex } p$ 3, EI, $(T_{4.5})$c.
5. $\rightarrow p \textit{ out } a$
6. $(z)(\exists p^\circ, a^\circ, b^\circ)(\{p/a;b!a\}/p;z!\{p;z!\{p^\circ/a^\circ;b^\circ!a^\circ\}$ $\text{.\&. } p/p;z!p^\circ \text{ .\&. } a/p;z!a^\circ \text{ .\&. } b/p;z!b^\circ)$ $(T_{4.13})$c, $(T_{4.11})$c, $*(s)$
7. $(z)(\exists p^\circ, a^\circ, b^\circ)(\{a/p;b!a\}/p;z!\{a^\circ/p^\circ;b^\circ!a^\circ\}$ $\text{.\&. } p/p;z!p^\circ \text{ .\&. } a/p;z!a^\circ \text{ .\&. } b/p;z!b^\circ)$ 1, $(T_{4.13})$c, $(T_{4.11})$c, $*(s)$.
8. $\{m/a;b^\circ!a\} = \{a/m;b^\circ!a\}$ 6, 7, ui, EI, (A_{13}), $(T_{4.3})$, $(T_{4.4})$, $*(=)$.
9. $m \textit{ out } a$ 4, df(ex), df(ind).
10. $a/m;b^\circ!a$ 9, df(*out*), ui.
11. $m/a;b^\circ!a$ 8, 10, $(T_{4.7})$.

12. $a\ out\ m$ 4, df(ex), df(ind).
13. $m = a$ 11, 12, $(T_{4.4})$.
14. __ F 4, 13, $(T_{4.5})$b, (8ei).
15. $p\ in\ a$ 14, reductio, df(*in*).
16. $a\ out\ p$ 2, 15, $(T_{4.1})$.
17. $(z)(\exists p^\circ, a^\circ, b^\circ)(\{p/a;b!a\}/a;z!\{p^\circ/a^\circ;b^\circ!a^\circ\}$
 $.\&.\ p/a;z!p^\circ\ .\&.\ a/a;z!a^\circ\ .\&.\ b/a;z!b^\circ)$
 1, $(T_{4.13})$c, $(T_{4.11})$c, $*(s)$.
18. $(z)(\exists p^\circ, a^\circ, b^\circ)(\{a/p;b!a\}/a;z!\{a^\circ/p^\circ;b^\circ!a^\circ\}$
 $.\&.\ p/a;z!p^\circ\ .\&.\ a/a;z!a^\circ\ .\&.\ b/a;z!b^\circ)$
 $(T_{4.13})$c, $(T_{4.11})$c, $*(s)$.
19. $\{p/m;b^\circ!m\} = \{m/p;b^\circ!m\}$ 17, 18, ui, EI, (A_{13}), $(T_{4.3})$, $(T_{4.4})$, $*(=)$.
20. $p\ out\ m$ 4, df(ex), df(ind).
21. $m/p;b^\circ!m$ 20, df(*out*), ui.
22. $p/m;b^\circ!m$ 19, 21, $(T_{4.7})$.
23. $m\ out\ p$ 4, df(ex), df(ind).
24. $p = m$ 22, 23, $(T_{4.4})$.
25. __ F 4, 24, $(T_{4.5})$b, (19ei), (4ei).
26. $2 \supset F$ cp.
27. __ $p = a$ 26, reductio, df($\neq$).
28. $1 \supset 27$ cp.
29. $(b)(p, a)28$ ug.

$(T_{4.18})$g $\vdash (b, q)(p, a)(\{p/a;b!q\} = \{a/p;b!q\}\ .\supset.\ p = a)$.

1. $\rightarrow \{p/a;b!q\} = \{a/p;b!q\}$
2. $\rightarrow p \neq a$
3. $(\exists m)(\ m\ \text{ex}\ \{p\ .\supset.\ a \supset q\})$ (q).
4. $m\ \text{ex}\ p\ .\&.\ m\ \text{ex}\ a\ .\&.\ m\ \text{ex}\ q$ 3, EI, $(T_{4.5})$c.
5. $\rightarrow p\ out\ a$
6. $(z)(\exists p^\circ, a^\circ, b^\circ, q^\circ)(\{p/a;b!q\}/p;z!\{p^\circ/a^\circ;b^\circ!q^\circ\}\ .\&.$
 $p/p;z!p^\circ\ .\&.\ a/p;z!a^\circ\ .\&.\ b/p;z!b^\circ\ .\&.\ q/p;z!q^\circ)$
 $(T_{4.13})$c, $(T_{4.11})$c, $*(s)$.
7. $(z)(\exists p^\circ, a^\circ, b^\circ, q^\circ)(\{a/p;b!q\}/p;z!\{a^\circ/p^\circ;b^\circ!q^\circ\}\ .\&.$
 $p/p;z!p^\circ\ .\&.\ a/p;z!a^\circ\ .\&.\ b/p;z!b^\circ\ .\&.\ q/p;z!q^\circ)$
 1, $(T_{4.13})$c, $(T_{4.11})$c, $*(s)$.
8. $\{m/a;b^\circ!q^\circ\} = \{a/m;b^\circ!q^\circ\}\ .\&.\ q/p;m!q^\circ\ .\&.\ b/p;m!b^\circ$
 6, 7, ui, EI, (A_{13}), $(T_{4.3})$, $(T_{4.4})$, $*(=)$.
9. $(z)(\exists a^{\circ\circ}, v, b^{\circ\circ}, q^{\circ\circ})(\{m/a;b^\circ!q^\circ\}/q^\circ;z!\{v/a^{\circ\circ};b^{\circ\circ}!q^{\circ\circ}\}\ .\&.$
 $m/q^\circ;z!v\ .\&.\ a/q^\circ;z!a^{\circ\circ}\ .\&.\ b^\circ/q^\circ;z!b^{\circ\circ}\ .\&.\ q^\circ/q^\circ; z!q^{\circ\circ})$
 $(T_{4.13})$c, $(T_{4.11})$e, $*(s)$.
10. $(z)(\exists a^{\circ\circ}, w, b^{\circ\circ}, q^{\circ\circ})(\{a/m;b^\circ!q^\circ\}/q^\circ;z!\{a^{\circ\circ}/w;b^{\circ\circ}!q^{\circ\circ}\}\ .\&.$
 $m/q^\circ;z!w\ .\&.\ a/q^\circ;z!a^{\circ\circ}\ .\&.\ b^\circ/q^\circ;z!b^{\circ\circ}\ .\&.\ q^\circ/q^\circ;z!q^{\circ\circ})$
 8, $(T_{4.13})$c, $(T_{4.11})$e, $*(s)$.
11. $\{v/a^{\circ\circ};b^{\circ\circ}!m\} = \{a^{\circ\circ}/v;b^{\circ\circ}!m\}\ .\&.\ m/q^\circ;m!v\ .\&.$
 $a/q^\circ;m!a^{\circ\circ}\ .\&.\ b^\circ/q^\circ;m!b^{\circ\circ}$
 8, 9, 10, ui, EI, (A_{13}), $(T_{4.3})$, $(T_{4.4})$, $*(=)$.

12. $m = v$ 4, 8, 11, $(T_{4.4})$, $(T_{4.18})$e.
13. $\{m/a^{\circ\circ};b^{\circ\circ}!m\} = \{a^{\circ\circ}/m;b^{\circ\circ}!m\}$ 11, 12, $*(=)$.
14. $a^{\circ\circ} = m$ 13, $(T_{4.18})$f.
15. $a/q^{\circ};m!m$ 11, 14, $*(=)$.
16. $q^{\circ} = a$.∨. $q^{\circ} out\ a$ 15, $(T_{4.2})$b.
17. $q^{\circ}\ out\ a$:⊃: $a = a^{\circ\circ}$.&. $a = m$ 11, 14, $(T_{4.4})$, $*(=)$.
18. $\sim (q^{\circ}\ out\ a)$ 4, 17, mt.
19. $q^{\circ} = a$ 16, 18, ds.
20. $\{m/a;b^{\circ}!a\} = \{a/m;b^{\circ}!a\}$ 8, 19, $*(=)$.
21. __ F 4, 20, $(T_{4.18})$f (8ei).
22. $5 \supset F$ cp.
23. $p\ in\ a$ 2, 26, reductio, df(*in*).
24. $a\ out\ p$ 2, 23, df(*in*).
25. $(z)(\exists p^{\circ}, a^{\circ}, b^{\circ}, q^{\circ})(\{p/a;b!q\}/a;z!\{p^{\circ}/a^{\circ};b^{\circ}!q^{\circ}\}$.&. $p/a;z!p^{\circ}$.&. $a/a;z!a^{\circ}$.&. $b/a;z!b^{\circ}$.&. $q/a;z!q^{\circ})$ 1, $(T_{4.13})$c, $*(T_{4.11})$d, $*(s)$.
26. $(z)(\exists p^{\circ}, a^{\circ}, b^{\circ}, q^{\circ})(\{a/p;b!q\}/a;z!\{a^{\circ}/p^{\circ};b^{\circ}!q^{\circ}\}$.&. $p/a;z!p^{\circ}$.&. $a/a;z!a^{\circ}$.&. $b/a;z!b^{\circ}$.&. $q/a;z!q^{\circ})$ $(T_{4.13})$c, $*(T_{4.11})$d, $*(s)$.
27. $\{p/m;b^{\circ}!q^{\circ}\} = \{m/p;b^{\circ}!q^{\circ}\}$.&. $q/a;m!q^{\circ}$.&. $b/a;m!b^{\circ}$ 24, 25, 26, ui, EI, (A_{13}), $(T_{4.3})$, $(T_{4.4})$, $*(=)$.
28. $(z)(\exists p^{\circ\circ}, v, b^{\circ\circ}, q^{\circ\circ})(\{p/m;b^{\circ}!q^{\circ}\}/q^{\circ};z!\{p^{\circ\circ}/v;b^{\circ\circ}!q^{\circ\circ}\}$.&. $m/q^{\circ};z!v$.&. $p/q^{\circ};z!p^{\circ\circ}$.&. $b^{\circ}/q^{\circ};z!b^{\circ\circ}$.&. $q^{\circ}/q^{\circ};z!q^{\circ\circ})$ $(T_{4.13})$c, $(T_{4.11})$e, $*(s)$.
29. $(z)(\exists p^{\circ\circ}, v, b^{\circ\circ}, q^{\circ\circ})(\{m/p;b^{\circ}!q^{\circ}\}/q^{\circ};z!\{v/p^{\circ\circ};b^{\circ\circ}!q^{\circ\circ}\}$.&. $m/q^{\circ};z!v$.&. $p/q^{\circ};z!p^{\circ\circ}$.&. $b^{\circ}/q^{\circ};z!b^{\circ\circ}$.&. $q^{\circ}/q^{\circ};z!q^{\circ\circ})$ $(T_{4.13})$c, $(T_{4.11})$e, $*(s)$.
30. $\{p^{\circ\circ}/v;b^{\circ\circ}!m\} = \{v/p^{\circ\circ};b^{\circ\circ}!m\}$.&. $m/q^{\circ};m!v$.&. $p/q^{\circ};m!p^{\circ\circ}$.&. $b^{\circ}/q^{\circ};m!b^{\circ\circ}$ 28, 29, ui, EI, (A_{13}), $(T_{4.3})$, $(T_{4.4})$, $*(=)$.
31. $v = m$ 4, 27, 30, $(T_{4.4})$, $(T_{4.18})$e.
32. $\{p^{\circ\circ}/m;b^{\circ\circ}!m\} = \{m/p^{\circ\circ};b^{\circ\circ}!m\}$ 30, 31, $*(=)$.
33. $p^{\circ\circ} = m$ 32, $(T_{4.18})$f.
34. $q^{\circ} = p$.∨. $q^{\circ}\ out\ p$ 30, 33, $(T_{4.2})$b.
35. $q^{\circ}\ out\ p$.⊃. $p = m$ 30, 33, $(T_{4.4})$.
36. $\sim (q^{\circ}\ out\ p)$ 4, 35, log.
37. $q^{\circ} = p$ 34, 36, ds.
38. $\{p/m;b^{\circ}!p\} = \{m/p;b^{\circ}!p\}$ 27, 37, $*(=)$.
39. $p = m$ 38, $(T_{4.18})$f.
40. __ F 4, 39, df(ex), log, (34ei), (31ei), (4ei).
41. $2 \supset F$ cp.
42. __ $p = a$ 40, reductio, df($\neq$).
43. $1 \supset 42$ cp.
44. $(b, q)(p, a)43$ ug.

Finally, we arrive at:

$(\mathrm{T}_{4.19})\mathrm{a} \quad \vdash \ (b, q)(p, a, r, c)(\ \{p/a;b!q\} = \{r/c;b!q\} \ .\supset.\ p = r).$

Proof:

1. $\rightarrow \{p/a;b!q\} = \{r/c;b!q\}$

2. $\rightarrow p \neq r$

3. $a \neq c$ 1, 2, $(\mathrm{T}_{4.18})$a.

4. $r \neq c$ 1, 2, $(\mathrm{T}_{4.18})$d.

5. $p \neq a$ 1, 2, $(\mathrm{T}_{4,18})$d.

6. $\rightarrow p$ *out* a

7. $(z)(\exists p^\circ, a^\circ, b^\circ, q)(\{p/a;b!q\}/p;z!\{p^\circ/a^\circ;b^\circ!q^\circ\}$.&. $p/p;z!p^\circ$.&. $a/p;z!a^\circ$.&. $b/p;z!b^\circ$.&. $q/p;z!q^\circ)$ $(\mathrm{T}_{4.13})$c, $*(\mathrm{T}_{4.11})$c, $*(s)$.

8. $(z)(\exists r^\circ, c^\circ, b^\circ, q^\circ)(\{r/c;b!q\}/p;z!\{r^\circ/c^\circ;b^\circ!q^\circ\}$.&. $r/p;z!r^\circ$.&. $c/p;z!c^\circ$.&. $b/p;z!b^\circ$.&. $q/p;z!q^\circ)$ 1, $(\mathrm{T}_{4.13})$c, $*(\mathrm{T}_{4.11})$c, $*(s)$.

9. $\{a/a;b^\circ!q^\circ\} = \{r^\circ/c^\circ;b^\circ!q^\circ\}$.&. $r/p;a!r^\circ$.&. $c/p;a!c^\circ$.&. $b/p;a!b^\circ$.&. $q/p;a!q^\circ$.&. $p/p;a!p^\circ$.&. $a/p;a!a^\circ$ 7, 8, ui, EI, (A_{13}), $(\mathrm{T}_{4.3})$, $(\mathrm{T}_{4.4})$, $*(=)$.

10. $a = r^\circ$ 9, $(\mathrm{T}_{4.18})$d.

11. $r/p;a!a$ 9, 10, $*(=)$.

12. p *out* r 2, 11, $(\mathrm{T}_{4.2})$b.

13. $r = a$ 11, 12, $(\mathrm{T}_{4.4})$.

14. $\{a/a;b^\circ!q^\circ\} = \{a/c^\circ;b^\circ!q^\circ\}$ 9, 10, $*(=)$.

15. $a = c^\circ$ 14, $(\mathrm{T}_{4.18})$b

16. $c/p;a!a$ 9, 15, $*(=)$.

17. $p = c$.$\vee$. p *out* c 16, $(\mathrm{T}_{4.2})$b.

18. $\rightarrow p = c$

19. $\{p/a;b!q\} = \{a/p;b!q\}$ 1, 13, 18, $*(=)$.

20. $\underline{\quad} F$ 5, 19, $(\mathrm{T}_{4.18})$g.

21. $18 \supset F$ cp.

22. $p \neq c$ 21, reductio.

23. p *out* c 17, 22, ds.

24. $c = a$ 16, 23, $(\mathrm{T}_{4.4})$.

25. $r = c$ 13, 24, $*(=)$.

26. $\underline{\quad} F$ 4, 25, conj, (9ei).

27. $6 \supset F$ cp.

28. p *in* a 27, reductio, df(*in*).

29. a *out* p 2, 28, $(\mathrm{T}_{4.1})$.

30. $(z)(\exists p^\circ, a^\circ b^\circ, q^\circ)(\{p/a;b!q\}/a;z!\{p^\circ/a^\circ;b^\circ!q^\circ\}$.&. $p/a;z!p^\circ$.&. $a/a;z!a^\circ$.&. $b/a;z!b^\circ$.&. $q/a;z!q^\circ)$ $(\mathrm{T}_{4.13})$c, $*(\mathrm{T}_{4.11})$d, $*(s)$.

31. $(z)(\exists r^\circ, c^\circ, b^\circ, q^\circ)(\{r/c;b!q\}/a;z!\{r^\circ/c^\circ;b^\circ!q^\circ\}$.&. $r/a;z!r^\circ$.&. $c/a;z!c^\circ b/a;z!b^\circ$.&. $q/a;z!q^\circ)$ 1, $(\mathrm{T}_{4.13})$c, $*(\mathrm{T}_{4.11})$d, $*(s)$.

32. $\{p/p;b^\circ!q^\circ\} = \{r^\circ/c^\circ;b^\circ!q^\circ\}$.&. $r/a;p!r^\circ$.&. $c/a;p!c^\circ$.&. $b/a;p!b^\circ$.&. $q/a;p!q^\circ$.&. $p/a;p!p^\circ$.&. $a/a;p!a^\circ$ 30, 31, ui, EI, (A_{13}), ($T_{4.3}$), ($T_{4.4}$), $*(=)$.
33. $p = r^\circ$ 32, ($T_{4.18}$)d.
34. $r/a;p!p$ 32, 33, $*(=)$.
35. $a = r$ 2, 34, ($T_{4.2}$)b.
36. $p = c^\circ$ 32, 33, ($T_{4.18}$)b.
37. $c/a;p!p$ 32, 36, $*(=)$.
38. $a = c$.∨. a *out* c 37, ($T_{4.2}$)b.
39. → $a = c$
40. $\{p/a;b!q\} = \{a/a;b!q\}$ 1, 35, 39, $*(=)$.
41. $p = a$ 40, ($T_{4.18}$)a.
42. __ F 5, 41, conj.
43. 39 ⊃ F cp.
44. $a \neq c$ 43, reductio.
45. a *out* c 38, 44, ds.
46. $c = p$ 37, 45, ($T_{4.4}$).
47. $\{p/a;b!q\} = \{a/p;b!q\}$ 1, 35, 46, $*(=)$.
48. __ F 5, 47, ($T_{4.18}$)g, (32ei).
49. 2 ⊃ F cp.
50. __ $p = r$ 49, reductio, df(≠).
51. 1 ⊃ 50 cp.
52. (p, a, r, c)51 ug.

Finally, from ($T_{4.19}$)a and ($T_{4.18}$)b, we arrive at (iv):

($T_{4.19}$)b ⊢ $(b, q)(p, a, r, c)(\ \{p/a;b!q\} = \{r/c;b!q\}$:⊃: $p = r$.&. $a = c)$.

As will be seen in Chapter 9, this theorem, as well as ($T_{4.17}$)b, plays a central role in the paradoxes of propositions plaguing the substitutional theory.[10]

Another proof is relevant to the paradoxes of propositions. One paradox that exercised Russell a good deal is the propositional *liar*. This will be examined in Part III, but for the present the following theorem is of interest.

($T_{4.20}$)
⊢ $(a)(a \neq \{(\exists p)(p = a$.&. $\sim p)\})$.

Moreover, the following is not a viable means of formulating a propositional *liar* paradox within substitution:

$a = \{(\exists p)(b = p$.&. $\sim p)\}$
$b = \{(\exists p)(a = p$.&. $p)\}$.

10. It should be possible to use the same techniques as above to render a proof of the more general:

$(q, b)(t, w)(p, a)(r, c)((\{p/a;b!q\} = \{r/c;w!t\}$:⊃: $p = r$.&. $a = c$.&. $b = w$.&. $q = t)$.

But this proof will certainly be tedious and so is avoided here.

The trouble here is that the entity b is a constituent of the entity a and also the entity a is a constituent of b. With Russell's axiom (A_9) of antisymmetry, we have that $a = b$. By theorem ($T_{4.20}$) this is impossible. The propositional *liar* is not formulable in the pure substitutional logic of propositions.

In closing, it should be noted that Russell's system can be formulated with every entity out of itself—replacing (A_{13}) with $(x, y)(x/x;y!x)$ and making appropriate adjustments. This simplifies the theory in some respects but complicates it in others; for instance $*(s)$ would require extensive modification. The present formulation stays with the basic substitutional theory as Russell formulated it in 1905.

CHAPTER FIVE

The "No Propositional Functions" Theory

5.0 By merging the calculus of substitution with the theory of definite descriptions the substitutional theory generates a proxy for a type stratified calculus which employs predicate variables (bound and free) in both subject and predicate positions.[1]

We must take care not to be misled. The plan of the theory was not to deny that there are properties and relations in *intension*. Russell is certainly committed to the subsistence of properties and relations in intension. He holds that we are immediately acquainted with many, not the least of which is 'implication' itself. Moreover, he maintains steadfastly that properties and relations have both an individual and a predicable nature—their predicable nature accounting for the unity of propositions. The substitutional theory simply endeavors to show how logicism can proceed without the blanket assumption that every open wff stands for a single entity with both a predicable and an individual nature. The blanket assumption leads to contradiction.

Another way of putting the point is to say that the theory proxies a type-theory of "propositional functions" (in an ontological sense). But this way of putting the matter is apt to mislead. Some say Russell held that a simple open wff comprehends a "universal" and a complex open wff comprehends a "propositional function." Now Russell's common illustrations of the universals of acquaintance were sense qualities such as 'hardness,' 'roundness,' 'redness,' and spatial relations such as 'left of' and 'similarity (of color)'.[2] The commonplace universals known by immediate acquaintance are sense qualities which we express through simple wffs. But surely other languages might express our

1. In modern parlance, it proxies a "second-order logic with nominalized predicates."
2. This is markedly so in his 1912–1914 writings.

sense qualities through syntactically complex wffs. The distinction between a simple "universal" and a complex universal ("propositional function") seems wholly unfounded. "Simple" and "complex" properly belong to formulae and universals are supposed to be independent of thought and linguistic representation.

The substitutional theory makes the whole debate moot. The theory does not adopt a predicational form for atomic propositions as a part of its syntax. Accordingly, the theory takes no stand whatsoever as to whether complex wffs can comprehend the universals which account for the unity of atomic propositions. The simple point is that substitutional logic is designed to get along without the assumption that every open wff comprehends a single entity with both a predicable and an individual nature. It can therefore enjoy complete agnosticism when it comes to whether there are universals comprehended by complex wffs.

In this chapter, we shall show how the logic of substitution was to make good on its agnosticism by proxying a calculus which employs type-stratified predicate variables (bound and free) in subject as well as predicate positions—a theory of "propositional functions" if you like. The 1905 theory of definite descriptions plays a central role as we shall see.

Substitution and Definite Descriptions

5.1 With the theory of definite descriptions at hand, December of 1905 found Russell with a calculus for propositions which allowed definite descriptions of propositions. In the manuscript "On Substitution" he has what is largely *Principia*'s apparatus for definite descriptions. Since the formal apparatus of definite descriptions was in place at the time of "On Substitution," it is not at all anachronistic to adopt the apparatus in our present discussion.

Principia, however, used the letters "ϕ", "ψ" in a dual way: at times they were to be predicate variables, and (when occurring free in definitions or theses) they were used as meta-linguistic schematic letters for wffs.[3] In the substitutional theory, "ϕ", and "ψ" are only used as schematic letters; there are no predicate variables. We shall therefore continue with our "A", "B", etc., for schematic letters and, where we cite principles from *Principia*'s $*14$ on definite descriptions, we rewrite accordingly. We shall also relax our practice of using Greek letters for terms in our definitions and schemata. Our metalanguage letters for terms shall look like our object language individual variables. There should be no confusion. In axiom and theorem schemata we shall continue our practice of tagging schemata with the $*$-label.

Principia's contextual definition for definite descriptions can be put as follows:

3. See Chapter 10 for a justification of this interpretation.

$$*14.01 \quad [\imath z Az]\ B(\imath z Az|v) = \text{df}$$
$$(\exists x)((y)(Ay|z\ .\equiv.\ y = x)\ \&\ Bx|v).$$

where y is free for z in Az and x is free for v in Bv. We shall, however, adopt the more convenient form:

df(desc)

$$[\imath z Az]\ B(\imath z Az|v)\ = \text{df}\ (\exists x)(Ax|z\ .\&.\ (y)(Ay|z\ .\supset.\ y = x)\ \&\ Bx|v),$$

where x and y are free for z in Az and x is free for v in Bv. To introduce definite descriptions for propositions, Russell introduces the expression "$p\frac{b}{a}$" (alternatively written as "$p/a;b$"). He puts (*S*, p. 4):

$$*12.12 \quad p\frac{b}{a} = \text{df}\ (\imath q)(p/a;b!q).$$

Recall that "$p/a;b!q$" is read "q results from substituting b for every occurrence of a *in* p." Thus "$p/a;b$" is a definite description of an entity q which is exactly like p except for containing b wherever p contains a. Now the following is an instance of df(desc),

df(desc)^1

$$[(\imath q)(p/a;b!q)]\ B((\imath q)(p/a;b!q)|v) = \text{df}$$
$$(\exists q)(p/a;b!q\ \&\ (r)(p/a;b!r\ .\supset.\ r = q)\ \&\ Bq|v)$$

df(desc.a)^1

$$[p/a;b]\ B(p/a;b) = \text{df}\ [(\imath q)(p/a;b!q)]\ B((\imath q)(p/a;b!q)).$$[4]

It is also convenient to have:

df($\exists^1$)

$$(\exists^1 q)(p/a;b!q\ .\&.\ Aq) = \text{df}$$
$$(\exists q)(p/a;b!q\ .\&.\ (r)(p/a;b!r\ .\supset.\ r = q)\ .\&.\ Aq).$$

We shall use this anon.

Russell adopts all the same scope conventions applicable to definite descriptions generally (*PM*, p. 173). The scope marker "$[p/a;b]$", or more formally "$[\imath q)(p/a;b!q)]$", is omitted where the narrowest possible scope is intended. For the situation where two or more definite descriptions occur in the same statement, we find (*ibid.*, p. 174):

$$*14.03 \quad [\imath x Ax,\ \imath x Bx]\ C(\imath x Ax,\ \imath x Bx) = \text{df}$$
$$[\imath x Ax]([\imath x Bx]\ C(\imath x Ax,\ \imath x Bx)).$$

This, however requires a slight modification in case the wff B is A. Russell should have put:

$$[\imath x Ax,\ \imath z Bz]\ C(\imath x Ax,\ \imath z Bz) = \text{df}\ [\imath x Ax]([\imath z Bz]\ C(\imath x Ax,\ \imath z Bz)).$$

4. There remain cases where the abbreviation is inappropriate. For example, to recover $[\imath x Ax][\imath z Az]B(\imath x Ax,\ \imath z Az)$ we shall need $[(\imath q)(p/a;x!q)][(\imath s)(p/a;x!s]\ B((\imath q)(p/a;x!q),\ (\imath s)\ (p/a;x!s))$.

Russell's convention has it that the leftmost description shall have greatest scope and then the next to the right in typographical order. But we may want to make a given description primary even if it is not leftmost. Accordingly, Russell has:

$$*14.04 \quad [\imath xBx]\ C(\imath xAx,\ \imath xBx) = \text{df}\ [\imath xBx,\ \imath xAx]\ C(\imath xAx,\ \imath xBx).$$

This too should be modified to that it applies when B is A.

$$[\imath zBz]\ C(\imath xAx, \imath zBz) = \text{df}\ [\imath zBz,\ \imath xAx]\ C(\imath xAx,\ \imath zBz).$$

These conventions and schemata enable Whitehead and Russell to avoid (in most cases) using scope markers.

To illustrate the conventions, consider the following context presented by Geach,

$$C(\imath xAx,\ \imath xAx).$$

This becomes,

$$[\imath xAx]\ C(\imath xAx,\ \imath xAx).$$

Applying ∗14.01 we have,

$$(\exists x)((y)(Ay\ .{\equiv}.\ y = x)\ \&\ C(x, x));$$

compare,

$$C(\imath xAx,\ \imath zAz),$$

which, on Russell's scope conventions abbreviates,

$$[\imath xAx,\ \imath zAz]\ C(\imath xAx,\ \imath zAz).$$

Applying ∗14.03 and ∗14.01 we get,

$$(\exists x)((y)(Ay\ .{\equiv}.\ y = x)\ .\&.\ [\imath zAz]C(x,\ \imath zAz)).$$

There is no problem of ambiguity.

It should be noted that there are no ambiguities when it comes to applying definitions. All too commonly it is argued that Russell's conventions on omitting scope markers leave him with no rules governing the order of elimination of defined signs. Consider,

$$\imath xAx \neq y.$$

Which should be applied first, *Principia*'s

$$*13.02 \quad x \neq y = \text{df} \sim (x = y)$$

or ∗14.01? There is no choice; it has to be ∗14.01. But it is *not* (some absent but implied) conventions on the omission of scope markers that ensure this! Definition schema ∗13.02 applies to *singular terms* x and y. Definite descriptions are not singular terms. The point is the same when it comes to definite descriptions of propositions. It would be an error to apply df(&) to

$p/a;b \ \& \ s$

to arrive at

$\sim(p/a;b \ .\supset. \ \sim s).$

The definition df(&) applies only to singular terms. Definite descriptions may occur in the positions of singular terms only by means of contextual definitions. Russell's conventions on omitting scope operators work smoothly and uniformly.[5]

Definite descriptions are not singular terms. Accordingly, *universal instantiation* does not apply to them. However, we do have *Principia*'s theorem (schema) $*14.18$ (*PM*, p. 180):

$$*(T_{5.1}) \quad \vdash \ E!(\iota x A x) \ .\supset. \ (x)Cx \ .\supset. \ [\iota x A x]C(\iota x A x).$$

In substitution we get an even stronger theorem schema for definite descriptions of propositions:

$$*(T_{5.1})a \quad \vdash \ (x)Cx \ .\supset. \ [p/a;b] \ C(p/a;b).$$

This follows because axiom (p) renders the antecedent $E!\iota(q)(p/a;b!q)$.

As we saw, in certain contexts a definite description of a proposition cannot have a secondary scope. For instance, using $*14.01$ we have,

$$\sim(p/a;b) = \text{df} \ (\exists q)((r)(p/a;b!r \ .\equiv. \ r = q) \ \& \sim q)$$
$$p/a;b \supset s = \text{df} \ (\exists q)((r)(p/a;b!r \ .\equiv. \ r = q) \ .\&. \ q \supset s)$$
$$p/a;b \ \& \ s = \text{df} \ (\exists q)((r)(p/a;b!r \ .\equiv. \ r = q) \ .\&. \ q \ \& \ s)$$
$$p/a;b = s = \text{df} \ (\exists q)((r)(p/a;b!r \ .\equiv. \ r = q) \ .\&. \ q = s).$$

There are, of course, secondary occurrences. Consider, for instance, the expression:

$$p/a;b \supset r \ .\supset. \ s.$$

By the convention on smallest scope this is to be

$$[p/a;b][p/a;b \ .\supset. \ r] \supset s.$$

The assumption of propositions realized in axiom (p), however, ensures that primary and secondary occurrences are always equivalent in truth-functional contexts. *Principia*'s $*14.3$ applies. We have:

$$*(T_{5.2})$$
$$\vdash \ (p)(q)(p \equiv q \ .\supset. \ Cp \equiv Cq) \ \& \ E!(\iota z A z) \ :\supset: \quad [\iota z A z]C(B(\iota z A z)) \ .\equiv. \ C([\iota z A z]B(\iota z A z)).$$

5. Geach claimed otherwise and went so far as to say that "Russell's contextual definitions of definite descriptions will not work as a symbolic device" (Geach 1950, p. 87). The problems Geach raises are trivial, however, as the above slight modifications of Russell's $*14.03$ and $*14.04$ show.

An instance of this schema is,

$$(p)(q)(p \equiv q \ .\supset.\ Cp \equiv Cq)\ \&\ \mathrm{E}!((\iota q)(p/a;b!q))\ :\supset:$$
$$[(\iota q)(p/a;b!q)]\ C(B((\iota q)(p/a;b!q)))\ :\supset:$$
$$C([(\iota q)(p/a;b!q)]\ B((\iota q)(p/a;b!q))).$$

Axiom (p) is just $\mathrm{E}!((\iota q)(p/a;b!q))$. So letting B be double tilde and eliminating the descriptions, we can see the validity of the following:

$$*\,\mathrm{T}(_{5.3})^1$$
$$\vdash\ (p, q)(p \equiv q \ .\supset.\ Cp \equiv Cq)\,.\supset.$$
$$(\exists q)(p/a;b!q \ .\&.\ (r)(p/a;b!r \ .\supset.\ r = q)\ .\&.\ Cq) \equiv$$
$$C((\exists q)(p/a;b!q \ .\&.\ (r)(p/a;b!r \ .\supset.\ r = q)\ .\&.\ q)).$$

We shall call this "$*(\mathrm{scope})^1$". The meta-proof is straightforward. The theorem schema $*(\mathrm{scope})^1$ is convenient. For instance we get:

$$\sim (\exists^1 q)(p/a;b!q \ .\&.\ q) \equiv \sim (p/a;b)$$

where C is the truth-functional context of tilde. The theorem schema is forthcoming because the assumption of propositions given by (p) ensures that the definite descriptions "$p/a;b$" behave as if they are genuine names of propositions in truth-functional contexts.

Universal instantiation is not applicable for definite descriptions, and this holds for such descriptions of propositions. Consider,

$$(x, y)(x \ .\supset.\ y \supset x).$$

Since definite descriptions are not singular terms of the formal language, one cannot use *Universal instantiation* to arrive at

$$p \ .\supset.\ t/w;b \supset p,$$

because "$t/w;b$", i.e., "$(\iota q)(t/w;b!q)$", is not a singular term. Nonetheless, we can get there another way. First get,

$$[t/w;b](\ p \ .\supset.\ t/w;b \supset p),$$

and apply $*(\mathrm{scope})^1$.

Multiple Substitutions

5.2 In order to proxy a type stratified language which allows predicate variables in subject as well as predicate positions, Russell goes on to what he calls "multiple substitutions." A multiple substitution is just a carefully crafted succession of single substitutions. At first blush, it may seem that one wants

$$p/a,b;x,y!q \ = \mathrm{df}\ (\exists h)(p/a;x!h \ .\&.\ h/b;y!q).$$

But this simple definition won't work. Consider, for example,

$\{a \supset b\}/a;x!h$.&. $h/b;y!q$.

We want q to be the proposition $\{x \supset y\}$. But if a is *in* b, then the result h of substituting an entity x for a may not contain b at all. The result, h, will be $\{x \supset b^\circ\}$ and we have $b/a;x!b^\circ$ and $b \neq b^\circ$. Now it may be that a is *out* of b, and the result, h, is $\{x \supset b\}$. But suppose b is *in* x; then substitution of an entity y for b will alter x. In this case, q is not $\{x \supset y\}$ but $\{x^\circ \supset y\}$ where $x/b;y!x^\circ$ and $x \neq x^\circ$.

In "On the Substantial Theory of Classes and Relations," Russell hoped to avoid these problems. He has the following definition of "simultaneous dual substitution" (*SCR*, p. 174):

$p/a,b;x,y!q$ = df
  b *in* a .⊃. $(\exists u)(u$ out p & $(u$ ind b .&. u ind $y)$ &
    $(\exists t)(p/a;u!t$ & $(\exists s)(t/b;y!s$ & $s/u;x!q)))$:&:
  b out a .⊃. $(\exists u)(u$ out p & $(u$ ind a .&. u ind $x)$ &
    $(\exists t)(p/b;u!t$ & $(\exists s)(t/a;x!s$ & $s/u;y!q)))$.

There is, however, a much simpler approach. It is not necessary to define "simultaneous" substitutions because there is no need to find a comprehensive definition covering all possible cases. We need not worry about a being a constituent of b (or vice versa) because we can find an a and a b appropriate to the outcome we want. Hence we put:

df(dual)
$p/a,b;x, y!q$ = df
  $(\exists h,e,t)(p/a;e!h$ & $h/b;y!t$ & $t/e;x!q)$.

The following proof demonstrates the adequacy of the definition. Analogous to $*(\mathrm{CP}_{\mathrm{sub}})^1$, we now arrive at $*(\mathrm{CP}_{\mathrm{sub}})^2$ for dual substitution.

$*(\mathrm{T}_{5.4})$
⊢ $(\exists p, a, b)(a$ *in* p .&. b *in* p .&. $(x, y)(p/a,b;x,y!\{A\})$,
where p, a, b are not free in A, and x, y are free in A.

Proof:
1. a ex $\{Amn\}$ (q), ui, EI.
2. $(\exists m, n)(a$ out $\{Amn\})$ 1, $(\mathrm{T}_{4.5})$b, EG.
3. b ex $\{Amn$.⊃. $a\}$ (q), ui, EI.
4. $(\exists m, n)(b$ out $\{Amn\})$ 3, $(\mathrm{T}_{4.5})$b, $*(\mathrm{A}_7)$, eg.
5. e ex $\{Amn$.⊃. $y \supset b\}$ (q), ui, EI.
6. $(\exists m, n)(e$ out $\{Amn\})$.&. e ex y .&. e ex b 5, $(\mathrm{T}_{4.5})$bc, $*(\mathrm{A}_7)$ eg.
7. $(z)(\exists b^\circ, a^\circ)(\{Aab\}/a;z!\{Aa^\circ b^\circ\}$
 .&. $a/a;z!a^\circ$.&. $b/a;z!b^\circ)$ 2, 3 $(\mathrm{T}_{4.5})$c, $*(\mathrm{A}_8)$, $*(s)$.
8. $(z)(\exists e^\circ, b^\circ)(\{Aeb\}/b;z!\{Ae^\circ b^\circ\}$
 .&. $e/b;z!e^\circ$.&. $b/b;z!b^\circ)$ 4, 5 $(\mathrm{T}_{4.5})$c, $*(\mathrm{A}_8)$ $*$ (s).
9. $(z)(\exists y^\circ, e^\circ)(\{Aey\}/e;z!\{Ae^\circ y^\circ\}$
 .&. $e/e;z!e^\circ$.&. $y/e;z!e^\circ)$ 5, 6, $(\mathrm{T}_{4.5})$c, $*(\mathrm{A}_8)$, $*(s)$.

10. $\{Aab\}/a;e!\{Aa^\circ b^\circ\}$
 $.\&.\ a/a;e!a^\circ\ .\&.\ b/a;e!b^\circ$ 7, ui, EI.
11. a out b 3, $(T_{4.5})$abc.
12. $b = b^\circ\ .\&.\ a^\circ = e$ 10, 11, $(T_{4.4})$, (A_{13}).
13. $\{Aab\}/a;e!\{Aeb\}$ 11, 12, $*(=)$.
14. $\{Aeb\}/b;y!\{Ae^\circ b^\circ\}\ .\&.\ e/b;y!e^\circ\ .\&.\ b/b;y!b^\circ$ 8, ui, EI.
15. b out e 5, $(T_{4.5})$abc.
16. $e = e^\circ\ .\&.\ b^\circ = y$ 14, 15, $(T_{4.4})$, (A_{13}).
17. $\{Aeb\}/b{:}y!\{Aey\}$ 14, 15, 16, simp, $*(=)$.
18. $\{Aey\}/e;x!\{Ae^\circ y^\circ\}\ .\&.\ e/e;x!e^\circ\ .\&.\ y/e;x!y^\circ$ 9, ui, EI.
19. e out y 6, simp, $(T_{4.5})$b.
20. $y = y^\circ\ .\&.\ e^\circ = x$ 18, 19, $(T_{4.4})$, (A_{13}).
22. $\{Aey\}/e;x!\{Axy\}$ 18, 20, $*(=)$.
23. $\{Aab\}/a;e!\{Aeb\}\ .\&.\ \{Aeb\}/b;y!\{Aey\}\ .\&.\ \{Aey\}/e;x!\{Axy\}$
 13, 17, 22, conj, (18ei), (14ei), (10ei).
24. $(\exists h, e, t)(\{Aab\}/a;e!h\ .\&.\ h/b;y!t\ .\&.\ t/e;x!\{Axy\})$
 23, eg, (5ei).
25. $\{Aab\}/a,b;x,y!\{Axy\}$ 24, df(dual).
26. $(x, y)(\{Aab\}/a,b;x,y!\{Axy\})$ 25, ug.
27. a *in* $\{Aab\}\ .\&.\ b$ *in* $\{Aab\}$ $*\,(A_7)$.
28. a *in* $\{Aab\}\ .\&.\ b$ *in* $\{Aab\}\ .\&.\ (x, y)(\{Aab\}/a,b;x,y!\{Axy\})$
 26, 27, conj.
29. $(\exists p, a, b)(a\ in\ p\ .\&.\ b\ in\ p\ .\&.\ p/a,b;x,y!\{Axy\})$
 28, eg, (3ei), (1ei).

As we can see, definition df(dual) is all that is needed.

No separate axioms analogous to (p) are therefore needed for dual substitutions. Axiom (p) and definitions yield the following theorem:

$(T_{5.5})$
$\vdash\ (p, a, b)(x, y)E!(\iota q)(p/a,b;x,y!q)$.

(We shall call this theorem "$(p)^2$".) In analogy with the case of single substitutions, the following is now an instance of $*14.01$,

df(desc)^2
$[(\iota q)(p/a,b;x,y!q)]\ A((\iota q)(p/a, b;x,y!q)|v) = \text{df}$
 $(\exists q)((r)(p/a,b;x,y!r\ .\equiv.\ r = q)\ \&\ Aq|v)$,

where q is free for v in Av. The expression "$p/a,b;x,y!q$", or its notational variant "$p\frac{x,y}{a,b}!q$" is read "q results from simultaneously substituting x for a and y for b in q." We can parallel single substitutions in having the convenient definition,

df(desc.a)^2 $[p/a,b;x,y]\ A(p/a,b;x,y) = \text{df}$
 $[(\iota q)(p/a,b;x,y!q)]\ A((\iota q)(p/a,b;x,y!q))$.

A scope theorem schema is forthcoming:

$*(\text{scope})^2$
$(p, q)(p \equiv q \ .\supset.\ Cp \equiv Cq)\ :\supset:$
$(\exists q)(p/a,b;x,y!q\ :\&:\ (r)(p/a,b;x,y!r\ .\supset.\ r = q)\ .\&.\ Cq) \equiv$
$C(\exists q)(p/a,b;x,y!q\ :\&:\ (r)(p/a,b;x,y!r\ .\supset.\ r = q)\ .\&.\ q).$

All the same laws apply to definite descriptions involving dual substitution as apply to definite descriptions generally.

In a similar way, definitions are possible for triple "simultaneous" substitutions and higher. Put:

df(triple)
$p/a,b,c;x,y,z!q\ = \text{df}\ (\exists h,e,t,m,s,l)(p/a;e!h\ .\&.\ h/b;m!t$
$.\&.\ t/c;z!s\ .\&.\ s/m;y!l\ .\&.\ l/e;x!q).$

This enables a proof of the schema:

$*(\text{CP}_{\text{sub}})^3$
$\vdash (\exists j, k, u, v)(k\ in\ j\ .\&.\ u\ in\ j\ .\&.\ v\ in\ j\ .\&.$
$(q,p,a)(\ j/k,u,v;q,p,a!\{A\}),$

where j,k,u,v are not free in A, and q,p,a are free in A. The process continues for any finite number of substitutions. We shall then arrive at a theorem $(p)^3$, in analogy with to $(p)^2$. One could generalize putting:

df(n-substitution)
$p/a_1, \ldots, a_n;b_1, \ldots, b_n!q\ = \text{df}$
$(\exists x_1, \ldots, x_{n-1}, y_1, \ldots, y_{n-1}, z_1, \ldots, z_{n-1})(p/a_1;x_1!y_1\ .\&.\ y_1/a_2;x_2!y_2$
$.\&., \ldots, .\&.\ y_{n-2}/a_{n-1};x_{n-1}!y_{n-1}\ :\&:\ y_{n-1}/a_n;b_n!z_1\ .\&.$
$z_1/x_{n-1};b_{n-1}!z_2\ .\&., \ldots, .\&.\ z_{n-1}/x_1;b_1!q)$

$*(\text{CP}_{\text{sub}})^n$
$(\exists p, a_1, \ldots, a_n)(b_1, \ldots, b_n)(a_1\ in\ p\ .\&., \ldots, .\&.\ a_n\ in\ p$
$:\&:\ p/a_1, \ldots, a_n;b_1, \ldots, b_n!\{A\})$

where $p,a_1, \ldots, a_n$ are not free in A, and $b_1, \ldots, b_n$ occur free in A.

New definite descriptions, scope theorem schemata, and the like, then follow. But the formulation of such general theorem schemata and definitions is not necessary. Any individual case can be introduced when needed.

Comprehension and Identity

5.3 From $*(\text{CP}_{\text{sub}})^1$ and (p) it is easy to see that the following theorem schema, which we call "$*(\text{CP})^1$", is valid.

$*(\text{T}_{5.6})^1$
$\vdash\ (\exists p, a)(z)(p/a;z\ .\equiv.\ A),$
where p, a are not free in A, and z is free in A.

Proof:

1. $(\exists p, a)(a \; in \; p \; .\&. \; (z)(p/a;z!\{Az\})$ $\quad * (CP_{sub})^1$.
2. $(z)(p/a;z!\{Az\})$ 1, EI, simp.
3. $p/a;z!\{Az\}$ 2, ui.
4. $p/a;z!q \;\&\; (s)(p/a;z!s \; .\supset. \; s = q)$ (p), ui, EI.
5. $(s)(p/a;z!s \; .\supset. \; s = q)$ 4, simp.
6. $p/a;z!\{Az\} \; .\supset. \; \{Az\} = q$ 5, ui.
7. $\{Az\} = q$ 3, 6, mp.
8. $Az \equiv q$ $(T_{4.7})$ 7, ui, mp.
9. $q \equiv Az$ 8, log.
10. 4 & 9 conj.
11. $(\exists q)10$ eg.
12. $p/a;z \; .\equiv. \; Az$ 12, df(desc)^1.
13. $(\exists p, a)12$ ug, eg.

Similarly, in virtue of $*(CP_{sub})^2$ and $(p)^2$, the following theorem schema called "$*(CP)^2$" is also valid:

$*(T_{5.6})^2$
$\vdash \; (\exists p, a, b)(x, y)(p/a,b;x, y \; .\equiv. \; A)$
where p, a, b are not free in A, and x, y are free in A.

From $*(CP_{sub})^3$ and $(p)^3$, we have

$*(CP)^3$
$\vdash \; (\exists j, k, u, v)(q, p, a)(j/k,u,v;q,p,a \; .\equiv. \; A)$,
where j, k, u, v are not free in A, and q, p, a are free in A.

The process continues for any finite number of substitutions.

Russell went on to introduce a number of convenient definitions so that substitutional grammar would emulate the grammar of type indexed predicate variables. To proxy

$$(x)(\phi x \equiv \psi x)$$

for the lowest type, Russell put (*SCR*, p. 163 fn):

$$p/a = r/c \quad = \text{df } (x)\left(p\frac{x}{a} \equiv r\frac{x}{c}\right);$$

then for the next type,

$$q/p,a = s/t,w \quad = \text{df } (r, c)\left(q\frac{r,c}{p,a} \equiv s\frac{r,c}{t,w}\right),$$

and so on. Russell's use of the identity sign here is unfortunate, however. The identity symbol is best reserved for a substitutional transcription of the identity sign flanked by predicate variables. Thus, we shall replace Russell's definition with the more appropriate triple bar. We have:

$\text{df}(\equiv.a)^1 \quad p/a \equiv r/c \quad = \text{df } (x)(p/a;x \; .\equiv. \; r/c;x)$,
$\text{df}(\equiv.a)^2 \quad q/p,a \equiv s/t,w = \text{df } (r, c)(q/p,a;r,c \; .\equiv. \; s/t,w;r,c)$,

and so on. From the transitivity, reflexivity, and symmetry of 'implication' the following theorems are immediately seen to be valid by $*(\text{scope})^1$.

$(T_{5.7})^1 \quad \vdash (p, a)(p/a \equiv p/a)$

Proof:
1. $(\exists^1 q)(p/a;x!q \;\&\; q) \;.{\equiv}.\; (\exists^1 q)(p/a;x!q \;\&\; q)$ taut.
2. $p/a;x \;.{\equiv}.\; p/a;x$ 1, $*(\text{scope})^1$, df(desc)^1, $\text{df(desc.}a)^1$, $\text{df}(\exists^1)$.
3. $(x)2$ ug.
4. $p/a \equiv p/a$ 3, $\text{df}(\equiv.a)$.
5. $(p, a)4$ ug.

$(T_{5.8})^1 \quad \vdash (p, a)(l, m)(p/a \equiv l/m \;.{\supset}.\; l/m \equiv p/a)$
$(T_{5.9})^1 \quad \vdash (p, a)(l, m)(r, c)(p/a \equiv l/m \;.\&.\; l/m \equiv r/c \;:{\supset}:\; p/a \equiv r/c)$
$(T_{5.10})^1 \quad \vdash (p, a)(x)({\sim}\,(p/a;x) \;.{\equiv}.\; {\sim}\,(\exists^1 q)(p/a;x!q \;\&\; q))$
$(T_{5.11})^1 \quad \vdash (p, a)(l, m)(p/a \equiv l/m \;.{\equiv}.\; (x)[(\exists^1 q)(p/a;x!q \;\&\; q) \equiv (\exists^1 q)(l/m;x!q \;\&\; q)]$.

The proofs are similar and left as an exercise. As expected, we have analogs of these theorems in higher types:

$(T_{5.7})^2 \quad \vdash (q, p, a)(q/p,a \equiv q/p,a)$
$(T_{5.8})^2 \quad \vdash (q, p, a)(s, t, w)(q/p,a \equiv s/t,w \;:{\supset}:\; s/t,w \equiv q/p,a)$
$(T_{5.9})^2 \quad \vdash (q, p, a)(s, t, w)(h, d, e)(q/p,a \equiv s/t,w \;.\&.\; s/t,w \equiv h/d,e \;:{\supset}:\; q/p,a \equiv h/d,e)$.
$(T_{5.10})^2 \quad \vdash (q, p, a)(r, c)({\sim}\,(q/p,a;r,c) \;.{\equiv}.\; {\sim}\,(\exists^1 t)(q/p,a;r,c!t \;\&\; t))$
$(T_{5.11})^2 \quad \vdash (q, p, a)(s, t, w)(q/p,a \equiv s/t,w \;.{\equiv}.\; (r, c)[(\exists^1 h)(q/p,a;r,c!h \;\&\; h) \;.{\equiv}.\; (\exists^1 h)(s/t,w;r,c!h \;\&\; h)]$.

These follow easily from $*(\text{scope})^2$. The process is similar as we increase the number of substitutions.

To proxy the identity sign flanking predicate variables, Russell needs a replacement for a statement such as "$\phi = \phi$". Excluding the issue of predicativity, *Principia* has the typically ambiguous:

$*13.01 \quad x = y \;=\text{df}\; (f)(fx \equiv fy)$.

So substitution needs to proxy,

$(f)(f\phi \equiv f\phi)$,

for each type in question. Accordingly, we shall allow the identity symbol to flank matrices by means of the following definitions:

$\text{df}(=.a)^1$
$p/a = r/c \;=\text{df}\; (s, t, w)(s/t,w;r,c \;.{\equiv}.\; s/t,w;p,a)$
$\text{df}(=.a)^2$
$s/t,w = q/p,a \;=\text{df}\; (\exists j, k, u, v)(j/k,u,v;s,t,w \;.{\equiv}.\; j/k,u,v;q,p,a)$,

and so on. We then have the expected theorems ensuring that "=" (when flanked by matrices) is (as it were) an equivalence relation—reflexive, symmetric, and transitive.

$(\mathrm{T}_{5.12})^1 \quad \vdash (1, m)(l/m = l/m)$
$(\mathrm{T}_{5.13})^1 \quad \vdash (l, m)(r, c)(l/m = r/c \ .\supset. \ r/c = l/m)$
$(\mathrm{T}_{5.14})^1 \quad \vdash (l, m)(r, c)(t, w)(l/m = r/c \ .\&. \ r/c = t/w \ .\supset. \ l/m = t/w)$

Similarly, we have:

$(\mathrm{T}_{5.12})^2 \quad \vdash (l, m, n)(l/m, n = l/m, n)$
$(\mathrm{T}_{5.13})^2 \quad \vdash (l, m, n)(s, t, w)(l/m,n = s/t,w \ .\supset. \ s/t,w = l/m,n)$
$(\mathrm{T}_{5.14})^2 \quad \vdash (l, m, n)(s, t, w)(q, p, a)(l/m,n = s/t,w \ .\&.$
$s/t,w = q/p,a \ :\supset: \ l/m,n = q/p,a).$

The pattern should be clear.

Now identity should imply coextensivity; and so indeed, it follows from our substitutional definitions.

$(\mathrm{T}_{5.15})^1 \quad \vdash (l, m)(r, c)(l/m = r/c \ .\supset. \ l/m \equiv r/c).$

Proof:

1. $\rightarrow \ l/m = r/c$
2. $(s, t, w)(s/t,w;l,m \ .\equiv. \ s/t,w;r,c)$ 1, df$(=.a)^1$.
3. $(\exists h, d, e)(q,b)(h/d,e;q,b \ .\equiv. \ l/m \equiv q/b)$ $*\,(\mathrm{CP})^2$.
4. $(q, b)(h/d,e;q,b \ .\equiv. \ l/m \equiv q/b)$ 3, EI.
5. $h/d,e;l,m \ .\equiv. \ h/d,e;r,c$ 2, ui.
6. $h/d,e;l,m \ .\equiv. \ l/m \equiv l/m$ 4, ui.
7. $h/d,e;r,c \ .\equiv. \ l/m \equiv r/c$ 4, ui.
8. $l/m \equiv l/m$ $(\mathrm{T}_{5.7})^1$, ui.
9. $l/m \equiv l/m \ .\supset. \ 1/m \equiv r/c$ 5, 6, 7, df$(\equiv.\mathrm{a})^1$, log.
10. __ $l/m \equiv r/c$ 8, 9, mp (4ei).
11. $1 \supset 10$ cp.
12. $(l,m,r,c)11$ ug.

As expected, we also have:

$(\mathrm{T}_{5.15})^2 \quad \vdash (l, m, n)(s, t, w)(l/m,n = s/t,w \ .\supset. \ l/m,n \equiv s/t,w),$

and so on. Recall that the definition of the identity sign is:

df(=)

$x = y = \mathrm{df} \ (p, a)(q, r)(p/a;x!q \ \& \ p/a;y!r \ .\supset. \ q \supset r).$

Since matrical expressions are not singular terms, it would be improper to attempt to apply df(=) to an expression of the form "$l/m = r/c$." However, we can prove the following:

$(\mathrm{T}_{5.16})^1$
$\vdash (p, a)(r, c)(p/a = r/c \ :\equiv: \ p = r \ .\&. \ a = c).$

Proof:

1. $(\exists l, m, o)(r, c)(l/m{,}o;r{,}c \;{:}{\equiv}{:}\; p = r\ .\&.\ a = c)$ $\quad * (\mathrm{CP})^2$.
2. $(r, c)(l/m{,}o;r{,}c \;{:}{\equiv}{:}\; p = r\ .\&.\ a = c)$ $\quad$ 1, EI.
3. $\rightarrow\ p/a = r/c$
4. $\quad (s, t, w)(s/t{,}w;p{,}a\ .{\equiv}.\ s/t{,}w;r.c)$ $\quad 3\mathrm{df}(=.a)^1$.
5. $\quad p = p\ .\&.\ a = a$ $\quad (\mathrm{T}_{4.8})$a, ui, conj.
6. $\quad l/m{,}o;p{,}a \;{:}{\equiv}{:}\; p = p\ .\&.\ a = a$ $\quad$ 2, ui.
7. $\quad l/m{,}o;p{,}a\ .{\equiv}.\ l/m{,}o;r{,}c$ $\quad$ 4, ui.
8. $\quad l/m{,}o;r{,}c \;{:}{\equiv}{:}\; p = r\ .\&.\ a = c$ $\quad$ 2, ui.
9. $\quad p = p\ .\&.\ a = a \;{:}{\equiv}{:}\; p = r\ .\&.\ a = c$ $\quad$ 6, 7, 8, df($\equiv$) log.
10. $\underline{\quad}\ p = r\ .\&.\ a = c$ $\quad$ 5, 9, (df $\equiv$), simp, mp.
11. $3 \supset 10$ $\quad$ cp.
12. $\rightarrow\ p = r\ .\&.\ a = c$
13. $\quad p/a = p/a$ $\quad (\mathrm{T}_{5.12})^1$
14. $\underline{\quad}\ p/a = r/c$ $\quad$ 12, 13, $*(=)$.
15. $12 \supset 14$ $\quad$ cp.
16. 11 & 15 $\quad$ conj, (2ei).
17. $p/a = r/c \;{:}{\equiv}{:}\; p = r\ .\&.\ a = c$ $\quad$ 16, df($\equiv$).
18. $(p, a)(r, c)17$ $\quad$ ug.

Similar theorems are possible for higher types. Thus, for example,

$$(\mathrm{T}_{5.16})^2 \quad \vdash\ (s, t, w)(q, p, a)(s/t{,}w = q/p{,}a \;{:}{\equiv}{:}\; s = q\ .\&.\ t = p\ .\&.\ w = a).$$

The nonpropositional formulations seem more natural, however, since our aim is to show the close relationship between the language of predicate variables and that of substitution.

It is interesting to recall that in Church's reconstruction of Russell's early intensional logic, propositional functions along with propositions were assumed. Church then sought an axiom schema which characterized the identity of propositional functions in terms of the identity of the propositions that are their values. For the lowest type, Church has an analog of the following (Church 1984, p. 514):

$$(x)(\{\phi x\} = \{\psi x\})\ .{\supset}.\ \phi = \psi.$$

Anderson (1986) objected to this schema, and rightly so. Consider the substitutional counterpart,

$$(p, a)(r, c)((x)(p/a;x = r/c;x)\ {:}{\supset}{:}\ p/a = r/c).$$

It is easy to find a counterexample. We find distinct a and c such that

$$(z)(\{\sim a\}/a;z!\{\sim z\})$$
$$(z)(\{\sim c\}/c;z!\{\sim z\}).$$

It follows that

$$(x)(\{\sim a\}/a;x = \{\sim c\}/c{:}x),$$

for the elimination of the descriptions yields

$$(x)((\exists q)[\{\sim a\}/a;x!q \ \&\ (s)(\{\sim a\}/a;x!s\ .\supset.\ s = q)\ \&\ (\exists t)(\{\sim c\}/c;x!t\ .\&.\ (s)(\{\sim c\}/c;x!s\ .\supset.\ s = t)\ .\&.\ q = t)]).$$

But since a and c are distinct, we know that

$$\sim (\{\sim a\}/a = \{\sim c\}/c).$$

Propositional functions (as proxied through the logic of propositions) can have all the same values for the same arguments and yet be quite distinct. By returning to *Principles*, Anderson hoped to find a new axiom schema. His amendments leave him with an analog of the following (for the lowest types):

$$(\exists x, y)(x \neq y)\ :\supset:\ (x, y)(\{\phi xy\} = \{\psi xy\})\ .\supset.\ \phi = \psi.$$

Substitutionally, this would be,

$$(p, a, r, c, b, e)[\ (\exists x, y)(x \neq y)\ .\supset.\ (x, y)(p/a,b;x,y = r/c,e;x,y)\ .\supset.\ p/a,b = r/c,e].$$

But from the perspective of the substitutional theory, Anderson's replacement is just as objectionable as was Church's schema. We find m, n, u, and v with m distinct from u and which are such that

$$(x, y)(\ \{m = n\}/m,n;x,y!\{x = y\})$$

$$(x, y)(\ \{u = v\}/u,v;x,y!\{x = y\}).$$

We have a counterexample to Anderson's schema. There is no point in searching further. A schema which uses the identity of the propositional values of propositional functions to form identity conditions for propositional functions will not be found in *Principles*. The work did not assume an ontology of propositional functions in the first place. In contrast, the substitutional theory, a theory which did grow out of *Principles*' concern to have only one style of unrestricted individual variable, shows how the identity of propositional functions is to be proxied in terms of the identity of propositions.

Types as Logical Grammar

5.4 The substitutional theory is entirely type-free. There are no types of *entities*, and the language of the theory has individual variables as its only variables. Nevertheless, we are beginning to realize that the substitutional theory proxies a type stratified calculus with predicate variables in subject as well as predicate positions. The hierarchy of types is built directly into the logical grammar. To see this clearly, we shall set out a formulation of the theory of "simple" types, show that every object-language formula in primitive notation of the language of simple types has a translation in the language of substitution, and show that for every theorem of the theory of simple types there is a theorem of the substitutional theory.

In setting out simple type theory, we begin with the following definition of a type index:

> A type index is any symbol which satisfies the following recursive definition (i) "o" is a type index; (ii) if $t_1, \ldots, t_n$ are type indices, then the expression $(t_1, \ldots, t_n)$ is a type index; (iii) these are the only type indices.

Where t is a type index, the variables of the language shall be "x^t", "y^t, "z^t", and any of these followed by one or more subscripted whole numerals. (For convenience, we shall informally use whole numeral subscripts instead of primes; and these will be omitted where clashes do not arise.) The predicate variables are any variables whose type index is not "o". Informally, we use as predicate variables "ϕ^t","ψ^t","Σ^t", where the superscripted t is not "o". Thus, "$\phi^{(o)}$" is a monadic predicate variable and "$\phi^{(o,o)}$" is a dyadic predicate variable. Similarly for the next type we have "$\phi^{((o))}$" and "$\phi^{((o),(o))}$". We also have predicate variables of mixed types, such as "$\phi^{(o,(o))}$".

The primitive signs are the variables, the brackets "(",")", the statement connectives "$\rightarrow$" and "$\neg$", our type indices, and the sign "$\forall$" used for the universal quantifier. The terms are just the variables. As usual, we use "μ", "ν" metalinguistically for any term, and "A", "B", "C", for wffs. Atomic formulae are of the form:

$$\mu^{(t_1,\ldots,t_n)}(\nu_1^{t_1}, \ldots, \nu_n^{t_n}),$$

where $\mu^{(t_1,\ldots,t_n)}$ is a predicate variable. The wffs are defined recursively as the smallest set K containing all the atomic wffs and such that $\ulcorner(A \rightarrow B)\urcorner$, $\ulcorner(\neg A)\urcorner$, and $\ulcorner(\forall\mu)C\mu\urcorner$ are in K whenever $\ulcorner A\urcorner$, $\ulcorner B\urcorner$ and $\ulcorner C\mu\urcorner$ are in K.

The formal calculus for the theory of simple types is as follows:

Ax1 $A \,.\!\rightarrow\!.\, B \rightarrow A$
Ax2 $A \,.\!\rightarrow\!.\, B \rightarrow C :\rightarrow: A \rightarrow B \,.\!\rightarrow\!.\, A \rightarrow C$
Ax3 $\neg B \rightarrow \neg A :\rightarrow: A \rightarrow B$
Ax4 $(\forall\mu^t)(A \rightarrow B\mu^t) :\rightarrow: A \rightarrow (\forall\mu^t)B\mu^t$, where μ^t is not free in A.
Ax5 $(\forall\mu^t)A\mu^t \rightarrow A\nu^t|\mu^t$, where ν^t is a term free for μ^t in A.
Ax6 (Comprehension)
$(\exists\phi^{(t_1,\ldots,t_n)})(\forall\mu_1^{t_1}), \ldots, (\forall\mu_n^{t_n})(\phi^{(t_1,\ldots,t_n)}(\mu_1^{t_1}, \ldots, \mu_n^{t_n}) \leftrightarrow A)$, where $\phi^{(t_1,\ldots,t_n)}$ is not free in A.
Rule 1 (modus ponens)
From A and $A \rightarrow B$, infer B
Rule 2 (universal generalization)
From $A\mu^t$ infer $(\forall\mu^t)A\mu^t$, where μ^t is a variable free in A.

For basic definitions we have:

$$\mu^t = \nu^t = \text{df } (\forall\phi^{(t)})(\phi^{(t)}(\mu^t) \leftrightarrow \phi^{(t)}(\nu^t))$$
$$A \,\&\, B = \text{df} \neg (A \rightarrow \neg B)$$

$A \vee B = \text{df} \neg A \rightarrow B$

$A \leftrightarrow B = \text{df} \ (A \rightarrow B) \ \& \ (B \rightarrow A)$

$(\exists \mu^t)A = \text{df} \neg (\forall \mu^t) \neg A.$

This completes the system.

Consider next the following pairs. The first is in the language of simple type theory, the second is its analog in the language of the substitutional theory.

$$
\begin{array}{ll}
\#1 & (\exists \phi^{(o)})(\forall x^{o})(\phi^{(o)}(x^{o}) \ .\!\leftrightarrow\!. \ x^{o} = x^{o}). \\
\#1S & (\exists p, a)(x)(p/a;x \ .\!\equiv\!. \ x = x). \\
\#2 & (\exists \psi^{((o))})(\forall \phi^{(o)})(\psi^{((o))}(\phi^{(o)}) \ .\!\leftrightarrow\!. \ \phi^{(o)} = \phi^{(o)}) \\
\#2S & (\exists q, p, a)(r, c)(q/p,a;r,c \ .\!\equiv\!. \ r/c = r/c) \\
\#3 & (\exists \psi^{((o))})(\forall \phi^{(o)})(\psi^{((o))}(\phi^{(o)}) \ .\!\leftrightarrow\!. \ (\forall x^{o}) \neg \phi^{(o)}(x^{o})) \\
\#3S & (\exists q, p, a)(r, c)(q/p,a;r,c \ .\!\equiv\!. \ (x) \sim (r/c;x)).
\end{array}
$$

For dyadic predicate variables, we have:

$$
\begin{array}{ll}
\#4 & (\exists \Sigma^{(o,o)})(\forall x^{o})(\forall y^{o})(\Sigma^{(o,o)}(x^{o}, y^{o}) \ .\!\leftrightarrow\!. \ x^{o} = y^{o}) \\
\#4S & (\exists q, p,a)(r, c)(q/p,a;r,c \ .\!\equiv\!. \ r = c) \\
\#5 & (\exists \Sigma^{((o),(o))})(\forall \phi^{(o)})(\forall \psi^{(o)})(\Sigma^{((o),(o))}(\phi^{(o)}, \psi^{(o)}) \\
 & \qquad .\!\leftrightarrow\!. \ \phi^{(o)} = \psi^{(o)}) \\
\#5S & (\exists q, p, a, r, c)(l, m)(d, e)(q/p,a,r,c;l,m,d,e \ .\!\equiv\!. \ l/m = d/e)
\end{array}
$$

For unequal-leveled relations, we have:

$$
\begin{array}{ll}
\#6 & (\exists \Sigma^{((o),o)})(\forall \phi^{(o)})(\forall x^{o})(\Sigma^{((o),o)}(\phi^{(o)}, x^{o}) \ .\!\leftrightarrow\!. \ \phi^{(o)}(x^{o})) \\
\#6S & (\exists q, p, a, x)(r, c)(y)(q/p,a,x;r,c,y \ .\!\equiv\!. \ r/c;y) \\
\#7 & (\exists \Sigma^{((o,o))})(\forall \phi^{(o,o)})(\Sigma^{((o,o))}(\phi^{(o,o)}) \ .\!\leftrightarrow\!. \ \phi^{(o,o)} = \phi^{(o,o)}) \\
\#7S & (\exists q, p, a, x)(r, c)(y)(q/p,a,x;r,c,y \ .\!\equiv\!. \ r/c,y = r/c,y).
\end{array}
$$

The first of the pairs are instances of the comprehension axiom schema (our Ax6 above) in the type stratified language of predicate variables, the second are their translations into the type-free substitutional language. They are instances of the substitutional theorem schemata $*(\text{CP})^1$–$*(\text{CP})^4$ respectively.

As we see, the translation is accomplished by the use of "matrices." A matrix is nothing but a sequence of symbols that forms a part of the expression of a substitution. For example, "*p/a*" is part of the expression "*p/a;b!q*" and so part of the definite description "(ɿ*q*)(*p/a;b!q*)". Both the symbol "*p*" (which is called the "prototype" of the matrix) and the symbol "*a*" (called the "argument") designate single entities. A matrix such as "*p/a*" will be used to proxy the use of a predicate variable of type (o). Here only one entity is substituted. Predicate variables of type ((o)) are proxied by using matrices such as "*q/p,a*", where double substitutions are involved. Similarly, predicate variables of type (((o))) are proxied by use of matrices of the form "*t/q,p,a*"; and so forth.

The type of the predicate variable proxied is related, therefore, to the number of the substitution (argument) places of the matrix used to form its proxy. The matrical form is determined by the number of its arguments. Observe, however, that the type of a predicate variable is not proxied solely by the number of argument places of a matrix. A matrical form such as "*q/p,a*" may be used to proxy the use of a predicate variable of type ((o)) but it may also be used to proxy use of a predicate variable of type (o, o). The matrical form is therefore not the sole substitutional counterpart of the type of a predicate variable. We must look at the comprehension principle governing the use of the matrix in question. We can see this clearly from our illustration above. Compare #4S and #2S, and also compare #6S and #7S. This is not a weakness of the substitutional language, however. Quite the contrary, it is a strength. That is, it is an immediate consequence of having only individual variables.

Another strength, and substitution's raison d'être, is that self-predication "$\phi(\phi)$" cannot be proxied in the substitutional grammar. The matrical forms used in proxying the use of predicate variables will be "p/a", or "$s/t,w$", or "$j/k,u,v$" and so on. Monadic predications are proxied in the substitutional grammar by using expressions "$p/a;x$", "$s/t,w;r,c$", "$j/k,u,v;q,p,a$", etc. So a predication such as "$\phi(\phi)$" cannot be proxied. A sequence such as "$p\frac{(p/a)}{a}$", as well as its negation, is ungrammatical because the symbol "p/a" is not a singular term of the substitutional language. The same is true as we ascend matrical forms. Type theory becomes part of the logical grammar of substitution.

We have given a few illustrations of translations from a type regimented language of predicate variables into the type-free language of substitution. The point can be made general. Any closed object language wff *in the primitive notation* of simple type theory can be translated into a wff in the type-free language of substitution. (We allow no predicate term forming operators such as Church's lambda operator.) We can see this by means of the recursive translation function, c, defined below.

(i) $c[(\forall x^{o})Ax^{o}] = (x)\ c[Ax|x^{o}]$

(ii) $c[\forall\phi^{(t_1,\ldots,t_n)})A\phi^{(t_1,\ldots,t_n)}]$
$= (p\vec{t}_1,\ldots,\vec{t}_n)\ c[\ A\ p/\vec{t}_1,\ldots,\vec{t}_n|\phi^{(t_1,\ldots,t_n)}]$,
where each $\vec{t}_i$, $1 \leq i \leq n$, is $z_1,\ldots,z_j$ for
$j =$ (the number of left-hand parentheses in t_i plus the number of occurrences of the symbol "o" in t_i).

(iii) $c[p/a(x)] = (\exists h)(p/a;x!h\ .\&.\ h)$

(iv) $c[p/x_1,\ldots,x_{n+1}(q/y_1,\ldots,y_n)] =$
$(\exists h)(p/x_1,\ldots,x_{n+1};q,y_1,\ldots,y_n!h\ .\&.\ h)$

(v) $c[\neg A] = \sim c[A]$

(vi) $c[A \rightarrow B] = c[A] \supset c[B]$.

It must be understood that new variables are to be chosen in case of any possible clash.

With our translation function at hand, we are in a position to see that any object-language thesis of the theory of simple types has an analog in the language of substitution that is a thesis of the substitutional theory. We have only to show:

(1) Any instance of the axiom schemata of simple type theory besides (Ax6) has an analog, under our translation function, c, that is an instance of the axiom schemata of the substitutional theory.
(2) Any object-language instance of (Ax6) of simple type theory has, under our translation function, c, an analog which is an instance of one of the comprehension theorem schemata of the substitutional theory.
(3) If B follows from A and $A \rightarrow B$, by Rule 1 of simple type theory, then $c[B]$ follows from $c[A]$ and $c[A \rightarrow B]$ of the substitutional theory.
(4) If $(\forall\mu^t)A\mu^t$ follows from $A\mu^t$ by Rule 2 of simple type theory, then $c[(\forall\mu^t)A\mu^t]$ follows from $c[A\mu^t]$ in the substitutional theory.

From the nature of our translation function, 1, 3, and 4 should be clear; and 2 is made clear from the above considerations concerning comprehension axioms. Accordingly, we can transform any proof in primitive notation of the object language of the theory of simple types into a proof within the substitutional theory.

5.5 Our translation function demonstrates that the type stratified language of predicate variables proxied in the substitutional theory solves the paradoxes in a way that is consistent with the doctrine of the unrestricted variable. All entities, propositions or otherwise, are treated on a par as individuals. The logical grammar is completely type-free. Indeed, there is not even a distinction between predicate variables and individual variables. There are no predicate variables. Nonetheless, the theory of definite descriptions extended (by means of substitution) to descriptions for propositions emulates a theory of types of "propositional functions"—that is, it proxies a simple type theory. The substitutional proxy builds type stratification into its logical grammar.

In this way, the formal solution of the paradoxes—viz., the theory of logical types–is reconciled with the demand of the universality of pure logic and the unrestricted variable essential to its calculus. It is instructive to recall Quine's diagnosis of the paradox of predication (and of classes). As Quine sees it, the calculus for logic must involve only individual variables and be type-free. The comprehension principle,

$$(\exists\phi)(x)(\phi x \leftrightarrow Ax).$$

seems straightforwardly won by *existential generalization* from

$$(x)(Ax \leftrightarrow Ax).$$

And yet comprehension generates Russell's paradox of predication. The error, in Quine's view, lies in "slighting the vital contrast between schematic letters and quantifiable variables" (Quine 1963, p. 258). This is precisely Russell's own view—although it was expressed with characteristic use–mention confusion as the doctrine that "ϕ must not be varied independently of x in ϕx". In agreement with Russell, Quine makes the presence of one style of individual variable definitive of a calculus for logic. Contrary to Quine, however, there is no reason to conclude that there is an important cleavage between logic and a (type stratified) second-order system with sets (or nominalized predicates). Quine overlooked an option, Russell's substitutional theory.

The lesson Russell drew was that since a calculus for logic must adopt one style of entity (individual) variable, we must follow—as it were—a sort of agnostic *eliminativism* with respect to the languages of classes and predicate variables. The calculus for logic must proceed without the assumption of classes or attributes (propositional functions). There are, for example, no entities that must be identified as Zeus's thunderbolts. That language and its ontology is to be supplanted by another. We can get along for explanatory purposes by appeal to electric discharges. Similarly, the language of substitution, which is just Russell's calculus for logic extended by the theory of definite descriptions, is to replace languages of classes and predicate variables. In the language of substitution, types become logical grammar, and so also does a type stratified theory of classes.

This is why Russell would later recollect that the theory of definite descriptions was "the first step towards overcoming the difficulties which had baffled me for so long" (*MPD*, p. 229). In "My Mental Development" he explains further (*MMD*, p. 14):

> What was of importance in this theory was the discovery that, in analyzing a significant sentence, one must not assume that each separate word or phrase has significance on its own account. "The golden mountain" can be part of a significant sentence, but is not significant in isolation. It soon appeared that class-symbols could be treated like descriptions, i.e., as non-significant parts of significant sentences. This made it possible to see, in a general way, how a solution of the contradictions might be possible.

Already skeptical about surface grammar, *Principles* had embraced Frege's analysis of number-words, accepted Cantor's theory of the infinite (which broke the hold of centuries of false antinomies based on the allure of "common sense" intuitions about the infinite), worked out the theory of relations, and provided a theory of denoting concepts adequate to avoid Meinongian impossibilia. But the theory of definite descriptions was needed before Russell could take the next step. Surface grammar was to be blamed for more than the conundrum of the existent golden mountain's being both existent and nonexistent. Armed with the new theory of definite descriptions (and the substitutional theory), Russell hoped to show that the antinomies are the result of embracing ontological assumptions imposed by surface grammar. An ontology of propositional functions, classes, and relations-in-extension, is unnecessary. The substitutional theory allows Russell to generate a proxy for a type regimented language which allows bindable predicate variables in subject and predicate positions. Through it, he will be able to go on to proxy a simple type-theory of classes and relations-in-extension from the pure type free calculus for the logic of propositions.

CHAPTER SIX

The "No-Classes" Theory

6.0 In the last chapter we discussed the theory of propositional functions proxied in the substitutional theory by the use of matrices. We now turn to classes. It not at all easy to recover the treatment of class symbols from Russell's writings and publications. Russell's published works are sketchy about the details. Many worknotes have been lost and one must piece together the theory of classes from the remaining fragments. There is a brief development of the theory of classes in "On the Substitutional Theory of Classes and Relations." This continues in the September 1906 "Les paradoxes de la logique." Nonetheless, in the April–May manuscript "On Substitution" there are hints of another approach which, if we can extrapolate from the scant remarks, offers a new treatment of the extensionality of classes and a correction of flaws in the earlier approach.

We know that the theory underwent modifications until reaching its mature stage in the 1907 "Mathematical Logic as Based on the Theory of Types." Indeed, we can recover the final stage because Chapter 5 taught us how to transform wffs in the type indexed language of predicate variables of "mathematical logic" into the substitutional. Thus, bracketing the matter of predicativity, we can recover what *Principia*'s theory of classes would look like if the substitutional notation had carried the day. The chapter closes by comparing the substitutional theory of class symbols with that of *Principia*.

Before we begin, it will be helpful to first make the following adjustment to notation. The reader can attest to the fact that the proliferation of commas in substitutional matrices is cumbersome. It is best to adopt the following remedy. We shall now write "$A(rc)$" instead of "$A(r,c)$" and "$q/pa;rc$" instead of "$q/p,a;r,c$". Similarly, we put "$(stw)(s/tw;rc\ .\supset.\ p)$" instead of "$(s, t, w)\ (s/t,w;r,c\ .\supset.\ p)$" and "$(\exists stw)(s/tw;rc\ .\supset.\ p)$" instead of "$(\exists s, t, w)\ (s/t,w;r,c\ .\supset.\ p)$", and so on for higher types of formulae.

Classes as Extensional Propositional Functions

6.1 A "class" is to be regarded as the extension of a monadic propositional function, though (of course) no ontology of propositional functions is to be assumed. A propositional function f of propositional functions is extensional when,

$$(\phi)(\psi)[(x)(\phi x \; .\!\equiv\!. \; \psi x) \; .\!\supset\!. \; f(\phi) \; .\!\equiv\!. \; f(\psi)].$$

Thus, if q, p, and a are to be used to proxy an extensional propositional function we should expect an analog.

In the last chapter we noted that Russell adopted the following definitions,

$$p/a \equiv r/c = \text{df } (x)(p/a;x \; .\!\equiv\!. \; r/c;x)$$
$$q/pa \equiv s/tw = \text{df } (rc)(q/pa;rc \; .\!\equiv\!. \; s/tw;rc),$$

and so on.[1] In manuscripts we also find,

$$x \in p/a = \text{df } p/a;x,$$

and extrapolating, there would also be:

$$p/a \in s/tw = \text{df } s/tw;pa$$
$$s/tw \in j/kuv = \text{df } j/kuv;stw,$$

and so on. Substitutional proxies for quantification over classes require restrictions to extensional propositional functions. To accomplish this, Russell introduces:[2]

$$\text{cls}^2(q, p, a) \; = \text{df}$$
$$(rc)(tw)(r/c \equiv t/w \; :\supset: \; q/pa;rc \; .\!\equiv\!. \; q/pa;tw)$$
$$\text{cls}^3(l, q, p, a) \; = \text{df}$$
$$(frc)(stw)(\text{cls}^2(f, r, c) \;\&\; \text{cls}^2(s, t, w) \; .:\supset:. $$
$$f/rc \equiv s/tw \; :\supset: \; l/qpa;stw \; .\!\equiv\!. \; l/qpa;frc).$$

Similar definitions are available for higher types. Russell uses "cls^2" to mean "class of classes of individuals"; "cls^3" means "class of classes of classes of individuals," etc.

For the algebra of classes, Russell's early writings suggest that he had something like the following in mind:

$$p/a \;\cup\; r/c = \text{df } \{e* \in p/a \; .\!\vee\!. \; e* \in r/c\}/e*$$
$$p/a \;\cap\; r/c = \text{df } \{e* \in p/a \; .\&. \; e* \in r/c\}e*$$
$$\breve{p}/a = \text{df } \{\sim p\}/a$$

1. Recall that we used "≡" instead of Russell's early use of "=", thereby freeing the identity sign for proxying the identity of the propositional functions in question. See Chapter 5.

2. See *OS*, p. 117 and also *SCR*, p. 176. I have altered Russell's original notation in accordance with our use of "≡".

Then a theorem such as

$$(\alpha)(\beta)(\overparen{\alpha \cup \beta} = \breve{\beta} \cap \breve{\alpha})$$

would take the form,

$$(pa)(rc)(\overparen{p/a \cup r/c} \;.\!\equiv\!.\; \breve{p}/a \cap \breve{r}/c)$$

A proxy for the schema of class abstraction within each type is forthcoming within this early approach. For instance, there would be:

$$\begin{array}{ll} *(\text{Abs})^1 & \vdash\ y \in \{Ac\}/c \;.\!\equiv\!.\; Ay \\ *(\text{Abs})^2 & \vdash \{Ac\}/c \in \{B(p, a)\}/p,a \;.\!\equiv\!.\; B(\{Ac\}, c) \\ *(\text{Abs})^3 & \vdash\ \{B(p, a)\}/p,a \in \{C(s, t, w)\}/s,t,w \;.\!\equiv\!.\; C(\{B(p, a)\}, p, a). \end{array}$$

In some ways this approach is certainly cumbersome, but not unduly so. Indeed, some definitions remain fairly normal. Consider, for example, those for "$\subseteq$" (subclass). We have:

$$\begin{aligned} p/a \subseteq r/c &= \text{df } (x)(x \in p/a \;.\!\supset\!.\; x \in r/c) \\ s/tw \subseteq q/pa &= \text{df } (rc)(r/c \in s/tw \;.\!\supset\!.\; r/c \in q/pa) \\ j/kuv \subseteq l/mno &= \text{df } (qpa)(\text{cls}^2(qpa) \;:\!\supset\!:\; \\ &\qquad\qquad q/pa \in j/kuv \;.\!\supset\!.\; q/pa \in l/mno) \end{aligned}$$

The notation for the algebra of classes largely tracks its usual formulation.

As we see, Russell's plan for class symbols is simply to choose matrices which proxy the use of predicate terms for extensional attributes (propositional functions). It is important, in this respect, not to be misled by Russell's remark that substitution is somehow "intermediate between intension and extension." He wrote (*SCR*, p. 175):

> Although a matrix is more akin to a class or relation *in extension* than in *intension*, it is not quite extensional; for even if p/a and p'/a' define the same class, they can still be distinguished if p is different from p' or a from a'. Thus the theory advocated is intermediate between that of intension and extension.

To be sure, many of the matrices used in the early substitutional treatment of class symbols proxy the use of predicate terms for extensional propositional functions. But we must not forget that substitution is an intensional calculus which proxies an intensional calculus of propositional functions whose grammar admits predicate variables in subject and predicate positions. Russell's early plan for classes was to simply choose substitutional proxies for predicate terms standing for extensional propositional functions and identify these as classes. But there remain a host of substitutional proxies for predicate terms that stand for intensional propositional functions.

The natural numbers are defined by following the general Fregean line of development. For instance, put:

$$V =\text{df } c* = c*$$
$$\Lambda =\text{df } c* \neq c*$$
$$\mathbf{0} =\text{df } (y) \sim (r*/c*;y).$$

Russell can then prove:

$$\vdash\ (y)(y \in V/c*)$$
$$\vdash\ (pa)(p/a \subseteq V/c*)$$
$$\vdash\ (pa)((y) \sim (y \in p/a)\ .\supset.\ p/a \equiv \Lambda/c*).$$
$$\vdash\ (y) \sim (y \in \Lambda/c*)$$
$$\vdash\ \Lambda/c* \in \mathbf{0}/r*c*.$$

Fregean definitions for "successor" and "natural number" are easily given as well:

$$S^{hde} =\text{df } (\exists y)(y \in r*/c*\ .\&.\ \{e* \in r*/c*\ .\&.\ e* \neq y\}/e* \in h/de)$$
$$N =\text{df } (lmno)(\text{cls}^3(lmno)\ .\&.\ \mathbf{0}/r*c* \in l/mno\ .\&.\ (hde)(\text{cls}^2(hde)\ \&\ h/de \in l/mno\ .\supset.\ S^{hde}/r*c* \in l/mno)\ :\supset:\ q*/p*a* \in l/mno).$$

The Peano Postulates would be formulated as follows:

Peano$_1$ $\vdash\ 0 \in \mathbb{N}$

$\vdash\ \mathbf{0}/r*c* \in N/q*p*a*$

Peano$_2$ $\vdash\ (\alpha) \sim (0 = S\alpha)$

$\vdash\ (stw)(\text{cls}^2(stw)\ :\supset:\ \sim(\mathbf{0}/r*c* \equiv S^{stw}/r*c*)\ )$

Peano$_3$ $\vdash\ (\alpha)(\alpha \in \mathbb{N}\ :\supset:\ S\alpha \in \mathbb{N})$

$\vdash\ (qpa)(\ \text{cls}^2(qpa)\ :\supset:\ q/pa \in N/q*p*a*\ \ .\supset.\ S^{qpa}/r*c* \in N/q*p*a*)$

Peano$_4$ $\vdash\ (\alpha\beta)(\alpha \in \mathbb{N}\ \&\ \beta \in \mathbb{N}\ :\supset:\ S\alpha = S\beta\ .\supset.\ \alpha = \beta)$

$\vdash\ (qpa)(stw)(\text{cls}^2(qpa)\ \&\ \text{cls}^2(stw)\ .:\supset:.\ q/pa \in N/q*p*a*\ .\&.\ s/tw \in N/q*p*a*\ :\supset:\ S^{qpa}/r*c* \equiv S^{stw}/r*c*\ \ .\supset.\ q/pa \equiv s/tw)$

Peano$_5$ $\vdash\ (\beta)(0 \in \beta\ \&\ (\alpha)(\alpha \in \beta\ \&\ \alpha \in \mathbb{N}\ .\supset.\ S\alpha \in \mathbb{N})\ .\supset.\ \mathbb{N} \subseteq \beta)$

$\vdash\ (lmno)(\text{cls}^3(lmno)\ .\&.\ \mathbf{0}/r*c* \in l/mno\ .\&.\ (hde)(\text{cls}^2(hde)\ .\&.\ h/de \in l/mno\ .\&.\ h/de \in N/q*p*a*\ :\supset:\ S^{hde}/r*c* \in l/mno)\ :\supset:\ N/q*p*a* \subseteq l/mno).$

The proofs go through without fuss—as if we were working with class symbols themselves.

Complex Prototypes and Extensionality

6.2 In the last section we discussed the way Russell originally planned to develop a theory of classes in the substitutional theory. This construction encounters a general problem in capturing the spirit of the view that for every formula (extensional or otherwise), with free variable μ, there is a class of just those entities satisfying the formula. This is the "problem of extensionality." Far worse, however, is a problem we shall call "the problem of complex prototypes." Russell solved both problems as we shall presently show.

Russell's use of complex prototypes in the early formulations of the substitutional treatment of classes is illegitimate. The problem is that in a class abstract $\hat{\mu}A\mu$ we have a bound variable μ. In the early proxy of classes, $\hat{\mu}A\mu$ is replaced by use of $\{A\alpha\}/\alpha$, and here α is not a bound variable but rather is a term occurring free in the matrix $\{A\alpha\}/\alpha$ whose complex prototype is $\ulcorner\{A\alpha\}\urcorner$. The problem of complex prototypes is that whatever singular term α is chosen, the entity it names would have to be *out* of every entity! To illustrate, consider finding a substitutional proxy for the class theoretical statement,

$$(y)(y \in \hat{z}(z = y)).$$

Russell's original intent was to simply put:

$$(y)(\exists h)(\{x = y\}/x;y!h \ \& \ h)$$

But universal instantiation yields:

$$(\exists h)(\{x = \{\sim x\}\}/x;\{\sim x\}!h \ \& \ h).$$

Now the proposition h, resulting from the substitution, is '$\{\sim x\} = \{\sim \{\sim x\}\}$', but this is false. Moreover, there is no entity which could do for x. Russell would need an entity out of *every* other entity. There is no escape here. Complex prototypes have to go.

The second problem is that of capturing the extensionality of classes. For all the simplicity of Russell's original approach to extensionality, it fails to capture the logicist view that every well-formed formula determines a class of just those entities (of appropriate type) satisfying that formula. In the substitutional proxy, Russell has classes for extensional conditions, and not for any formula, extensional or otherwise.

In the April–May 1906 manuscript "On Substitution$_2$," a different substitutional notation for class symbols appears. The manuscript's theme is not the theory of classes, but rather a paradox of propositions which we shall take up in Chapter 9. The new notation, and the idea behind it, is submerged in Russell's discussion of the paradox of propositions. Little more than hints of how this new notation was to work occur. Nonetheless, the notation is very suggestive and when followed out reveals that Russell was coming upon a new treatment of extensionality which shows how to proceed without complex prototypes. In a 1906 manuscript "On Substitution$_2$," he introduces the notation

$$\ulcorner\iota(p/a)(p/a \equiv \{\phi x\}/x)\urcorner$$

as a proxy for the more familiar "$\hat{x}\phi x$". Further work in the manuscript leads him to worry about the free variable "x" in this notation. Eventually, he comes to see that $\ulcorner\iota(p/a)[(x)(p/a;x \ .\!\equiv\!. \ \phi x)]\urcorner$ avoids the problem of the occurrence of the free variables. To proxy the class theoretic

$$(y)(y \in \hat{z}(z = y)\ ),$$

Russell can now put:

$$(y)(y \in \imath(p/a)[(x)(p/a;x\ .\!\equiv\!.\ x = y)]\ ),$$

with appropriate supporting definitions. The problem of complex prototypes is solved.

Russell's new approach to class symbols also reveals that he had developed a new way to handle the extensionality of classes. Classes of classes (of individuals) are identical because they are defined by functions coextensive with respect to classes (of individuals). That two propositional functions (of functions of individuals) are coextensive with respect to all classes (of individuals) can be proxied with the definition:

$$l/mn \approx h/de = \text{df}$$
$$(rc)(\imath(p/a)[p/a \equiv r/c] \in l/mn\ .\!\equiv\!.\ \imath(p/a)[p/a \equiv r/c] \in h/de).$$

This would be taken with the supporting definition:

$$\imath(p/a)[p/a \equiv r/c] \in l/mn = \text{df}$$
$$(\exists pa)(p/a \equiv r/c\ .\&.\ p/a \in l/mn).$$

The following is then a theorem of extensionality for classes of classes (of individuals):

$$(lmn)(hde)(\imath(s/tw)[s/tw \approx l/mn] = \imath(s/tw)[s/tw \approx h/de]\ .\!\equiv\!.$$
$$(rc)(\imath(p/a)[p/a \equiv r/c] \in \imath(s/tw)[s/tw \approx l/mn]\ .\!\equiv\!.$$
$$\imath(p/a)[p/a \equiv r/c] \in \imath(s/tw)[s/tw \approx h/de])).$$

Identical classes (of classes of individuals) have all the same classes as members.

The idea behind this treatment of extensionality owes its origin to Russell's solution of Frege's *belief paradox* set out in "On Denoting." Let $\ulcorner Ax \urcorner$ abbreviate the context:

"John believes that . . . ."

We may form the extensional context,

$$(\exists q)[q \equiv x\ .\&.\ Aq],$$

based on the original. It is clear that this is extensional for we know that:

$$(r)(s)(r \equiv s\ :\supset:\ (\exists q)[q \equiv r\ .\&.\ Aq]\ .\!\equiv\!.\ (\exists q)[q \equiv s\ .\&.\ Aq]).$$

Thus, in order to capture the extensionality of classes through the use of matrices, Russell proceeds analogously. Consider an intensional two-placed context $\ulcorner A(uv) \urcorner$. Now form the extensional context,

$$(\exists qb)[q/b \equiv x/y\ .\&.\ A(qb)].$$

The context is clearly extensional for we have:

$$(rc)(lm)(r/c \equiv l/m \ .\supset.$$
$$(\exists qb)[q/b \equiv r/c \ .\&.\ A(qb)] \ .\equiv.\ (\exists qb)[q/b \equiv l/m \ .\&.\ A(qb)]).$$

A similar result can be obtained in higher types. The new symbols for classes set forth (admittedly only in sketch) in "On Substitution$_2$" achieves just this end. In the next section, we set out the substitutional theory of classes anew, attending now to Russell's mature approach.

The General Theory of Classes

6.3 We now begin afresh, developing the substitutional theory of classes by introducing new definitions appropriate to Russell's solutions to the problem of extensionality and the problem of complex prototypes.

The contextual definitions for class symbols now proceed from class abstracts (if you will). We have:

df$(C)^1$

$$x \in \imath(p/a)[p/a \approx_z Az] \ = \text{df}\ (\exists pa)[p/a \approx_z Az \ .\&.\ x \in p/a]$$

df$(C.a)^1$

$$x \in p/a \ = \text{df}\ (\exists q)(p/a;x!q \ \&\ q)$$

df$(C)^2$

$$\imath(p/a)[p/a \approx_z Az] \in \imath(s/tw)[s/tw \approx_{rc} B(rc)] \ = \text{df}$$
$$(\exists stw)(s/tw \approx_{rc} B(rc) \ .\&.\ (\exists pa)(p/a \approx_z Az \ .\&.\ p/a \in s/tw))$$

df$(C.a)^2$

$$p/a \in s/tw \ = \text{df}\ (\exists q)(s/tw;pa!q \ \&\ q)$$

df$(C)^3$

$$\imath(s/tw)[s/tw \approx_{rc} A(rc)] \in \imath(j/kuv)[j/kuv \approx_{qpa} B(qpa)] \ = \text{df}$$
$$(\exists jkuv)(j/kuv \approx_{qpa} B(qpa) \ .\&.$$
$$(\exists stw)(s/tw \approx_{rc} A(rc) \ .\&.\ s/tw \in j/kuv))$$

df$(C.a)^3$

$$s/tw \in j/kuv \ = \text{df}\ (\exists q)(j/kuv;stw!q \ \&\ q).$$

The definitions may be extended to higher types. For the context of the identity sign, we have:

df$(C =)^1$

$$\imath(p/a)[p/a \approx_z Az] = \imath(p/a)[p/a \approx_z Bz] = \text{df}$$
$$(\exists pa)(p/a \approx_z Az \ .\&.\ (\exists qb)(q/b \approx_z Bz \ .\&.\ p/a = q/b))$$

df$(C =)^2$

$$\imath(s/tw)[s/tw \approx_{rc} A(rc)] = \imath(s/tw)[s/tw \approx_{rc} B(rc)] \ = \text{df}$$
$$(\exists stw)(s/tw \approx_{rc} A(rc) \ .\&.$$
$$(\exists lmn)(l/mn \approx_{rc} B(rc) \ .\&.\ s/tw = l/mn))$$

and so on.

In an effort to make the notations convenient, the following supporting definitions are employed:

$\mathrm{df}(A)^1$
$$[r/c]\ A(rc) \ = \mathrm{df}\ (\exists pa)(p/a \approx r/c \ .\&.\ A(pa))$$
$\mathrm{df}(A)^2$
$$[q/pa]\ A(qpa) \ = \mathrm{df}\ (\exists stw)(s/tw \approx q/pa \ .\&.\ A(stw))$$

and so forth as we ascend types. Next,

$\mathrm{df}(\approx)^1$
$$p/a \approx r/c \ = \mathrm{df}\ (z)(z \in p/a \ .\equiv.\ z \in r/c)$$
$\mathrm{df}(\approx .a)^1$
$$p/a \approx_z \ Az \ = \mathrm{df}\ (z)(z \in p/a \ .\equiv.\ Az)$$
$\mathrm{df}(\approx .b)^1$
$$Az \approx_z \ Bz \ = \mathrm{df}\ (z)(Az \equiv Bz)$$
$\mathrm{df}(\approx)^2$
$$s/tw \approx l/mn \ = \mathrm{df}\ (rc)([r/c][r/c \in s/tw] \ .\equiv.\ [r/c][r/c \in l/mn])$$
$\mathrm{df}(\approx .a)^2$
$$s/tw \approx_{rc} \ A(rc) \ = \mathrm{df}\ (rc)([r/c][r/c \in s/tw] \ .\equiv.\ [r/c]A(rc))$$
$\mathrm{df}(\approx .b)^2$
$$A(rc) \approx_{rc} \ B(rc) \ = \mathrm{df}\ (rc)([r/c]A(rc) \ .\equiv.\ [r/c]B(rc))$$
$\mathrm{df}(\approx)^3$
$$j/kuv \approx l/mno \ = \mathrm{df}$$
$$(qpa)([q/pa][q/pa \in j/kuv] \ .\equiv.\ [q/pa][q/pa \in l/mno])$$
$\mathrm{df}(\approx .a)^3$
$$j/kuv \approx_{qpa} \ A(qpa) \ = \mathrm{df}\ (qpa)([q/pa][q/pa \in j/kuv] \ .\equiv.\ [q/pa]A(qpa))$$
$\mathrm{df}(\approx .b)^3$
$$A(qpa) \approx_{qpa} \ B(qpa) \ = \mathrm{df}\ (qpa)([q/pa]A(qpa) \ .\equiv.\ [q/pa]B(qpa))$$

and so on as needed.

As we can see, "$\approx$" flanking matrices has all the usual properties of an equivalence relation:

$(\mathrm{T}_{6.1})^1 \quad \vdash (pa)(p/a \approx p/a)$
$(\mathrm{T}_{6.2})^1 \quad \vdash (pa)(lm)(p/a \approx l/m \ .\supset.\ l/m \approx p/a)$
$(\mathrm{T}_{6.3})^1 \quad \vdash (pa)(lm)(rc)(p/a \approx l/m \ .\&.\ l/m \approx r/c \ :\supset:\ p/a \approx r/c).$

We get similar theorems for higher types. Also we have:

$(\mathrm{T}_{6.4})^1 \quad \vdash (pa)(rc)(p/a \equiv r/c \ .\supset.\ p/a \approx r/c)$
$(\mathrm{T}_{6.4})^2 \quad \vdash (stw)(lmn)(s/tw \equiv l/mn \ .\supset.\ s/tw \approx l/mn)$
$(\mathrm{T}_{6.4})^3 \quad \vdash (jkuv)(lmno)(j/kuv \equiv l/mno \ .\supset.\ j/kuv \approx l/mno).$

Since the identity of propositional functions implies their mutual exemplification, we have:

$(\mathrm{T}_{6.5})^1 \quad \vdash (pa)(rc)(p/a = r/c \ .\supset.\ p/a \approx r/c)$
$(\mathrm{T}_{6.5})^2 \quad \vdash (stw)(lmn)(s/tw = l/mn \ .\supset.\ s/tw \approx l/mn)$
$(\mathrm{T}_{6.5})^3 \quad \vdash (jkuv)(lmno)(j/kuv = l/mno \ .\supset.\ j/kuv \approx l/mno),$

and so forth. Moreover, from the definitions we get:

$$*(\mathrm{T}_{6.6})^1 \quad \vdash \; Az \approx_z \; Bz \;:\supset: \quad (\exists pa)(p/a \approx_z \; Az \;.\&.\; C(pa)) \;.\equiv.\; (\exists pa)(p/a \approx_z \; Bz \;.\&.\; C(pa))$$

$$*(\mathrm{T}_{6.6})^2 \quad \vdash \; A(rc) \approx_{rc} \; B(rc) \;:\supset: \quad (\exists stw)(s/tw \approx_{rc} \; A(rc) \;.\&.\; C(stw)) \;.\equiv.\; (\exists stw)(s/tw \approx_{rc} \; B(rc) \;.\&.\; C(stw))$$

Again, similar theorem schemata come in higher types. Now putting the wff, "$z \in l/m$", for Az, and applying (ug) on the first, and similarly putting "$r/c \in s/tw$" for $A(rc)$, and applying (ug) on the second, we get:

$$*(\mathrm{T}_{6.7})^1 \quad \vdash \; (lm)(l/m \approx_z \; Bz \;:\supset: \quad (\exists pa)(p/a \approx l/m \;.\&.\; C(pa)) \;.\equiv.\; (\exists pa)(p/a \approx_z \; Bz \;.\&.\; C(pa)))$$

$$*(\mathrm{T}_{6.7})^2 \quad \vdash \; (hde)(h/de \approx_{rc} \; B(rc) \;:\supset: \quad (\exists stw)(s/tw \approx h/de \;.\&.\; C(stw)) \;.\equiv. \quad (\exists stw)(s/tw \approx_{rc} \; B(rc) \;.\&.\; C(stw))).$$

Existence theorem schemata for classes are readily available. That is, paralleling $*(\mathrm{CP})^1$, $*(\mathrm{CP})^2$, and so on, we have:

$*(\mathrm{T}_{6.8})^1 \quad \vdash (\exists pa)(p/a \approx_z \; A)$, where p, a are not free in A.
$*(\mathrm{T}_{6.8})^2 \quad \vdash (\exists stw)(s/tw \approx_{rc} \; A)$, where s,t,w are not free in A.
$*(\mathrm{T}_{6.8})^3 \quad \vdash (\exists jkuv)(j/kuv \approx_{qpa} \; A)$, where j,k,u,v are not free in A.

and so forth. We shall call these "$*(\mathrm{CP}_\in)^1$", "$*(\mathrm{CP}_\in)^2$", "$*(\mathrm{CP}_\in)^3$", etc. We then can put:

$$\mathrm{E!}\imath(p/a)[p/a \approx_z \; Az] \; = \mathrm{df} \; (\exists pa)(p/a \approx_z \; Az)$$
$$\mathrm{E!}\imath(s/tw)[s/tw \;.\!\approx_{rc}.\; A(rc)] = \mathrm{df} \; (\exists stw)(s/tw \approx_{rc} \; A(rc))$$
$$\mathrm{E!}\imath(j/kuv)[j/kuv \approx_{qpa} \; A(qpa))] = \mathrm{df} \; (\exists jkuv)(j/kuv \approx_{qpa} A(qpa))$$

and so on.

6.4 The new definitions for "class abstracts" allow us to go on to prove substitutional analogs for the basic theorems for classes of *Principia Mathematica*. We have:

$*20.14 \quad \vdash \hat{z}Az = \hat{z}Bz \;.\supset.\; (x)(Ax \;.\equiv.\; Bx)$
$*20.15 \quad \vdash (x)(Ax \;.\equiv.\; Bx) \;.\supset.\; \hat{z}Az = \hat{z}Bz$
$*20.18 \quad \vdash \hat{z}Az = \hat{z}Bz \;.\supset.\; C(\hat{z}Az) \supset C(\hat{z}Bz).$
$*20.19 \quad \vdash \hat{z}Az = \hat{z}Bz \;.\equiv.\; (f)(f(\hat{z}Az) \;.\equiv.\; f(\hat{z}Bz))$
$*20.2 \quad \vdash \hat{z}Az = \hat{z}Az$
$*20.21 \quad \vdash \hat{z}Az = \hat{z}Bz \;.\supset.\; \hat{z}Bz = \hat{z}Az$
$*20.22 \quad \vdash \hat{z}Az = \hat{z}Bz \;.\&.\; \hat{z}Bz = \hat{z}Cz \;:\supset:\; \hat{z}Az = \hat{z}Cz$
$*20.3 \quad \vdash x \in \hat{z}Az \;.\equiv.\; Ax$
$*20.31 \quad \vdash \hat{z}Az = \hat{z}Bz \;.\equiv.\; (x)(x \in \hat{z}Az \;.\equiv.\; x \in \hat{z}Bz).$

Principia sets out the theorems with typical ambiguity. The idea is that one can restore type indices in any way consistent with the rules for forming wffs. Now substitution cannot employ typical ambiguity since types are part of its grammar. Nonetheless, having revealed the pattern the reader can easily go on to higher types. In substitutional language, the pattern for higher types commonly emerges at classes of classes (of individuals), and not at classes of individuals. So we shall state the theorem schemata for the first two types, proving it in the second.

The following are very useful theorem schemata. They number $*(T_{6.9})$–$*(T_{6.14})$, but we shall give each a more convenient name. Let us begin with class abstraction.

$*20.3 \quad \vdash x \in \hat{z}Az .\equiv. Az$
$*(\text{Abs})^1 \quad \vdash x \in \imath(p/a)[p/a \approx_z Az] .\equiv. Ax$
$*(\text{Abs})^2 \quad \vdash \imath(p/a)[p/a \approx_z Az] \in \imath(s/tw)[s/tw \approx_{rc} B(rc)] .\equiv.$
$(\exists pa)(p/a \approx_z Az .\&. B(pa))$

Proof:
1. $\rightarrow \imath(p/a)[p/a \approx_z Az] \in \imath(s/tw)[s/tw \approx_{rc} B(rc)]$
2. $(\exists stw)(s/tw \approx_{rc} B(rc) .\&. (\exists pa)(p/a \approx_z Az .\&. p/a \in s/tw))$ 1, df$(C)^2$.
3. $s/tw \approx_{rc} B(rc) .\&. (\exists pa)(p/a \approx_z Az .\&. p/a \in s/tw)$ 2, EI.
4. $s/tw \approx_{rc} B(rc)$ 3, simp
5. $(\exists pa)(p/a \approx_z Az .\&. p/a \in s/tw)$ 3, simp.
6. $q/b \approx_z Az .\&. q/b \in s/tw$ 5, EI.
7. $(\exists pa)(p/a \approx q/b .\&. p/a \in s/tw))$ 6, eg, $*(T_{6.1})^1$.
8. $(\exists pa)(p/a \approx q/b .\&. p/a \in s/tw) \equiv$ $(\exists pa)(p/a \approx q/b .\&. B(pa))$ 4, df$(\approx.a)^2$, df$(A)^1$, ui.
9. $(\exists pa)(p/a \approx q/b .\&. B(pa))$ 7, 8, df$(\equiv)$, simp, mp.
10. __ $(\exists pa)(p/a \approx_z Az .\&. B(pa))$ 6, 9, $(T_{6.7})^1$, log, (6ei), (3ei).
11. $1 \supset 10$ cp.
12. $\rightarrow (\exists pa)(p/a \approx_z Az .\&. B(pa))$
13. $(\exists stw)(s/tw \approx_{rc} B(rc))$ $*(CP_\in)^2$
14. $s/tw \approx_{rc} B(rc)$ 13, EI.
15. $q/b \approx_z Az .\&. B(qb)$ 12, EI.
16. $(\exists pa)(p/a \approx q/b .\&. B(pa))$ 15, eg, $*(T_{6.7})^1$.
17. $(\exists pa)(p/a \approx q/b .\&. p/a \in s/tw) \equiv$ $(\exists pa)(p/a \approx q/b .\&. B(pa))$ 14, df$(\approx.a)^2$, df$(A)^1$, ui.
18. $(\exists pa)(p/a \approx q/b .\&. p/a \in s/tw)$ 16, 17, df$(\equiv)$, simp, mp.
19. 14 & 18 conj.
20. $(\exists stw)19$ eg, (15ei), (14ei).
21. __ $\imath(p/a)[p/a \approx_z Az] \in \imath(s/tw)[s/tw \approx_{rc} B(rc)]$ 20, df$(C)^2$.
22. $12 \supset 21$ cp.
22. 11 & 22 conj.
23. $\imath(p/a)[p/a \approx_z Az] \in \imath(s/tw)[s/tw \approx_{rc} B(rc)] \equiv$ $(\exists pa)(p/a \approx_z Az .\&. B(pa))$ 22, df$(\equiv)$.

$*20.15 \quad \vdash (x)(Ax \equiv Bx) .\equiv. \hat{z}Az = \hat{z}Bz$
$*(V)^1 \quad \vdash (z)(Az .\equiv. Bz) .\equiv.$
$\imath(p/a)[p/a \approx_z Az] = \imath(p/a)[p/a \approx_z Bz]$
$*(V)^2 \quad \vdash A(rc) \approx_{rc} B(rc) .\equiv.$
$\imath(s/tw)[s/tw \approx_{rc} A(rc)] = \imath(s/tw)[s/tw \approx_{rc} B(rc)]$

Proof:

1. $\rightarrow A(rc) \approx_{rc} B(rc)$
2. $(\exists stw)(s/tw \approx_{rc} A(rc))$ $*(\mathrm{CP}_{\in})^2$.
3. $s/tw \approx_{rc} A(rc)$ 2, EI.
4. $s/tw \approx_{rc} B(rc)$ 1, 3, df$(\approx.\mathrm{a})^2$, df$(A)^1$, log.
5. $s/tw = s/tw$ $*(\mathrm{T}_{5.12})^2$, ui.
6. $s/tw \approx_{rc} B(rc)$.&. $s/tw = s/tw$ 4, 5, conj.
7. $(\exists lmn)(l/mn \approx_{rc} B(rc)$.&. $l/mn = s/tw)$ 6, eg, (3ei).
8. 3 & 7 conj.
9. $(\exists stw)8$ eg.
10. __ $\imath(s/tw)[s/tw \approx_{rc} A(rc)] = \imath(s/tw)[s/tw \approx_{rc} B(rc)]$ 9, df$(\mathrm{C} =)^2$.
11. $1 \supset 10$ cp.
12. $\rightarrow \imath(s/tw)[s/tw \approx_{rc} A(rc)] = \imath(s/tw)[s/tw \approx_{rc} B(rc)]$
13. $(\exists stw)(s/tw \approx_{rc} A(rc)$.&. $(\exists lmn)(l/mn \approx_{rc} B(rc)$.&. $l/mn = s/tw))$ 12, df$(\mathrm{C} =)^2$.
14. $s/tw \approx_{rc} A(rc)$.&. $(\exists lmn)(l/mn \approx_{rc} B(rc)$.&. $l/mn = s/tw)$ 13, EI.
15. $l/mn \approx_{rc} B(rc)$.&. $l/mn = s/tw$ 14, simp, EI.
16. $s/tw \approx_{rc} B(rc)$ 15, $(\mathrm{T}_{5.16})^2$, $*(=)$.
17. __ $A(rc) \approx_{rc} B(rc)$ 14, 16, simp, df$(\approx.a)^2$, df$(\approx.b)^2$, (15ei), (14ei).
18. $12 \supset 17$ cp.
19. 11 & 18 conj.
20. $*(\mathrm{V})^2$ 19, df$(\equiv)$.

Next consider the theorem schemata $*20.31$ of extensionality. It is useful in the substitutional theory to have a schematic version. We prove a lemma.

$$*(\mathrm{Ext.a})^1 \vdash \imath(p/a)[p/a \approx_z Az] = \imath(p/a)[p/a \approx_z Bz] :\supset: y \in \imath(p/a)[p/a \approx_z Az] .\equiv. y \in \imath(p/a)[p/a \approx_z Bz]$$

$$*(\mathrm{Ext.a})^2 \vdash \imath(s/tw)[s/tw \approx_{rc} A(rc)] = \imath(s/tw)[s/tw \approx_{rc} B(rc)] :\supset: \imath(p/a)[p/a \approx_z Dz] \in \imath(s/tw)[s/tw \approx_{rc} A(rc)] .\equiv. \imath(p/a)[p/a \approx_z Dz] \in \imath(s/tw)[s/tw \approx_{rc} B(rc)]$$

Proof:

1. $\rightarrow \imath(s/tw)[s/tw \approx_{rc} A(rc)] = \imath(s/tw)[s/tw \approx_{rc} B(rc)]$
2. $A(rc) \approx_{rc} B(rc)$ 1, $*(\mathrm{V})^2$, df$(\equiv)$, simp, mp.
3. $A(rc) \approx_{rc} B(rc) :\supset:$
   $(\exists stw)(s/tw \approx_{rc} A(rc)$.&. $(\exists pa)(p/a \approx_z Dz$.&. $p/a \in s/tw))$
   $.\equiv. (\exists stw)(s/tw \approx_{rc} B(rc)$.&. $(\exists pa)(p/a \approx_z Dz$.&. $p/a \in s/tw))$
   $*(\mathrm{T}_{6.6})^2$.
4. $(\exists stw)(s/tw \approx_{rc} B(rc)$.&. $(\exists pa)(p/a \approx_z Dz$.&. $p/a \in s/tw))$
   $.\equiv. (\exists stw)(s/tw \approx_{rc} B(rc)$.&. $(\exists pa)(p/a \approx_z Dz$.&. $p/a \in s/tw))$
   2, 3, mp.
5. __ $\imath(p/a)[p/a \approx_z Dz] \in \imath(s/tw)[s/tw \approx_{rc} A(rc)] .\equiv.$
   $\imath(p/a)[p/a \approx_z Dz] \in \imath(s/tw)[s/tw \approx_{rc} B(rc)]$ 4, df$(\mathrm{C})^2$
6. $1 \supset 5$cp.

Then we prove:

$*20.31 \vdash \hat{z}Az = \hat{z}Bz .\equiv. (y)(y \in \hat{z}Az .\equiv. y \in \hat{z}Bz)$
$*(\mathrm{Ext})^1 \vdash \imath(p/a)[p/a \approx_z Az] = \imath(p/a)[p/a \approx_z Bz] .\equiv.$
$(y)(y \in \imath(p/a)[p/a \approx_z Az] .\equiv. y \in \imath(p/a)[p/a \approx_z Bz])$
$*(\mathrm{Ext})^2 \vdash \imath(s/tw)[s/tw \approx_{rc} A(rc)] = \imath(s/tw)[s/tw \approx_{rc} B(rc)] .\equiv.$
$(rc)(\imath(p/a)[p/a \approx r/c] \in \imath(s/tw)[s/tw \approx_{rc} A(rc)] .\equiv.$
$\imath(p/a)[p/a \approx r/c] \in \imath(s/tw[s/tw \approx_{rc} B(rc)]).$

Proof:

1. $\imath(s/tw)[s/tw \approx_{rc} A(rc)] = \imath(s/tw)[s/tw \approx_{rc} B(rc)]$:⊃:
$\imath(p/a)[p/a .\approx_z. z \in r/c] \in \imath(s/tw)[s/tw \approx_{rc} A(rc)] .\equiv.$
$\imath(p/a)[p/a .\approx_z. z \in r/c] \in \imath(s/tw)[s/tw \approx_{rc} B(rc)]$
$*(\mathrm{Ext.a})^2$
2. $\imath(s/tw)[s/tw \approx_{rc} A(rc)] = \imath(s/tw)[s/tw \approx_{rc} B(rc)]$:⊃:
$\imath(p/a)[p/a \approx r/c] \in \imath(s/tw)[s/tw \approx_{rc} A(rc)] .\equiv.$
$\imath(p/a)[p/a \approx r/c] \in \imath(s/tw)[s/tw \approx_{rc} B(rc)]$
1, $\mathrm{df}(\approx)^1$, $\mathrm{df}(\approx.\mathrm{a})^1$.
3. $(rc)2$ ug.
4. $\imath(s/tw)[s/tw \approx_{rc} A(rc)] = \imath(s/tw)[s/tw \approx_{rc} B(rc)]$:⊃:
$(rc)(\imath(p/a)[p/a \approx r/c] \in \imath(s/tw)[s/tw \approx_{rc} A(rc)] .\equiv.$
$\imath(p/a)[p/a \approx r/c] \in \imath(s/tw)[s/tw \approx_{rc} B(rc)])$
3, $*(\mathrm{A}_6)$, mp.
5. → $(rc)(\imath(p/a)[p/a \approx r/c] \in \imath(s/tw)[s/tw \approx_{rc} A(rc)] .\equiv.$
$\imath(p/a)[p/a \approx r/c] \in \imath(s/tw)[s/tw \approx_{rc} B(rc)])$
6. $(rc)((\exists pa)(p/a \approx r/c .\&. A(rc)) .\equiv.$
$(\exists pa)(p/a \approx r/c .\&. B(rc)))$ 5, $*(\mathrm{Abs})^2$.
7. $A(rc) \approx_{rc} B(rc)$ 6, $\mathrm{df}(\mathrm{A})^1$, $\mathrm{df}(\approx.b)^2$.
8. __ $\imath(s/tw)[s/tw \approx_{rc} A(rc)] = \imath(s/tw)[s/tw \approx_{rc} B(rc)]$
7, $*(\mathrm{V})^2$.
9. $5 \supset 8$ cp.
10. 4 & 9 conj.
11. $*(\mathrm{Ext})^2$ 10, df(≡).

We next show the parallel for *Principia*'s ∗20.19. First we have the schema ∗20.18,

$*20.18 \vdash \hat{z}Az = \hat{z}Bz .\supset. f(\hat{z}Az) .\equiv. f(\hat{z}Bz)$
$*(\mathrm{E.a})^1 \vdash \imath(p/a)[p/a \approx_z Az] = \imath(p/a)[p/a \approx_z Bz] .\supset.$
$(\exists pa)(p/a \approx_z Az .\&. C(pa)) .\equiv.$
$(\exists pa)(p/a \approx_z Bz .\&. C(pa))$
$*(\mathrm{E.a})^2 \vdash \imath(s/tw)[s/tw \approx_{rc} A(rc)] = \imath(s/tw)[s/tw \approx_{rc} B(rc)] .\supset.$
$(\exists stw)(s/tw \approx_{rc} A(rc) .\&. C(stw)) .\equiv.$
$(\exists stw)(s/tw \approx_{rc} B(rc) .\&. C(stw)).$

The proofs are straightforward from $*(\mathrm{V})^1$ and $*(\mathrm{T}_{6.6})^1$, and $*(\mathrm{V})^2$ and $*(\mathrm{T}_{6.6})^2$. From here we have:

$*20.19 \vdash \hat{z}Az = \hat{z}Bz .\equiv. (f)(f(\hat{z}Az) .\equiv. f(\hat{z}Bz))$

$*(\mathrm{E})^1 \quad \vdash \ \iota(p/a)[p/a \approx_z Az] = \iota(p/a)[p/a \approx_z Bz] \ .\!\equiv\!.$
$(stw)((\exists pa)(p/a \approx_z Az \ .\&.\ p/a \in s/tw)$
$.\!\equiv\!.\ (\exists pa)(p/a \approx_z Bz \ .\&.\ p/a \in s/tw))$

$*(\mathrm{E})^2 \ \vdash \ \iota(s/tw)[s/tw \approx_{rc} A(rc)] = \iota(s/tw)[s/tw \approx_{rc} B(rc)] \ .\!\equiv\!.$
$(lmno)(\ (\exists stw)(s/tw \approx_{rc} A(rc) \ .\&.\ s/tw \in l/mno) \ .\!\equiv\!.$
$(\exists stw)(s/tw \approx_{rc} B(rc) \ .\&.\ s/tw \in l/mno))$

Proof:

1. $\iota(s/tw)[s/tw \approx_{rc} A(rc)] = \iota(s/tw)[s/tw \approx_{rc} B(rc)] \ :\!\supset\!:$
   $(\exists stw)(s/tw \approx_{rc} A(rc) \ .\&.\ s/tw \in l/mno) \ .\!\equiv\!.$
   $(\exists stw)(s/tw \approx_{rc} A(rc) \ .\&.\ s/tw \in l/mno)$ $\quad *(\mathrm{E.a})^2$.
2. $(lmno)1$ ug.
3. $\iota(s/tw)[s/tw \approx_{rc} A(rc)] = \iota(s/tw)[s/tw \approx_{rc} B(rc)] \ :\!\supset\!:$
   $(lmno)(\ (\exists stw)(s/tw \approx_{rc} A(rc) \ .\&.\ s/tw \in l/mno) \ .\!\equiv\!.$
   $(\exists stw)(s/tw \approx_{rc} A(rc) \ .\&.\ s/tw \in l/mno)$ $\quad$ 2, $*(\mathrm{A}_6)$.
4. $\rightarrow$ $(lmno)(\ (\exists stw)(s/tw \approx_{rc} A(rc) \ .\&.\ s/tw \in l/mno) \ .\!\equiv\!.$
   $(\exists stw)(s/tw \approx_{rc} B(rc) \ .\&.\ s/tw \in l/mno))$
5. $s/tw \approx_{rc} A(rc)$ $\quad *(\mathrm{CP}_{\in})^2$, EI.
6. $(\exists lmno)(qpa)(l/mno;qpa \ .\!\equiv\!.\ q/pa = s/tw)$ $\quad *(\mathrm{CP}_{\mathrm{sub}})^3$.
7. $(qpa)(l/mno;qpa \ .\!\equiv\!.\ q/pa = s/tw)$ $\quad$ 6, EI.
8. $(qpa)(q/pa \in l/mno \ .\!\equiv\!.\ q/pa = s/tw)$ $\quad$ 7, $\mathrm{df(C.a)}^3$, $*(\mathrm{scope})^3$.
9. $s/tw = s/tw$ $\quad *(\mathrm{T}_{5.12})^2$, ui.
10. $s/tw \in l/mno$ $\quad$ 8, 9, log.
11. 5 & 10 conj.
12. $(\exists stw)11$ eg.
13. $(\exists stw)(s/tw \approx_{rc} B(rc) \ .\&.\ s/tw \in l/mno)$ $\quad$ 4, 12, ui, df($\equiv$), simp, mp.
14. $h/de \approx_{rc} B(rc) \ .\&.\ h/de \in l/mno$ $\quad$ 13, EI.
15. $h/de = s/tw$ $\quad$ 8, 14, ui, simp, df($\equiv$), mp.
16. $h/de \approx_{rc} B(rc) \ .\&.\ h/de = s/tw$ $\quad$ 14, 15, simp, conj.
17. $(\exists hde)16$ eg.
18. $s/tw \approx_{rc} A(rc) \ .\&.\ (\exists hde)(h/de \approx_{rc} B(rc) \ .\&.\ h/de = s/tw)$
    5, 17 conj.
19. $(\exists stw)18$ eg, (14ei), (7ei), (5ei).
20. __ $\iota(s/tw)[s/tw \approx_{rc} A(rc)] = \iota(s/tw)[s/tw \approx_{rc} B(rc)]$ $\quad$ 19, $\mathrm{df(C{=})}^2$.
21. $4 \supset 20$ cp.
27. 3 & 21 conj.
28. $*(E)^2$ $\quad$ 27, df($\equiv$).

We should note that from $*(\mathrm{E}.a)^1$ we can get the following schema:

$*(\mathrm{T}_{6.15})^1 \ \iota(p/a)[p/a \approx_z Az] = \iota(p/a)[p/a \approx_z Bz] \ :\!\supset\!:$
$\iota(p/a)[p/a \approx_z Az] \in \iota(s/tw)[s/tw \approx_{rc} C(rc)] \ .\!\equiv\!.$
$\iota(p/a)[p/a \approx_z Bz] \in \iota(s/tw)[s/tw \approx_{rc} C(rc)].$

Putting the wff $\ulcorner(\exists stw)(s/tw \approx_{rc} D(rc) \ .\&.\ p/a \in s/tw)\urcorner$ for $\ulcorner C(pa)\urcorner$ in $*(\mathrm{E.a})^1$, we arrive at the proof. Similar theorems $*(\mathrm{T}_{6.15})^2$, $*(\mathrm{T}_{6.15})^3$, and so on, follow the pattern.

Finally, the usual properties of identity apply to classes constructed in the substitutional theory. We have:

$*20.2 \quad \vdash \hat{z}Az = \hat{z}Az$

$*(T_{6.16})^1 \quad \vdash \iota(p/a)[p/a \approx_z Az] = \iota(p/a)[p/a \approx_z Az]$

$*20.21 \quad \vdash \hat{z}Az = \hat{z}Bz \;.\!\supset\!.\; \hat{z}Bz = \hat{z}Az$

$*(T_{6.17})^1 \quad \vdash \iota(p/a)[p/a \approx_z Az] = \iota(p/a)[p/a \approx_z Bz] \;.\!\supset\!.$
$\iota(p/a)[p/a \approx_z Bz] = \iota(p/a)[p/a \approx_z Az]$

$*20.22 \quad \vdash \hat{z}Az = \hat{z}Bz \;.\&.\; \hat{z}Bz = \hat{z}Cz \;:\!\supset\!:\; \hat{z}Az = \hat{z}Cz$

$*(T_{6.18})^1 \quad \vdash \iota(p/a)[p/a \approx_z Az] = \iota(p/a)[p/a \approx_z Bz] \;.\&.$
$\iota(p/a)[p/a \approx_z Bz] = \iota(p/a)[p/a \approx_z Cz] \;:\!\supset\!:$
$\iota(p/a)[p/a \approx_z Az] = \iota(p/a)[p/a \approx_z Cz].$

This completes the needed schemata.

Scope is, of course, important in the substitutional treatment of class symbols. The following are very useful theorem schema:

$*(T_{6.19})^1 \quad \vdash C(B(\iota(p/a)[p/a \approx_z Az])) \equiv$
$(\exists q, b)(q/b \approx_z Az \;.\&.\; C(B(\iota(p/a)[p/a \approx q/b]))),$
where C is truth-functional, and B is a wff with $\ulcorner\iota(p/a)[p/a \approx_z Az]\urcorner$ flanking a membership sign or an identity sign.

$*(T_{6.19})^2 \quad \vdash C(B(\iota(s/tw)[s/tw \approx_{rc} A(rc)])) \equiv$
$(\exists hde)(h/de \approx_{rc} Arc \;.\&.\; C(B(\iota(s/tw)[s/tw \approx h/de]))),$
where C is truth-functional, and B is a wff with $\ulcorner\iota(s/tw)[s/tw \approx_{rc} A(rc)]\urcorner$ flanking a membership sign or an identity sign.

The proofs are straightforward. We also get a proxy for universal instantiation to class symbols.

$*(T_{6.20})^1 \quad \vdash (lm)(C(B(\iota(p/a)[p/a \approx l/m]))) \;.\!\supset\!.\; C(B(\iota(p/a)[p/a \approx_z Az])),$
where C is truth-functional, and B is a wff with $\ulcorner\iota(p/a)[p/a \approx_z Az]\urcorner$ flanking a membership sign or an identity sign.

$*(T_{6.20})^2 \quad \vdash (hde)(C(B(\iota(s/tw)[s/tw \approx h/de]))) \;.\!\supset\!.$
$C(B(\iota(s/tw)[s/tw \approx_{rc} A(rc)]))$
where C is truth-functional, and B is a wff with $\ulcorner\iota(s/tw)[s/tw \approx_{rc} A(rc)]\urcorner$ flanking a membership sign or an identity sign.

We call these "$*(\text{uic})^1$, "$*(\text{uic})^2$", and so on. From these we have:

$*(T_{6.20}\text{a})^1 \quad \vdash C(B(\iota(p/a)[p/a \approx_z Az])) \;.\!\supset\!.\; (\exists lm)(C(B(\iota(p/a)[p/a \approx l/m]))$
where C is truth-functional, and B is a wff with $\ulcorner\iota(p/a)[p/a \approx_z Az]\urcorner$ flanking a membership sign or an identity sign.

$*(T_{6.20}\text{a})^2 \quad \vdash C(B(\iota(s/tw)[s/tw \approx_{rc} A(rc)])) \;.\!\supset\!.$
$(\exists hde)(C(B(\iota(s/tw)[s/tw \approx h/de]))$
where C is truth-functional, and B is wff with $\ulcorner\iota(s/tw)[s/tw \approx_{rc} A(rc)]\urcorner$ flanking a membership sign or an identity sign.

We shall call these "$*(\text{egc})^1$", "$*(\text{egc})^2$", and so on. We can get along without these theorem schemata, of course. But the price is great tedium.

6.5 Let us now go on to some simple theorems in the type-indexed theory of classes. We first express the theorems in the familiar notion of classes, and then render its substitutional proxy. To do this, we shall use lower-case Greek letters "α", "β", etc., with appropriate type indices, for classes. All the theorems repeat as one ascends types. The following theorems are in lowest type.

The following definitions prove convenient and help remind us of the class-theoretical statements being proxied in the language of substitution.

$\text{df}(V)^1 \quad Vz = \text{df}\ z = z.$

$\text{df}(\Lambda)^1 \quad \Lambda z = \text{df}\ z \neq z.$

The usual theorems follow. For instance,

$(\mathrm{T}_{6.21})^1$
$\vdash (\forall x^{\mathrm{o}})(x^{\mathrm{o}} \in V^{(\mathrm{o})})$
$\vdash (x)(x \in \imath(p/a)[p/a \approx_z Vz])$

Proof:
1. $x \in \imath(p/a)[p/a \approx_z Vz]$.≡. $Vx \quad *(\text{Abs})^1$.
2. $Vx \quad (\mathrm{T}_{4.8})a$, ui, $\text{df}(V)^1$.
3. $x \in \imath(p/a)[p/a \approx_z Vz] \quad$ 1, 2, df(≡), simp, mp.
4. $(x)3 \quad$ ug.

$(\mathrm{T}_{6.22})^1$
$\vdash (\forall \alpha^{(\mathrm{o})})((\forall x^{\mathrm{o}})(x^{\mathrm{o}} \in \alpha^{(\mathrm{o})})\ .\supset.\ \alpha^{(\mathrm{o})} = V^{(\mathrm{o})})$
$\vdash (lm)((x)(x \in \imath(p/a)[p/a \approx l/m])\ .\supset.$
$\imath(p/a)[p/a \approx l/m] = \imath(p/a)[p/a \approx_z Vz])$

Proof:
1. $\rightarrow (x)(x \in \imath(p/a)[p/a \approx 1/m])$
2. $x = x \quad (\mathrm{T}_{4.8})$a, ui.
3 $Vx \quad 2, \text{df}(V)^1$.
4. $x \in l/m\ .\supset.\ Vx \quad$ 3, log.
5. $x \in \imath(p/a)[p/a \approx l/m] \quad$ 1, ui.
6. $x \in l/m \quad 5, *(\text{Abs})^1$.
7. $Vx\ .\supset.\ x \in l/m \quad$ 6, log.
8. $x \in l/m$.≡. $Vx \quad$ 4, 7, conj, df(≡).
9. $(x)8 \quad$ ug.
10. $\_\_\ \imath(p/a)[p/a \approx 1/m] = \imath(p/a)[p/a \approx_z Vz] \quad 9, *(V)^1$.
11. $1 \supset 10 \quad$ cp.
12. $(lm)11 \quad$ ug.

$(\mathrm{T}_{6.23})^1$
$\vdash (\forall x^{\mathrm{o}}) \sim (x^{\mathrm{o}} \in \Lambda^{(\mathrm{o})})$,
$\vdash (x) \sim (x \in \imath(p/a)[p/a \approx_z \Lambda z])$

Proof:
1. $x \in \imath(p/a)[p/a \approx_z \Lambda z]$.≡. $\Lambda x \quad *(\text{Abs})^1$.
2. $\sim (x \neq x) \quad (\mathrm{T}_{4.8})$a.

3. $\sim \Lambda x$ 2, df(Λ)[1].
4. $\sim (x \in \imath(p/a)[p/a \approx_z \Lambda z])$ 1, 3, df($\equiv$), simp, mt.
5. $(x)4$ ug.

The substitutional theory of classes does not require adjustments for "urelements," i.e., entities which, because they are not classes, have no members. In the naive theory of classes such adjustments are needed to ensure, for example, that the empty class is unique. But in the substitutional theory there really are no entities which are classes. The distinction between classes and non-classes is clear in the formal grammar.

$(T_{6.24})^1$
$\vdash (\forall \alpha^{(o)})((\forall y^o) \sim (y^o \in \alpha^{(o)}) .\equiv. \alpha^{(o)} = \Lambda^{(o)})$
$\vdash (lm)((y) \sim (y \in \imath(p/a)[p/a \approx l/m)]) .\equiv.$
$\imath(p/a)[p/a \approx l/m] = \imath(p/a)[p/a \approx_z \Lambda z])$

Proof:
1. $\rightarrow (y) \sim (y \in \imath(p/a)[p/a \approx l/m])$
2. $x \in \imath(p/a)[p/a \approx l/m] .\equiv. x \in l/m$ $*(\text{Abs})^1$.
3. $\sim (x \in \imath(p/a)[p/a \approx l/m])$ 1, ui.
4. $\sim (x \in l/m)$ 2, 3, df($\equiv$), simp, mt.
5. $\sim (x \in l/m) \vee \Lambda x$ 4, add.
6. $x \in l/m .\supset. \Lambda x$ 5, imp.
7. $\sim \Lambda x$ $(T_{4.8})$a, ui, dneg, df($\neq$), df(Λ)[1].
8. $\sim \Lambda x .\vee. x \in l/m$ 7, add.
9. $\Lambda x .\supset. x \in l/m$ 8, imp.
10. $x \in l/m .\equiv. \Lambda x$ 6, 9, conj, df($\equiv$).
11. $(x)9$ ug.
12. __ $\imath(p/a)[p/a \approx l/m] = \imath(p/a)[p/a \approx_z \Lambda z]$
11, $*(V)^1$, ui, df($\approx$)[1], simp, mp.
13. $1 \supset 12$ cp.
14. $\rightarrow \imath(p/a)[p/a \approx l/m] = \imath(p/a)[p/a \approx_z \Lambda z]$
15. $(x)(x \in l/m .\equiv. \Lambda x)$ $*(V)^1$, ui, df($\approx$.a)[1].
17. $\sim \Lambda y$ $T_{(4.8)}$a, ui, dneg, df($\neq$), df(Λ)[1].
18. $\sim (y \in l/m)$ 15, 17, ui, df($\equiv$), simp, mt.
19. $\sim (y \in \imath(p/a)[p/a \approx l/m])$
18, $*(\text{Abs})^1$, df($\equiv$), simp, mt.
20. __ $(y)19$ ui.
21. $14 \supset 20$ cp.
22. 13 & 21 conj.
23. $1 \equiv 14$ 22, df($\equiv$).
24. $(lm)23$ ug.

It is instructive to remind ourselves that any matrix "x/y", where the entity named by "x" is not a proposition, will proxy a propositional function which is necessarily not exemplified. Recall that all entities, propositions or otherwise, are on a par as "individuals" in the substitutional theory; nothing requires the prototype of a matrix to express a proposition. All that is required is that it be a genuine name for an entity (*SCR*, p. 168). Thus, consider the matrix "a/b", where "a" and "b" name Socrates and Plato, respectively. Now for every entity x, the result of substituting x for Plato in Socrates will be Socrates—since Plato does not occur in Socrates. Hence, for each x the result of the substitution, q, is Socrates. Since Socrates is nontrue, the matrix "a/b" proxies an unexemplified propositional function of individuals. Consequently, Russell is quite justified in saying that "any two entities p, a define a class p/a" (*SCR*, p. 180). That is, they define a propositional function whose extension is a class.

We can go on to define subclass and proper subclass in the now familiar way.

$$\alpha^{(o)} \subseteq \beta^{(o)} \;=\text{df}\; (\forall x^{o})(x^{o} \in \alpha^{(o)} \;.\supset.\; x^{o} \in \beta^{(o)})$$
$$\alpha^{(o)} \subset \beta^{(o)} \;=\text{df}\; \alpha^{(o)} \subseteq \beta^{(o)} \;.\&.\; \sim (\alpha^{(o)} = \beta^{(o)})$$

df$(\subseteq)^1$

$$\imath(p/a)[p/a \approx_z Az] \subseteq \imath(p/a)[p/a \approx_z Bz] = \text{df}$$
$$(y)(y \in \imath(p/a)[p/a \approx_z Az] \;.\supset.\; y \in \imath(p/a)[p/a \approx_z Bz])$$

df$(\subset)^1$

$$\imath(p/a)[p/a \approx_z Az] \subset \imath(p/a)[p/a \approx_z Bz] \;=\text{df}$$
$$\imath(p/a)[p/a \approx_z Az] \subseteq \imath(p/a)[p/a \approx_z Bz] \;.\&.$$
$$\sim (\imath(p/a)[p/a \approx_z Az] = \imath(p/a)[p/a \approx_z Bz])$$

Of course the definition repeats at higher types.

$$\alpha^{((o))} \subseteq \beta^{((o))} \;=\text{df}\; (\forall \delta^{(o)})(\delta^{(o)} \in \alpha^{((o))} \;.\supset.\; \delta^{(o)} \in \beta^{((o))})$$
$$\alpha^{((o))} \subset \beta^{((o))} \;=\text{df}\; \alpha^{((o))} \subseteq \beta^{((o))} \;.\&.\; \sim (\alpha^{((o))} = \beta^{((o))})$$

df$(\subseteq)^2$

$$\imath(s/tw)[s/tw \approx_{rc} A(rc)] \subseteq \imath(s/tw)[s/tw \approx_{rc} B(rc)] \;=\text{df}$$
$$(lm)(\imath(p/a)[p/a \approx l/m] \in \imath(s/tw)[s/tw \approx_{rc} A(rc)] \;.\supset.$$
$$\imath(p/a)[p/a \approx l/m] \in \imath(s/tw)[s/tw \approx_{rc} B(rc)])$$

df$(\subset)^2$

$$\imath(s/tw)[s/tw \approx_{rc} A(rc)] \subset \imath(s/tw)[s/tw \approx_{rc} B(rc)] \;=\text{df}$$
$$\imath(s/tw)[s/tw \approx_{rc} A(rc)] \subseteq \imath(s/tw)[s/tw \approx_{rc} B(rc)] \;.\&.$$
$$\sim (\imath(s/tw)[s/tw \approx_{rc} A(rc)] = \imath(s/tw)[s/tw \approx_{rc} B(rc)]).$$

To illustrate the use of the definition for subclass, let us take a simple theorem.

$(\text{T}_{6.25})^1$

$\vdash (\forall \alpha^{(o)})(\alpha^{(o)} \subseteq V^{(o)})$

$\vdash (lm)(\imath(p/a)[p/a \approx l/m] \subseteq \imath(p/a)[p/a \approx_z Vz]$

Proof:

1. $x \in \imath(p/a)[p/a \approx_z Vz]$ $\quad * (\text{T}_{6.21})^1$, ui.
2. $x \in \imath(p/a)[p/a \approx l/m] \;.\supset.\; x \in \imath(p/a)[p/a \approx_z Vz]$ $\quad$ 1, log.
3. $(x)2$ $\quad$ ug.
4. $\imath(p/a)[p/a \approx l/m] \subseteq \imath(p/a)[p/a \approx_z Vz]$ $\quad$ 3, df$(\subseteq)^1$.
5. $(lm)4$ $\quad$ ug.

Definitions for subclass (and proper subclass) operate normally.

Unfortunately, the same cannot be said for the usual algebraic class operations such as union, intersection, complement, and singleton. Using class abstract terms, ⌜$\{\mu\colon A\mu\}$⌝ instead of Whitehead and Russell's ⌜$\hat{\mu}(A\mu)$⌝, they would normally be defined (for lowest type) as follows:

$$\alpha^{(o)} \cup \beta^{(o)} = \text{df } \{z^o\colon z^o \in \alpha^{(o)} \ .\!\vee\!.\ z^o \in \beta^{(o)}\}^{(o)}$$
$$\alpha^{(o)} \cap \beta^{(o)} = \text{df } \{z^o\colon z^o \in \alpha^{(o)} \ .\&.\ z^o \in \beta^{(o)}\}^{(o)}$$
$$\bar{\alpha}^{(o)} = \text{df } \{z^o\colon \sim (z^o \in \alpha^{(o)})\}^{(o)}$$
$$\iota x^{(o)} = \text{df } \{z^o\colon z^o = x^o\}^{(o)}$$
$$\alpha^{(o)} - \iota y^o = \text{df } \{z^o\colon z^o \in \alpha^{(o)} \ .\&.\ z^o \neq y^o\}.$$

These cannot be replicated in the language of substitution without introducing for each a battery of auxiliary definitions. Consider putting:

df(∪)[1]
$\imath(p/a)[p/a \approx_z Az] \cup \imath(p/a)[p/a \approx_z Bz)] = \text{df}$
$\quad \imath(p/a)[p/a :\approx_z\colon z \in \imath(p/a)[p/a \approx_z Az] \ .\!\vee\!.\ z \in \imath(p/a)[p/a \approx_z Bz]]$
df(∩)[1]
$\imath(p/a)[p/a \approx_z Az] \cap \imath(p/a)[p/a \approx_z Bz)] = \text{df}$
$\quad \imath(p/a)[p/a :\approx_z\colon z \in \imath(p/a)[p/a \approx_z Az] \ .\&.\ z \in \imath(p/a)[p/a \approx_z Bz]].$

With usual definitions, class symbols are single letters and this allows the above definition of intersection to apply to $(\alpha^{(o)} \cup \beta^{(o)}) \cap \delta^{(o)}$. Not so in substitution. A new definition would be needed together with supporting definitions to ensure its elimination from all contexts. This makes substitution exceedingly inconvenient when compared to the ordinary notation for classes.

The best way to get on in substitution is to avoid the introduction of operation signs flanked by the substitutional "class symbols." For the practical purposes of the next chapter we shall employ the following definitions:

df$(-\iota y)$ $\quad (r/c - \iota y)(z) = \text{df } z \in \imath(p/a)[p/a \approx r/c] \ .\&.\ z \neq y$
df$(+\iota y)$ $\quad (r/c + \iota y)(z) = \text{df } z \in \imath(p/a)[p/a \approx r/c] \ .\!\vee\!.\ z = y$
df$(+\iota y - \iota x)$ $\quad (r/c + \iota y - \iota x)(z) = \text{df}$
$\qquad z \in \imath(p/a)[p/a \ .\!\approx\! z.\ (r/c + \iota y)(z)] \ .\&.\ z \neq x$
df$(-\iota y + \iota x)$ $\quad (r/c - \iota y + \iota x)(z) = \text{df}$
$\qquad z \in \imath(p/a)[p/a \ .\!\approx\! z.\ (r/c - \iota y)(z)] \ .\!\vee\!.\ z = x$
df$(-\iota y - \iota x)$ $\quad (r/c - \iota y - \iota x)(z) = \text{df}$
$\qquad z \in \imath(p/a)[p/a \ .\!\approx\! z.\ (r/c - \iota y)(z)] \ .\&.\ z \neq x.$

These definitions simplify the needed notation. It is not necessary to have a special symbol for singleton. Substitution already has a straightforward proxy.

$(\text{T}_{6.26})^1$
$\vdash (\forall x^o)(x^o \in \iota x^{(o)})$
$\vdash (x)(x \in \imath(p/a)[p/a \ .\!\approx\! z.\ z = x])$

Proof:

1. $x = x$ ($T_{4.8}$)a, ui.
2. $x \in \imath(p/a)[p/a .\approx z.\ z = x]$ 1, $*(\text{Abs})^1$, log.
3. $(x)2$ ug.

$(T_{6.27})^1$

$\vdash (\forall y^o)(\forall \alpha^{(o)})(y^o \in \alpha^{(o)} \cup \iota y^{(o)})$

$\vdash (y)(lm)(y \in \imath(p/a)[p/a .\approx z.\ (l/m + \iota y)(z)])$

Proof:

1. $y = y$ ($T_{4.8}$)a, ui.
2. $y \in \imath(p/a)[p/a \approx l/m] .\vee.\ y = y$ 1, add.
3. $(l/m + \iota y)(y)$ 3, df$(+\iota y)$,
4. $y \in \imath(p/a)[p/a .\approx z.\ (lm + \iota y)(z)]$ 4, $*(\text{Abs})^1$.
5. $(y)(lm)4$ ug.

$(T_{6.28})^1$

$\vdash (\forall^o y)(\forall \alpha^{(o)})(y^o \in \alpha^{(o)}\ :\supset:\ \alpha^{(o)}\ .=.\ \alpha^{(o)} - \iota y^{(o)} \cup \iota y^{(o)})$

$\vdash (y)(lm)(y \in \imath(p/a)[p/a \approx l/m]\ :\supset:$
$\quad \imath(p/a)[p/a \approx l/m]\ .=.\ \imath(p/a)[p/a .\approx z.\ (l/m - \iota y + \iota y)(z)])$

Proof:

1. $\rightarrow\ y \in \imath(p/a)[p/a \approx l/m]$
2. $\quad \rightarrow\ w \in \imath(p/a)[p/a \approx l/m]$
3. $\quad\quad w \in \imath(p/a)[p/a \approx l/m]\ \&\ w \neq y\ .\vee.\ w = y$ 2, log.
4. $\quad\quad (l/m - \iota y + \iota y)(w)$ 3, df$(-\iota y + \iota y)$, df$(\iota - \iota y)$, $*(\text{Abs})^1$.
5. $\quad\quad$ __ $w \in \imath(p/a)[p/a .\approx z.\ (l/m - \iota y + \iota y)(z)]$ 4, $*(\text{Abs})^1$.
6. $\quad 2 \supset 5$ cp.
7. $\quad \rightarrow\ w \in \imath(p/a)[p/a .\approx z.\ (l/m - \iota y + \iota y)(z)]$
8. $\quad\quad (l/m - \iota y + \iota y)(w)$ 7, $*(\text{Abs})^1$.
9. $\quad\quad w \in \imath(p/a)[p/a \approx l/m]\ .\&.\ w \neq y\ :\vee:\ w = y$ df$(-\iota y + \iota y)$, df$(-\iota y)$, $*(\text{Abs})^1$.
10. $\quad\quad$ __ $w \in \imath(p/a)[p/a \approx l/m]$ 1, 9, log.
11. $\quad 7 \supset 10$ cp.
12. $w \in \imath(p/a)[p/a \approx l/m]\ .\equiv.$
    $w \in \imath(p/a)[p/a .\approx z.\ (l/m - \iota y + \iota y)(z)]$
    6 & 11 conj, df$(\equiv)$.
13. $(w)12$ ug.
14. $\imath(p/a)[p/a \approx l/m] = \imath(p/a)[p/a .\approx z.\ (l/m - \iota y + \iota y)(z)]$
    13, $*(\text{Ext})^1$.

$*(T_{6.29})^1$

$\vdash (\forall \alpha^{(o)})(\forall x^o)(\forall y^o)((\alpha^{(o)} - \iota x^o) - \iota y^o\ .=.\ (\alpha^{(o)} - \iota y^o) - \iota x^o)$

$\vdash (lm)(x)(y)(\imath(p/a)[p/a .\approx.\ (l/m - \iota x - \iota y)(z)] =$
$\quad \imath(p/a)[p/a .\approx.\ (l/m - \iota y - \iota x)(z)])$

Proof:

1. $z \in \imath(p/a)[p/a \approx l/m]\ \&\ z \neq x\ .\&.\ z \neq y\ :\equiv:$
   $z \in \imath(p/a)[p/a \approx l/m]\ \&\ z \neq y\ .\&.\ z \neq x$ log.

2. $z \in \text{ı}(p/a)[p/a\ .\!\approx\!.\ (l/m - \iota x)(z)]\ .\&.\ z \neq y\ :\equiv:$
   $z \in \text{ı}(p/a)[p/a\ .\!\approx\!.\ (l/m - \iota y)(z)]\ \&\ z \neq x$
   1, df$(-\iota x)$, df$(-\iota y)$, $*(\text{Abs})^1$.
3. $(l/m - \iota x - \iota y)(z) \equiv (l/m - \iota y - \iota x)(z)$ 2, df$(-\iota x - \iota y)$, df$(-\iota y - \iota x)$.
4. $(z)((l/m - \iota x - \iota y)(z) \equiv (l/m - \iota y - \iota x)(z))$ 3, ug.
5. $\text{ı}(p/a)[p/a\ .\!\approx\!.\ (l/m - \iota x - \iota y)(z)] = \text{ı}(p/a)[p/a\ .\!\approx\!.\ (l/m - \iota y - \iota x)(z)]$
   4, $*(V)^1$.
6. $(lm)(x)(y)5$ ug.

The Boolean operations of the calculus of classes are thus transformed when articulated within the language of substitution. Nonetheless, it is clear that analogs of the theorems of the calculus of classes are always forthcoming.

Comparison with *Principia Mathematica*

6.6 Let us now compare the substitutional treatment of class symbols with the treatment of class symbols in *Principia Mathematica*. As it is tangential to our present concerns, the matter of predicativity will be omitted here and taken up separately in Part III. Moreover, we shall omit type indices where convenient. We begin by setting out the salient features of the treatment of class symbols in *Principia*, responding to some common criticisms. There is no need to become mired in ontological issues surrounding the question of whether an ontology "propositional functions" as values of the predicate variables is assumed in *Principia*. My intent is simply to compare two distinct languages.

Whitehead and Russell give the following definition schemata for class symbols (*PM*, p. 190):

$*20.01\quad f(\hat{z}\psi z) = \text{df}\ (\forall\phi)(\phi x\ .\!\equiv_x\!.\ \psi x\ :\&:\ f(\phi\hat{z}))$.
$*20.02\quad x \in \phi\hat{z} = \text{df}\ \phi x$.

The scope indicator "$[\hat{z}\phi z]$" discussed in their introduction is omitted and should be added (*PM*, p. 80). Moreover, schematic letters should be distinguished from *Principia*'s predicate variables "ϕ", "ψ" "Γ", and "Σ", etc. We shall rewrite the definitions accordingly. Thus, $*20.01$ should be:

$[\hat{z}Az]\ B(\hat{z}Az|v) = \text{df}\ (\exists\phi)(\phi x\ .\!\equiv_x\!.\ A(x|z)\ .\&.\ B(\phi\hat{z}|v))$,
where x is free for z in A, and $\phi\hat{z}$ is free for v in B.

Definition $*20.02$ can stay as it is. With the scope indicator added, one can go on to adopt *Principia*'s convention of omitting scope indicators where the narrowest scope is intended.

The use of the two schemata can be illustrated by considering the following process of elimination:

i. $x \in \hat{z}Az$
ii. $(\exists\phi)(\phi z\ .\!\equiv_z\!.\ Az\ :\&:\ x \in \phi\hat{z})$ i, $*20.01$
iii. $(\exists\phi)(\phi z\ .\!\equiv_z\!.\ Az\ :\&:\ \phi x)$ ii, $*20.02$

To take another example, consider:

i. $\hat{z}Az = \hat{z}Bz$.
ii. $[\hat{z}Az]([\hat{z}Bz][\hat{z}Az = \hat{z}Bz])$ i, scope convention
iii. $(\exists\phi)(\phi z .\equiv_z. Az :\&: (\exists\psi)(\psi z .\equiv_z. Bz :\&: \phi\hat{z} = \psi\hat{z}))$ ii, $*20.01$

It would not have been possible to apply,

$$*13.01 \quad x = y \;.{=}.\; (\phi)(\phi x \equiv \phi y) \quad \text{Df}$$

before $*20.01$. As we saw in the case of definite descriptions, definitions applicable to singular terms cannot be applied to incomplete symbols.

The similarities between the treatment of class symbols and the treatment of definite descriptions in *Principia* are many. But there are important differences as well. Comprehension in *Principia* (ignoring the issue of predicativity) is:

$$(\exists\phi^{(t_1,\ldots,t_n)})(x_1^{t_1},\ldots,x_n^{t_n})(\phi^{(t_1,\ldots,t_n)}(x_1^{t_1},\ldots,x_n^{t_n}) \;.{\equiv}.\; A),$$
where $\phi^{(t_1,\ldots,t_n)}$ is not free in A.

This ensures that

$$\text{E!}(\hat{z}Az),$$

since this is just

$$(\exists\phi)(x)(\phi x \equiv Ax).$$

Accordingly,

$$\hat{z}Az = \hat{z}Az,$$

is a theorem schema, for by $*20.01$ this is,

$$(\exists\phi)(\phi x \equiv_x Ax \;.\&.\; \phi\hat{z} = \phi\hat{z}).$$

Its analog in the theory of definite descriptions, however, is

$$\imath zAz = \imath zAz,$$

and this is not a theorem schema. Similarly, in the theory of descriptions one finds,

$$\text{E!}(\imath zAz) \;.{\supset}.\; [\imath zAz]C(B(\imath zAz)) \equiv C([\imath zAz]B(\imath zAz)),$$

where C is any truth-functional context. The direct analog,

$$E!(\hat{z}Az) \;.{\supset}.\; [\hat{z}Az]C(B(\hat{z}Az)) \equiv C([\hat{z}Az]B(\hat{z}Az))),$$

for C a truth-functional context, is not theorem schema.[3] Let C be the context $\sim(\ldots)$, and B the context $(\ldots = \Gamma\hat{z})$. We get:

$$(\exists\phi)(x)(\phi x \equiv Ax) \;.{\supset}.$$
$$(\exists\phi)(\phi x \equiv_x Ax \;.\&.\; \sim(\phi\hat{z} = \Gamma\hat{z})) \equiv$$
$$\sim(\exists\phi)(\phi x \equiv_x Ax \;.\&.\; \phi\hat{z} = \Gamma\hat{z}).$$

3. Linsky errs in attributing this to *Principia* (1983, p. 160).

The left-hand side of the biconditional is true; the right side is false. The proper theorem schema for the theory of classes requires that C be truth-functional and that B be extensional. That is, we get:

$$\text{E!}[\hat{z}Az] \;\&\; (\phi)(\psi)(\phi x \equiv_x \psi x \;.\!\supset\!.\; B(\phi\hat{z}) \equiv B(\psi\hat{z})) \;.\!\supset\!. \\ [\hat{z}Az]C(B(\hat{z}Az)) \equiv C([\hat{z}Az]B(\hat{z}Az)),$$

where C is truth-functional.

These differences are significant and one must be on the look out for them. The omission of scope markers for class symbols offers important cases. For example, Carnap objects to Whitehead and Russell's claim that their contextual definitions permit class symbols to behave *as if* they were genuine singular terms. Carnap finds trouble with this because statements of the form,

(1) $\hat{z}\phi z = \phi\hat{z}$
(2) $\hat{z}\phi z \neq \phi\hat{z}$

"look like" contradictories when they are not. The first expands to

(1′) $(\exists\Gamma)[\Gamma x \;.\!\equiv_x\!.\; \phi x \;:\!\&\!:\; \Gamma\hat{z} = \phi\hat{z}]$,

the second to

(2′) $(\exists\Gamma)[\Gamma x \;.\!\equiv_x\!.\; \phi x \;:\!\&\!:\; \Gamma\hat{z} \neq \phi\hat{z}]$.

Both may be true. To avoid the problem, Carnap suggests a departure from *Principia*'s contextual definitions for class symbols, offering

$$f(\hat{z}\psi z) \;=\text{df}\; (\phi)[\phi x \;.\!\equiv_x\!.\; \psi x \;:\!\supset\!:\; f(\phi\hat{z})],$$

as the replacement for $*20.01$ (Carnap 1947, p. 146).

In his *Oblique Contents*, Linsky was eager to come to Russell's rescue. Taking issue with Carnap's replacement, Linsky arrives at

(3) $(\psi)(\phi)[\phi x \;.\!\equiv_x\!.\; \psi x \;:\!\supset\!:\; \phi\hat{z} = \psi\hat{z}]$,

which is clearly false, by applying Carnap's contextual definition to

(4) $(\Gamma)(\hat{z}\Gamma z = \hat{z}\Gamma z)$,

which, he says, is clearly true (Linsky, 1983, p. 145). Carnap's contextual definition, Linsky explains, transforms (4) into,

(5) $(\Gamma)(\phi)[\phi x \;.\!\equiv_x\!.\; \Gamma x \;:\!\supset\!:\; (\psi)(\psi x \;.\!\equiv_x\!.\; \Gamma x \;:\!\supset\!:\; \phi\hat{z} = \psi\hat{z})]$.

From this (3) follows. Linsky concludes that we must abandon Carnap's suggestion. Linsky's own solution is to keep the original $*20.01$ and embellish the conventions on the omission of scope operators (*ibid.*).

Linsky's argument against Carnap is wanting. In considering (4), Linsky has omitted scope markers. But we must choose between the following:

(4)a $(\Gamma)(\,[\hat{z}\Gamma z][\hat{x}\Gamma x][\hat{z}\Gamma z = \hat{x}\Gamma x])$
(4)b $(\Gamma)(\,[\hat{z}\Gamma z][\hat{z}\Gamma z = \hat{z}\Gamma z])$.

With Russell's existence axioms for classes and his original contextual definition of class symbols, the two are logically equivalent. But with Carnap's suggested contextual definition, this is no longer true. As we see, (4)a is false because it expands to (5) on Carnap's definition. On the other hand, (4)b is true; it expands to,

(6) $(\Gamma)(\phi)[\phi x \mathrel{.{\equiv}_x.} \Gamma x :\supset: \phi\hat{z} = \phi\hat{z}]$,

on Carnap's definition.

The difficulty with Carnap's proposal is that while hoping to avoid one problem (i.e., that (1) and (2) look like contradictories when they are not) he introduced another—(4a) and (4b) look like equivalences when they are not. Linsky was wrong about Carnap. But Carnap has constructed a pseudo-problem in the first place. The wffs (1) and (2) will look like contradictories only to those readers of *Principia* that misapply its definitions. One cannot apply the definition,

$*13.02 \quad x \neq y = \text{df} \sim (x = y)$

to the clause

$\hat{z}\phi z \neq y.$

It is not some unstated but intended scope convention governing the omission of scope markers the prevents the application. Its application is prevented by the fact that class expressions are not terms of the formal system. Surely, the fact that there may be readers who will misapply the definitions is no reason to regard *Principia*'s treatment of class symbols as a failure.

6.7 There are, however, substantive questions concerning the introduction and elimination of class symbols in *Principia*. Definitions $*20.01$ and $*20.02$ are not the only schemata for the contextual definition of such symbols. Whitehead and Russell introduce lower-case Greek symbols as well:

$*20.07 \quad (\alpha)f(\alpha) = \text{df } (\phi)\, f\{\hat{z}\phi z\}$
$*20.071 \quad (\exists\alpha)\, f(\alpha) = \text{df } (\exists\phi)\, f\{\hat{z}\phi z\}.$
$*20.08 \quad f(\hat{\alpha}\psi\alpha) = \text{df } (\exists\phi)(\phi\alpha \mathrel{.{\equiv}_\alpha.} \psi\alpha :\&: f(\phi\hat{\alpha})).$
$*20.081 \quad \alpha \in \phi\hat{\alpha} = \text{df } \phi\alpha.$

Rewriting $*20.07$, $*20.071$, and $*20.08$ with schematic letters and scope operators we have:

$(\alpha)B(\alpha) = \text{df } (\phi)\ B(\hat{z}\phi z|\alpha)$
$(\exists\alpha)B(\alpha) = \text{df } (\exists\phi)B(\hat{z}\phi z|\alpha),$
$[\hat{\alpha}A\alpha]\ B(\hat{\alpha}A\alpha|v) = \text{df } (\exists\phi)(\phi\alpha \mathrel{.{\equiv}_\alpha.} A\alpha :\&: B(\phi\hat{\alpha}|v)).$

where $\phi\hat{\alpha}$ is free for v in B.

Many definitions are introduced via lower-case Greek letters. For instance, we find:

$*22.03 \quad \alpha \cup \beta \mathrel{.=.} \hat{z}(z \in \alpha \vee z \in \beta) \quad \text{Df.}$

Now the letters "α" and "β" here are really meta-linguistic letters for the object-language expressions for classes—that is, lower-case Greek or class abstract symbols. This, however, raises questions of the order of elimination again. It is clear that one must apply $*22.03$ to

"$\hat{z}\phi z \cup \hat{z}\psi z$",

before applying $*20.01$, for definition $*22.03$ does not permit "$\phi \cup \psi$" as an expression. So the order of elimination here is determined by the fact that definition $*22.03$ is rendered with lower-case Greek symbols (used as meta-linguistic schematic letters).

The situation is now the inverse of that found with definitions such as $*13.01$, definitions introduced with meta-linguistic letters for genuine variables of the system. *Principia* intends that class terms be eliminated in the narrowest scope possible. For this to be secured, one must always eliminate lower-case Greek first. Consider the following:

i. $(\alpha)(\beta)(\alpha \cup \beta = \beta \cup \alpha)$.
ii. $(\alpha)(\beta)(\hat{z}(z \in \alpha \vee z \in \beta) = \hat{z}(z \in \beta \vee z \in \alpha))$ i, $*22.03$
iii. $(\phi)(\psi)(\hat{z}(z \in \hat{x}\phi x \ .\vee.\ z \in \hat{x}\psi x) = \hat{z}(z \in \hat{x}\psi x \ .\vee.\ z \in \hat{x}\phi x))$ ii, $*20.07$

It is only after the application of definition $*22.03$ that we see the narrowest context of scope—the context to which definition $*20.01$ and $*20.02$ are to be applied. Similarly, we have

i. $\Sigma\alpha \ .\equiv_\alpha.\ (\beta)(0 \in \beta \ .\&.\ (\delta)(\delta \in \beta \ .\supset.\ \delta + 1 \in \beta) \ :\supset:\ \alpha \in \beta)$
ii. $(\phi)([\hat{z}\phi z][\Sigma(\hat{z}\phi z)] \ .\equiv.$
$(\beta)(0 \in \beta \ .\&.\ (\delta)(\delta \in \beta \ .\supset.\ \delta + 1 \in \beta) \ :\supset:\ [\hat{z}\phi z][\hat{z}\phi z \in \beta])$

The definition $*20.07$ is to be employed with narrowest possible scope.

Scope arises in another way as well. Consider the following elimination:

i. $\hat{z}Az \in \hat{\alpha}B\alpha$
ii. $(\exists\Sigma)(\Sigma\alpha \ .\equiv_\alpha.\ B\alpha \ :\&:\ \hat{z}Az \in \Sigma\hat{\alpha})$ i, $*20.08$
iii. $(\exists\Sigma)(\Sigma\alpha \ .\equiv_\alpha.\ B\alpha \ :\&:\ \Sigma(\hat{z}\phi z))$ ii, $*20.081$
iv. $(\exists\Sigma)(\Sigma\alpha \ .\equiv_\alpha.\ B\alpha \ :\&:\ (\exists\Gamma)(\Gamma x \ .\equiv_x.\ Ax \ :\&:\ \Sigma(\Gamma\hat{z})))$ iii, $*20.01$

All is well. But can one go another way? Starting from ii, put:

iii$*$. $(\exists\Sigma)(\Sigma\alpha \ .\equiv_\alpha.\ B\alpha \ :\&:\ (\exists\Gamma)(\Gamma x \ .\equiv_x.\ Ax \ :\&:\ \Gamma\hat{z} \in \Sigma\hat{\alpha}))$ ii, $*20.01$.

Trouble arises with the clause "$\Gamma\hat{z} \in \Sigma\hat{\alpha}$". We cannot apply $*20.081$; and neither, it seems, can we apply $*20.02$ for we have a circumflex lower-case Greek letter involved. A similar problem arises if one begins as follows:

ii°. $(\exists\Gamma)(\Gamma x \ .\equiv_x.\ Ax \ :\&:\ \Gamma\hat{z} \in \hat{\alpha}B\alpha)$ i, $*20.01$
iii°. $(\exists\Gamma)(\Gamma x \ .\equiv_x.\ Ax \ :\&:\ (\exists\Sigma)(\Sigma\alpha \ .\equiv_\alpha.\ B\alpha \ :\&:\ \Gamma\hat{z} \in \Sigma\hat{\alpha})))$ ii°, $*20.08$.

Again we arrive at "$\Gamma\hat{z} \in \Sigma\hat{\alpha}$". The presence of the circumflexed lower-case Greek is the trouble.

Now one might regard this as showing that there is a determinate order of elimination here. But in their *Introduction*, Whitehead and Russell discuss the following definition (*PM*, p. 80):

$$f\alpha = \text{df } (\exists\phi)(\phi x \equiv_x \psi x \text{ .\&. } f(\phi\hat{z})).$$

From this, they say, one arrives at:

$$f\hat{\alpha} = \text{df } (\exists\phi)(\phi x \equiv_x \hat{\psi} x \text{ .\&. } f(\phi\hat{z})).$$

This would avoid our problem, for we could use such definitions to eliminate the circumflex occurrence of the lower-case Greek "α" in the clause "$\Gamma\hat{z} \in \Sigma\hat{\alpha}$". Put:

$$\Gamma\hat{z} \in (\exists\phi)(\phi x \equiv_x \hat{\psi} x \text{ .\&. } \Sigma(\phi\hat{z})).$$
$$(\exists\phi)(\phi x \equiv_x \Gamma x \text{ .\&. } \Sigma(\phi\hat{z})) \quad \text{by } {*}20.02.$$

If the definitions of the Introduction are allowed, then the different orders of elimination above would not matter. Unfortunately, the discussions defining free and cimcumflexed occurrences of lower-case Greek letters set out in the Introduction are formally improper. As Whitehead and Russell admit, the definiens contains a free variable, viz., "ψ", not contained in the definiendum. Perhaps the best plan is to replace $*20.08$ with:

$$[\hat{\alpha}A\alpha]B(\hat{\alpha}A\alpha) = \text{df } (\exists\phi)[\phi\alpha \text{ .}{\equiv_\alpha}\text{. } A\alpha \text{ :\&: } B(\phi\hat{z}|v)),$$

where $\phi\hat{z}$ is free for v in B. This simply drops the circumflexed Greek in definition $*20.08$.

There is a related problem. Whitehead and Russell claim to have demonstrated the following theorem (schema),

$$*20.61 \quad (\alpha)f\alpha \text{ .}{\supset}\text{. } f\beta,$$

noting that this is the analog for class symbols of

$$*10.1 \quad (x).\phi x. \supset \phi y.$$

The trouble is that *Principia* makes make no formal provisions for free occurrences of lower-case Greek symbols. In characteristic fashion, Whitehead and Russell muddled the distinction between object-language Greek and metalinguistic Greek, writing that "a Greek letter, such as α, is merely an abbreviation for an expression of the form $\hat{z}\phi z$" (*PM*, p. 194). There are two sorts of such "abbreviations"; those introduced into the object language by definitions such as $*20.07$ and those which are meta-linguistic expressions for object-language expressions. The muddle makes $*20.61$ improper, unless "β" is just intended to stand in for a class expression. In any event, $*20.61$ should be replaced by:

$$(\alpha)B\alpha \supset B(\hat{z}Az|\alpha),$$
where $\ulcorner \hat{z}Az \urcorner$ is free for α in A.

There are no provisions for free or circumflex occurrences of object-language lower-case Greek symbols. But this is not serious; such occurrences are never needed in any case.

6.8 With the salient features of *Principia*'s treatment of class symbols before us, we are in a position to compare the substitutional treatment.

Consider the following class-theoretical statement expressed in the language of predicate variables,

$$x \in \hat{z}\phi z.$$

Principia defines this as

$$(\exists\Gamma)[\Gamma z \mathrel{.{\equiv_z}.} \phi z \mathrel{:\&:} x \in \Gamma\hat{z}].$$

Substitutionally, we have

$$x \in \imath(p/a)[p/a \approx l/m].$$

This becomes

$$(\exists pa)[p/a \approx l/m \mathrel{.\&.} x \in p/a].$$

There is an exact parallel. For classes of classes (of individuals) we saw that *Principia* has

$$\Sigma\alpha \equiv_\alpha \psi\alpha.$$

Eliminating the lower-case Greek, we arrive at:

$$(\phi)([\hat{z}\phi z]\Sigma(\hat{z}\phi z) \mathrel{.{\equiv}.} [\hat{z}\phi z]\psi(\hat{z}\phi z)).$$

Substitution has

$$s/tw \approx h/de$$

which by definition was

$$(rc)([r/c][r/c \in s/tw] \mathrel{.{\equiv}.} [r/c][r/c \in h/de]).$$

We can see the parallel. Next, consider eliminating the class symbols from

$$\hat{z}\phi z \in \hat{\alpha}\psi\alpha.$$

Principia arrives at

$$(\exists\Sigma)[\Sigma\alpha \mathrel{.{\equiv_\alpha}.} \psi\alpha \mathrel{:\&:} (\exists\Gamma)[\Gamma x \mathrel{.{\equiv_x}.} \phi x \mathrel{:\&:} \Sigma(\Gamma\hat{z})]].$$

Avoiding the class variable "α", this yields

$$(\exists\Sigma)[(\phi)([\hat{z}\phi z]\Sigma(\hat{z}\phi z) \mathrel{.{\equiv}.} [\hat{z}\phi z]\psi(\hat{z}\phi z)) \mathrel{\&} (\exists\Gamma)[\Gamma x \mathrel{.{\equiv_x}.} \phi x \mathrel{:\&:} \Sigma(\Gamma\hat{z})]].$$

The substitutional language has

$$\imath(p/a)[p/a \approx l/m] \in \imath(s/tw)[s/tw \approx h/de].$$

Eliminating the class symbols, one gets,

$$(\exists stw)[(rc)([r/c][r/c \in s/tw] \;.{\equiv}.\; [r/c][r/c \in h/de]) \;\&\; (\exists pa)[(x)(x \in p/a \;.{\equiv}.\; x \in l/m) \;\&\; p/a \in s/tw]].$$

Again we find a close parallel.

In virtue of this parallel, we can see the direct evolutionary link from the substitutional treatment of classes to the treatment (bracketing predicativity for now) that would appear in the 1907 "Mathematical Logic as Based on the Theory of Types," and later in *Principia*. After some eighty-seven years, Russell can finally be believed when, in "Mathematical Logic," he maintained that the language of predicate variables (the so-called language of "propositional functions") was but a "convenience" replaceable by the language of substitution. Every closed wff in Russell's primitive type-indexed notation of predicate variables—every closed wff which is the result of eliminating class symbols—has an analog in the type-free language of substitution. This is, of course, just what we should have expected from the transformation function set forth in our Chapter 5.

There are, however, important differences in the languages. For instance, in the language of predicate variables, type indices may be suppressed so that the axioms and theorems have "typical ambiguity." Obviously, typical ambiguity is impossible in the substitutional notation. The very *raison d'être* of the theory was to make types part of logical grammar. This is its great asset. The asset is also its practical weakness. The language of predicate variables has the important advantage of allowing the introduction of the commonplace class symbol $\hat{z}Az$ directly in its language. From here the usual definitions and the calculus of classes takes over.

Interestingly, Russell originally envisioned definitions similar to $*20.01$ and $*20.02$ for substitution. In the manuscript, "On Substitution$_2$" Russell wrote (*OS*, p. 110):

$$\hat{x}\phi x \text{ is to be } \imath(p/a)\{p/a = \phi x/x\}^4;$$

and later he puts (*OS*, p. 117):

$$x \in p/a = \text{df } p/a;x.$$

The plan was to proxy a class abstract symbol such as $\hat{x}Ax$, typed for a class of individuals, with the definite description-like expression, $\ulcorner\imath(p/a)[p/a = \phi x/x]\urcorner$. To be sure, the notation

$$\imath(p/a)[p/a \equiv \{Ax\}/x]$$

still employs complex prototypes. But the remedy is at hand as we saw. Russell goes on to put (*ibid.*):

$$\psi\{\imath(p/a)\phi(p,a);x\} \;.{=}:\; (\exists q,b){:}\; \phi(p,a) \;.{\equiv_{p,a}}.\; p/a = q/b : \psi(q/b;x) \quad \text{Df.}$$

4. Once again note that Russell used "=" flanking matrices to capture coexemplification. We altered this to "≡".

He is coming close to what we would find with *Principia*'s $*20.01$. It is but a short step to:

$$B(\imath(p/a)[p/a;x \equiv_x Ax]) = \text{df}$$
$$(\exists pa)(p/a;x \equiv_x Ax \;.\&.\; B(p/a)).$$

The context B would be either that of membership or identity. For example, an expression of the form

$$x \in \imath(p/a)[p/a;x \;.{\equiv_x}.\; x = x],$$

would become

$$(\exists pa)[p/a;x \;.{\equiv_x}.\; x = x \;:\&:\; x \in p/a].$$

Then applying the definition of "$x \in p/a$", this becomes,

$$(\exists pa)(p/a;x \;.{\equiv_x}.\; x = x \;:\&:\; p/a;x).$$

The intended result, of course, is to achieve

$$x \in \imath(p/a)[p/a;x \;.{\equiv_x}.\; x = x] \;.{\equiv}.\; x = x,$$

which parallels the following instance of the naive class theoretic abstraction schema:

$$x \in \hat{z}(z = z) \;.{\equiv}.\; x = x.$$

We see the general idea.

Unfortunately, class symbols cannot be introduced into the language of substitution in a way that parallels their introduction into the language of predicate variables. A definition schema such as

$$B(\imath(p/a)[p/a \equiv \{Ac\}/c]) = \text{df}$$
$$(\exists pa)(p/a \equiv \{Ac\}/c \;.\&.\; B(p/a)),$$

is not well-formed. The context B is really a dyadic context masquerading as if it were monadic. We must not forget that "p/a" is not a singular term. In *Principia*'s $*20.01$, on the other hand, the context f is monadic. In the language of substitution, one must contextually define substitutional proxies for class symbols for each context in which they occur. There can be no one contextual definition covering all contexts.

Class symbols are enormously practical. As we remarked in our introduction, Russell's original intent in 1906 was to write the whole of *Principia* using the substitutional theory. Whitehead would hear nothing of it! On 21 February 1906 he wrote to Russell criticizing the symbolism of the theory:[5]

> I look upon these remarks as a protest against your excessive formalism. (There's a change of heart for you!)

Mind you, Whitehead was not objecting to the substitutional theory. He was prodding Russell to find a way to preserve class symbols in the notation.

5. See manuscript #210.057392 in the Russell Archives.

Whitehead wrote that he was delighted by Russell's work on the substitutional theory in a letter of 25 December 1905. (Probably he had seen Russell's 22 December 1905 axiomatization of the theory.) But he was steadfastly against abandoning class symbols. On 30 April 1905, he wrote Russell:

> This extreme rigour must be tempered by practical considerations. *Classes* can be kept in use by the consideration that our object is to systematize the actual reasoning of mathematics, and this actual reasoning does in fact habitually employ classes when it need not do so. Thus our object is to systematize the reasoning concerning classes, even when it is a primitive idea which might be avoided.

At Whitehead's suggestion, Russell seems to have tried to get ordinary class symbols into the language of substitution. His first step is to allow a single letter "α" for a matrix. In a letter to Jourdain of 15 March 1906 this becomes clear:

> $u \in$ Cls' is not *nonsense* on my present view, only the u should be replaced by p/a, where p/a is a 'matrix'. I use Greek letters for 'matrices', Latin letters for entities; thus
>
> $\alpha \in \text{Cls} \ .{=}.\ (\exists p,a).\ \alpha = p/a \quad \text{Df.}$
>
> (roughly speaking). And
>
> $\alpha = p/a \ .{=}.\ \alpha;x \equiv_x \ .\ p/a;x \quad \text{Df,}$
>
> where $p/a;x$ means the same as $p\frac{x}{a}$. α is nothing but a typographical abbreviation for something of the form q/b; it has no more reality than [sic] $\frac{\mathrm{d}}{\mathrm{d}x}$ or ∇^2. But symbolically, my no-classes theory makes very little difference—almost everything remains much as it was, but with a new interpretation of the symbols (Quoted from Grattan-Guinness 1977b, p. 77).

There are manuscripts showing Whitehead grappling with this approach.[6] But single letters for substitutional matrices cannot work.

It is, in fact, tedious to the point of practical impossibility to work in the substitutional notation. But the tedium of working within the substitutional language is not significantly greater than that which is required to work with predicate variables. The difference between the two languages is that the language of predicate variables (in subject as well as predicate positions) permits the introduction of class symbols proper. When in 1907 Whitehead saw Russell's definition *20.01 he wrote that it is "beyond all praise."[7] The substitutional theory was demoted. The theory, it was admitted, offers the best philosophical explanation of types, and one can translate from the language of

6. See the manuscript "Miscellaneous notes on PM," #230.031230f1 in the Russell Archives; there is also #210.057393.

7. See manuscript #710.057402 in the Russell Archives.

predicate variables into the language of substitution. So early in 1907 it was decided that the substitutional theory might properly belong in a philosophical appendix to *Principia*, the practical language being that of predicate variables so that class symbols could be introduced.[8]

Of course, the substitutional theory never made it even to an appendix of *Principia*. The reasons why will be the subject of Part III of this book. But it is important to realize that tedium is not among them. In fact, the option of translation is one that Whitehead and Russell did not fully explore. Practically, all that is required is a type-indexed language of class symbols plus a translation manual. If one thinks substitution holds the proper solution to the paradoxes plaguing logicism, then type-theory can be set forth largely as it appears in what Quine calls the "modern type-theory of classes" (Quine 1963, p. 259). It is straightforward to transform any closed statement (in primitive notion) in Quine's formulation of the type-indexed language of classes directly into the type-free language of substitution. After the translation there is only one style of variable—viz., the individual variable of the pure calculus for the substitutional logic of propositions.

8. See Russell's "On Types" (*T*, p. 4).

CHAPTER SEVEN

The "No-Relations$_e$" Theory

7.0 The substitutional proxy for the language of propositional functions—that is, a type-indexed language which allows predicate variables in subject and predicate positions—was discussed in Chapter 5. The substitutional treatment for symbols for relations-in-extension (hereafter "relations$_e$") is a natural continuation of the technique applied to class symbols examined in Chapter 6.

In dealing with relations$_e$, one might think it natural to adopt the 1912/4 Wiener–Kuratowski definition of ordered pair. This would be to have:

$$\langle x, y \rangle = \text{df } \{\{x\}, \{x, y\}\}.$$

Relations$_e$ would then be classes of classes. We shall not do this, however. The technique is out of sorts with Russell's idea of type-stratifying logicism's tenet that every n-place open wff determines a relation$_e$ of just those (ordered) entities satisfying the wff. When set within type-theory, the Wiener-Kuratowski technique makes this impossible. Moreover, heterogeneous relations$_e$ are lost. Cardinal similarity of classes of different type is the immediate casualty. We would have to raise the type of the one class (by taking its singletons, or the singletons of its singletons, etc.) to recover a form of comparability.

The Wiener-Kuratowski definition certainly simplifies types by ruling out heterogeneous relations$_e$. But in the substitutional theory types are part of logical form and (as we shall see) this has its own elegance and simplicity where heterogeneous relations$_e$ are concerned. In what follows, we set forth the theory of relations$_e$, and illustrate it by proving Cantor's power-class theorem. We demonstrate how substitution solves Cantor's paradox of the greatest cardinal.

Relations-in-Extension in *Principia Mathematica*

7.1 *Principia*'s treatment of the symbols for relations$_e$ follows the same lines as its treatment of class symbols. We find:

$$*21.01 \quad f(\hat{z}\hat{w}\ \phi zw) = \text{df} \ (\exists\psi)[(xy)(\psi xy \ .\!\equiv\!. \ \phi xy) \ \&\ f(\psi\hat{z}\hat{w})].$$
$$*21.02 \quad m(\psi\hat{z}\hat{w})n = \text{df} \ \psi mn.$$

Of course, $*21.01$ leaves out the needed scope operator. It also seems better to use the symbol "$\langle m, n\rangle$" so that $*21.02$ becomes:

$$\langle m, n\rangle \in \psi\hat{z}\hat{w} = \text{df} \ \psi mn.$$

To illustrate, consider

$$\langle a, b\rangle \in \hat{z}\hat{w}\phi zw.$$

Applying $*21.01$ we have:

$$(\exists\psi)[(xy)(\psi xy \ .\!\equiv\!. \ \phi xy) \ .\&. \ \langle a, b\rangle \in \psi\hat{z}\hat{w}].$$

Then using $*21.02$ we shall arrive at:

$$(\exists\psi)[(xy)(\psi xy \ .\!\equiv\!. \ \phi xy) \ .\&. \ \psi ab].$$

The parallel with the substitutional treatment (where relations$_e$ of individuals are concerned) should be clear.

The definitions $*21.01$ and $*21.02$ apply to relations$_e$ of individuals. Through typical ambiguity they also apply for relations$_e$ of entities of whatever type. But they do not apply to relations$_e$ of other relations$_e$. Just as in the treatment of class symbols, separate contextual definitions are needed for such cases. By typical ambiguity these definitions will apply for all types of relations$_e$ of relations$_e$. Whitehead and Russell introduce special quantifiers:

$$*21.07 \quad (R)fR = \text{df} \ (\phi)f(\hat{z}\hat{w}\phi zw)$$
$$*21.071 \quad (\exists R)fR = \text{df} \ (\exists\phi)f(\hat{z}\hat{w}\phi zw).$$

Then they put:

$$*21.08 \quad f\{\hat{R}\hat{S}\psi(R, S)\} = \text{df} \ (\exists\psi^\circ)[(R, S)(\psi^\circ(R, S) \ .\!\equiv\!. \ \psi(R, S)) \ \&\ f\{\psi^\circ(\hat{R}, \hat{S})\}].$$
$$*21.081 \quad R\{\psi(\hat{R}, \hat{S})\}S = \text{df} \ \psi(R, S).$$

As before, the scope operator should be added; and it is better to have

$$\langle R, S\rangle \in \psi(\hat{R}, \hat{S}) = \text{df} \ \psi(R, S)$$

as the expression of $*21.081$.

To illustrate, let us consider a two-placed relation between relations of individuals. Consider the following:

$$\langle \hat{z}\hat{w}\phi zw \ , \ \hat{z}\hat{w}\Gamma zw\rangle \in \hat{R}\hat{S}\{\psi(RS)\}$$

By ∗21.08 Russell gets:

$$(\exists\psi^\circ)[(RS)(\psi^\circ(RS) \,.{\equiv}.\, \psi(RS)) \,.\&.\, \langle \hat{z}\hat{w}\phi z\mathrm{w},\ \hat{z}\hat{w}\Gamma zw\rangle \in \psi^\circ(\hat{R}\hat{S})].$$

Now by ∗21.07, the clause

$$(RS)(\psi^\circ(RS) \,.{\equiv}.\, \psi(RS))$$

becomes:

$$(\phi^\circ)(\Gamma^\circ)([\hat{z}\hat{w}\phi^\circ zw][\hat{z}\hat{w}\Gamma^\circ zw]\ \psi^\circ(\hat{z}\hat{w}\phi^\circ zw, \hat{z}\hat{w}\Gamma^\circ zw) \,.{\equiv}.\, [\hat{z}\hat{w}\phi^\circ zw][\hat{z}\hat{w}\Gamma^\circ zw]\ \psi(\hat{z}\hat{w}\phi^\circ zw, \hat{z}\hat{w}\Gamma^\circ zw)).$$

Then applying ∗21.081 we arrive at:

$$(\exists\psi^\circ)[(\phi^\circ)(\Gamma^\circ)([\hat{z}\hat{w}\phi^\circ zw][\hat{z}\hat{w}\Gamma^\circ zw]\ \psi^\circ(\hat{z}\hat{w}\phi^\circ zw, \hat{z}\hat{w}\Gamma^\circ zw) \,.{\equiv}.\, [\hat{z}\hat{w}\phi^\circ zw][\hat{z}\hat{w}\Gamma^\circ zw]\ \psi(\hat{z}\hat{w}\phi^\circ zw, \hat{z}\hat{w}\Gamma^\circ zw)) \,.\&.\, \psi^\circ(\hat{z}\hat{w}\phi zw, \hat{z}\hat{w}\Gamma zw)].$$

The last clause can be treated in accordance with ∗21.01 and ∗21.02. What results is finally:

$$(\exists\psi^\circ)[(\phi^\circ)(\Gamma^\circ)([\hat{z}\hat{w}\phi^\circ zw][\hat{z}\hat{w}\Gamma^\circ zw]\ \psi^\circ(\hat{z}\hat{w}\phi^\circ zw, \hat{z}\hat{w}\Gamma^\circ zw). \,.{\equiv}.\, [\hat{z}\hat{w}\phi^\circ zw][\hat{z}\hat{w}\Gamma^\circ zw]\ \psi(\hat{z}\hat{w}\phi^\circ zw, \hat{z}\hat{w}\Gamma^\circ zw)) \,.\&.\, (\exists\phi^\circ)[(xy)(\phi^\circ xy \,{:}{\equiv}{:}\, \phi xy) \,.\&.\, (\exists\Gamma^\circ)[(xy)(\Gamma^\circ xy \,.{\equiv}.\, \Gamma xy) \,.\&.\, \psi^\circ(\phi^\circ\hat{z}, \Gamma^\circ\hat{z})]]].$$

Again the reader cannot miss the parallel with the substitutional treatment.

Whitehead and Russell do not give separate definitions for relations$_e$ of classes, or for classes of relations$_e$, etc. They simply write: "The results obtained in ∗20 and ∗21 for classes and relations whose members or terms are neither classes nor relations can be extended, by mere repetition of the proofs, to classes of classes, classes of relations, relations of classes, relations of relations, and so on" (*PM*, p. 204). Presumably then, dyadic relations$_e$ of classes would be defined by means of the following:

$$f\{\hat{\alpha}\hat{\beta}\psi\alpha\beta\} = \mathrm{df}\ (\exists\Sigma)[(\alpha\beta)(\Sigma\alpha\beta \,.{\equiv}.\, \psi\alpha\beta) \,\&\, f(\Sigma\hat{\alpha}\beta)]$$
$$\langle\alpha, \beta\rangle \in \psi\hat{\alpha}\beta = \mathrm{df}\ \psi\alpha\beta.$$

Consider the following example:

$$\langle\hat{z}\phi z, \hat{z}\Gamma z\rangle \in \hat{\alpha}\hat{\beta}\psi\alpha\beta.$$

Applying the contextual definition, we get:

$$(\exists\Sigma)[(\alpha\beta)(\Sigma\alpha\beta \,.{\equiv}.\, \psi\alpha\beta) \,\&\, \langle\hat{z}\phi z,\ \hat{z}\Gamma z\rangle \in \psi\hat{\alpha}\beta].$$

Then by ∗20.07 the clause

$$(\alpha\beta)(\Sigma\alpha\beta \,.{\equiv}.\, \psi\alpha\beta),$$

becomes:

$$(\phi^\circ)(\Gamma^\circ)([\hat{z}\phi^\circ z][\hat{z}\Gamma^\circ z]\Sigma(\hat{z}\phi^\circ z, \hat{z}\Gamma^\circ z) .\equiv. [\hat{z}\phi^\circ z][\hat{z}\Gamma^\circ z]\, \psi(\hat{z}\phi^\circ z, \hat{z}\Gamma^\circ z)).$$

Finally, eliminating the class symbols of the last conjunct by means of $*20.01$ and $*20.02$ we arrive at:

$$(\exists\Sigma)[(\phi^\circ)(\Gamma^\circ)([\hat{z}\phi^\circ z][\hat{z}\Gamma^\circ z]\Sigma(\hat{z}\phi^\circ z, \hat{z}\Gamma^\circ z) .\equiv. [\hat{z}\phi^\circ z][\hat{z}\Gamma^\circ z]\psi(\hat{z}\phi^\circ z, \hat{z}\Gamma^\circ z)) .\&. (\exists\phi^\circ)[(z)(\phi^\circ z .\equiv. \phi z) \;\&\; (\exists\Gamma^\circ)[(z)(\Gamma^\circ z .\equiv. \Gamma z) \;\&\; \Sigma(\phi^\circ\hat{z}, \Gamma^\circ\hat{z})]]].$$

We now see more clearly how the plan was to work.

Relations-in-Extension in the Substitutional Theory

7.2 The substitutional treatment of relations$_e$ parallels the functional, just as we found was the case in the theory of classes. Recall, however, that the substitutional theory treats all entities, propositions or otherwise, on a par. No distinction between propositional and non-propositional substitutions is made. A two-place propositional function of individuals is proxied by using matrices of the form "$p/a, b$", triatic functions of individuals are proxied through matrices of the form "$p/a,b,c$"; and so on. On the other hand, Russell typically uses "$q/p,a$" for a matrix which (in use) proxies a function of functions of individuals. Russell's use of letters "q" and "p", although they easily lend themselves to the interpretation that they are to refer only to propositions, actually are variables for all entities, propositions or otherwise. This means, however, that the matrical form for propositional functions of functions is the same as that for dyadic propositional functions of individuals—viz., "$x/y,z$". Similarly, the matrical form for propositional functions of functions of functions of individuals would be the same as that for triadic, and so forth.

How then are the relevant distinctions in the proxies to be made? Now Russell sometimes highlighted differences between proxying relations of individuals and propositional functions of functions (of individuals) by using the matrix "$x/(y,z)$" for the former as opposed to "$x/(y/z)$". Similarly, one would use "$w/(x/v, y/z)$" when the intent is to proxy a dyadic propositional function of two propositional functions of individuals; the matrix "$w/(x,v,y,z)$" would be used to proxy a four-place function of individuals. Of course, this solves nothing by itself. Moreover, putting, "$q/(x/z);(y/w)$" as opposed to "$q/xz;yw$" may lead the unwary into misunderstanding—as if "x/z" designates a single entity after all.

Grattan-Guinness (1974, 1977a) has called attention to this issue in the substitutional theory on a number of occasions. He suggests that a solution might be found if the theory would distinguish propositional and individual substitutions. But such a distinction would achieve little by itself. The difference between a propositional function of functions of individuals and a dyadic propositional function of individuals is fixed solely by the proposition in which the substitution is to take place. Even if "p" has to indicate a proposition and "a" an individual, the matrix "$q/p,a$" can still be used to proxy a function of functions (of individuals) or a dyadic function between a proposition and an individual. It all depends on the nature of the substitution performed in the proposition indicated by the prototype.

Now the problem does not arise when it comes to proxying propositional functions (no matter the adicity). In such cases there is no need for the matrical expression to syntactically mark differences between, say, a two-placed propositional function of individuals and a propositional function of functions (of individuals). Grattan-Guinness does have an important point, however, when it comes to relations$_e$. The substitutional theory must be able to capture extensionality when it comes to, say, relations$_e$ of individuals, relations$_e$ of classes, relations$_e$ between classes and individuals, and so on. Failure to do so, would destroy the theory.

The substitutional theory can face up to this challenge. The technique I shall employ is to use type superscripts. The superscripts are not appended to variables, however. The substitutional language is type-free. Rather, they appear in definitions. Let us begin with homogeneous dyadic relations$_e$.

$$\text{df}(R{\approx})^{(\text{oo})}$$
$$q/ce \approx^{(\text{oo})} r/ab \;=\text{df}\; (xy)(\langle x, y\rangle \in q/ce \;.{\equiv}.\; \langle x, y\rangle \in r/ab)$$
$$\text{df}(R{\approx}.\text{a})^{(\text{oo})}$$
$$q/ce \approx^{(\text{oo})}_{xy} B(xy) = \text{df}\; (xy)(\langle x, y\rangle \in q/ce \;.{\equiv}.\; B(xy))$$
$$\text{df}(R{\approx}.\text{b})^{(\text{oo})}$$
$$A(xy) \approx^{(\text{oo})}_{xy} B(xy) = \text{df}\; (xy)(A(xy) \;.{\equiv}.\; B(xy))$$
$$\text{df}(R)^{(\text{oo})}$$
$$\langle a, b\rangle \in \iota(q/ce)[q/ce \approx^{(\text{oo})}_{xy} B(xy)] \;=\text{df}$$
$$(\exists qce)(q/ce \approx^{(\text{oo})}_{xy} B(xy) \;.\&.\; \langle a, b\rangle \in q/ce)$$
$$\text{df}(R.\text{a})^{(\text{oo})}$$
$$\langle a, b\rangle \in q/ce = \text{df}\; (\exists h)(q/ce;ab!h \;\&\; h)$$

For identity we shall need:

$$\text{df}(R =)^{(\text{oo})}$$
$$\iota(q/ce)[q/ce \approx^{(\text{oo})}_{xy} A(xy)] = \iota(r/ab)[r/ab \approx^{(\text{oo})}_{xy} B(xy)] = \text{df}$$
$$(\exists qce)(q/ce \approx^{(\text{oo})} A(xy) \;.\&.\; (\exists rab)(r/ab \approx^{(\text{oo})}_{xy} B(xy) \;.\&.\; q/ce = r/ab)).$$

For homogeneous dyadic relations$_e$ between classes (of individuals) we have new definitions:

$$\text{df}(R{\approx})^{((\text{o})(\text{o}))}$$
$$s/twrc \approx^{((\text{o})(\text{o}))} q/lmpa \;=\text{df}$$
$$(xyzu)([x/y][z/u][\langle x/y, z/u\rangle \in s/twrc] \;.{\equiv}.$$
$$[x/y][z/u][\langle x/y, z/u\rangle \in q/lmpa])$$
$$\text{df}(R{\approx}.\text{a})^{((\text{o})(\text{o}))}$$
$$s/twrc \approx^{((\text{o})(\text{o}))}_{xyzu} B(xyzu) \;=\text{df}$$
$$(xyzu)([x/y][z/u][\langle x/y,z/u\rangle \in s/twrc \;.{\equiv}.\; [x/y][z/u]B(xyzu))$$

$\mathrm{df}(R{\approx}.b)^{((o)(o))}$

$$A(xyzu) \approx^{((o)(o))}_{xyzu} B(xyzu) \ = \mathrm{df}$$
$$(xyzu)([x/y][z/u]A(xyzu) \ .{\equiv}.\ [x/y][z/u]B(xyzu))$$

$\mathrm{df}(R)^{((o)(o))}$

$$\langle \imath(p/a)[p/a \approx_z Az], \imath(l/m)[l/m \approx_z Bz]\rangle \in$$
$$\imath(s/twrc)[s/twrc \approx^{((o)(o))}_{xyzu} C(xyzu)] = \mathrm{df}$$
$$(\exists stwrc)(s/twrc \approx^{((o)(o))}_{xyzu} C(xyzu)\ .\&.$$
$$\langle \imath(p/a)[p/a \approx_z Az], \imath(p/a)[p/a \approx_z Bz]\rangle \in s/twrc)$$

$\mathrm{df}(R.\mathrm{a})^{((o)(o))}$

$$\langle \imath(p/a)[p/a \approx_z Az], \imath(p/a)[p/a \approx_z Bz]\rangle \in s/twrc = \mathrm{df}$$
$$(\exists pa)(p/a \approx_z Az\ .\&.\ (\exists lm)(l/m \approx_z Bz\ .\&.\ \langle p/a, l/m\rangle \in s/twrc))$$

$\mathrm{df}(R.\mathrm{b})^{((o)(o))}$

$$\langle p/a, l/m\rangle \in s/twrc = \mathrm{df}(\exists h)(s/twrc;palm!h \ \&\ h)$$

It is convenient to amalgamate at times. This allows us to go from

$$\langle \imath(p/a)[p/a \approx_z Az], \imath(l/m)[l/m \approx_z Bz]\rangle \in$$
$$\imath(s/twrc)[s/twrc \approx^{((o)(o))}_{xyzu} C(xyzu)]$$

directly to

$$(\exists stwrc)(s/twrc \approx^{((o)(o))}_{xyzu} C(xyzu)\ .\&.$$
$$(\exists pa)(p/a \approx_z Az\ .\&.\ (\exists lm)(l/m \approx_z Bz\ .\&.\ \langle p/a, l/m\rangle \in s/twrc))$$

citing only "$\mathrm{df}(R)^{((o)(o))}$". For identity we need:

$\mathrm{df}(R=)^{((o)(o))}$

$$\imath(s/twrc)[s/twrc \approx^{((o)(o))}_{xyzu} A(xyzu)] =$$
$$\imath(q/lmpa)[q/lmpa \approx^{((o)(o))}_{xyzu} B(xyzu)] = \mathrm{df}$$
$$(\exists stwrc)(s/twrc \approx^{((o)(o))}_{xyzu} A(xyzu)\ .\&.$$
$$(\exists qlmpa)(q/lmpa \approx^{((o)(o))}_{xyzu} B(xyzu)\ .\&.\ s/twrc = q/lmpa))$$

These definitions should make the pattern for higher types of homogeneous dyadic relations$_e$ of classes clear. Polyadic homogeneous relations$_e$ follow the pattern as well.

7.3 Nothing prevents heterogeneous relations$_e$. We just need new definitions. For example, we have dyadic heterogeneous relations$_e$ between entities and classes of entities. We put:

$\mathrm{df}(R{\approx})^{(o(o))}$

$$q/clm \approx^{(o(o))} r/atw = \mathrm{df}$$
$$(xyz)([y/z][\langle x, y/z\rangle \in q/clm]\ .{\equiv}.\ [y/z][\langle x, y/z\rangle \in r/atw])$$

$\mathrm{df}(R{\approx}.\mathrm{a})^{(o(o))}$

$$q/clm \approx^{(o(o))}_{xyz} B(xyz) = \mathrm{df}$$
$$(xyz)([y/z][\langle x, y/z\rangle \in q/clm\ .{\equiv}.\ [y/z]B(xyz))$$

$\mathrm{df}(R{\approx}.\mathrm{b})^{(o(o))}$

$$A(xyz) \approx^{(o(o))}_{xyz} B(xyz) = \mathrm{df}$$
$$(xyz)([y/z]A(xyz)\ .{\equiv}.\ [y/z]B(xyz))$$

$\mathrm{df}(R)^{(o(o))}$

$$\langle e, \imath(p/a)[p/a \approx_z Az]\rangle \in \imath(q/clm)[q/clm \approx^{(o(o))}_{xyz} B(xyz)] = \text{df}$$
$$(\exists qclm)(q/clm \approx^{(o(o))}_{xyz} B(xyz) \text{ .\&. } \langle e, \imath(p/a)[p/a \approx_z Az]\rangle \in q/clm)$$

df$(R.\text{a})^{(o)(o))}$

$$\langle e, \imath(p/a)(p/a \approx_z Az]\rangle \in q/clm = \text{df}$$
$$(\exists pa)(p/a \approx_z Az \text{ .\&. } \langle e, p/a\rangle \in q/clm)$$

df$(R.\text{b})^{(o(o))}$

$$\langle e, p/a\rangle \in q/clm \ = \text{df } (\exists h)(q/clm;epa!h \ \& \ h)$$

For identity we have the expected clause:

df$(R =)^{(o(o))}$

$$\imath(q/clm)[q/clm \approx^{(o(o))}_{xyz} A(xyz)] = \imath(r/atw)[r/atw \approx^{(o(o))}_{xyz} B(xyz)] = \text{df}$$
$$(\exists qclm)(q/clm \approx^{(o(o))}_{xyz} A(xyz) \text{ .\&. }$$
$$(\exists ratw)(r/atw \approx^{(o(o))}_{xyz} B(xyz) \text{ .\&. } q/lcm = r/atw)).$$

Of course, we may also have heterogeneous relations$_e$ between classes of entities and entities. The needed definitions should now be clear.

The reader can see how dyadic heterogeneous relations$_e$ of whatever type could be proxied. For instance, there is a relation$_e$ between an individual and an heterogeneous relation$_e$ between individuals. To facilitate this, we just need new definitions. Among them will be:

df$(R{\approx})^{(o(oo))}$

$$s/twrc \approx^{(o(oo))}_{xyzu} q/lmpa = \text{df}$$
$$(xyzu)((\exists bde)(b/de \approx^{(oo)} y/zu \text{ .\&. } \langle x, b/de\rangle \in s/twrc) \text{ .}{\equiv}\text{.}$$
$$(\exists bde)(b/de \approx^{(oo)} y/zu \text{ .\&. } \langle x, b/de\rangle \in q/lmpa))$$

df$(R{\approx}.\text{a})^{(o(oo))}$

$$s/twrc \approx^{(o(oo))}_{xyzu} B(xyzu) = \text{df}$$
$$(xyzu)((\exists bde)(b/de \approx^{(oo)} y/zu \text{ .\&. } \langle x, b/de\rangle \in s/twrc) \text{ .}{\equiv}\text{.}$$
$$(\exists bde)(b/de \approx^{(oo)} y/zu \text{ .\&. } B(xbde))).$$

Here we could not employ df(A). This shall be the usual case for relations$_e$ involving relations$_e$ as arguments. To take another example, the following will be used for a dyadic relation$_e$ between an individual and a dyadic relation$_e$ between a monadic property (of individuals) and an individual:

df$(R{\approx}.\text{a})^{(o((o)o))}$

$$s/twrcn \approx^{(o((o)o))}_{xyzwv} B(xyzwv) = \text{df}$$
$$(xyzwv)((\exists abde)(a/bde \approx^{((o)o)} y/zwv \text{ .\&. } \langle x, a/bde\rangle \in s/twrcn) \text{ .}{\equiv}\text{.}$$
$$(\exists abde)(a/bde \approx^{((o)o)} y/zwv \text{ .\&. } B(xabde)).$$

There are no problems in generating such definitions. Indeed, definitions enabling proxies for polyadic heterogeneous relations$_e$ are easily generated as well. The following is used to help proxy a triadic relation$_e$ between an individual and two classes of individuals:

df$(R{\approx}.\text{a})^{(o(o)(o))}$

$$s/twrcn \approx^{(o(o)(o))}_{xyzwv} A(xyzwv) = \text{df}$$
$$(xyzwv)([y/z][w/v][\langle x, y/z, w/v\rangle \in s/twrcn] \text{ .}{\equiv}\text{. } [y/z][w/w]A(xyzwv)).$$

Observe that the difference between df$(\approx.\text{a})^{(o((o)o))}$ and df$(R\approx.\text{a})^{(o(o)(o))}$ above does not show up in the matrical form itself. In both case we have "*s/twrcn*". Further definitions are readily available as needed to proxy polyadic relations$_e$ of whatever type.

The above definitions show that the problem raised by Grattan-Guinness is solved without making any distinction between propositional and non-propositional substitutions. We can see how the definitions make clear when the use of a matrix "*q/lmc*" proxies a class of classes of classes (of individuals) and when it shall proxy a heterogeneous relation$_e$ between classes of individuals and individuals. Of course, we might have made distinctions in the matrical forms by building them into the definitions. That is, we could have put "*q/(l/m,c)*" for the former and "*q/(l/(m/c))*" for the latter. This would avoid the superscripts. But these are notationally inconvenient; and as already said, they are apt to mislead one into thinking the difference lies in the nature of the entities substituted.

Our definitions for relations$_e$ show that the treatment of the extensions of polyadic propositional functions parallels the treatment of classes. The schema $*(\text{Abs})^{((o)(o))}$, for example, is:

$$
\begin{aligned}
&*(\text{Abs})^{((o)(o))}\\
&\vdash\ \langle \imath(p/a)[p/a \approx_z Az],\ \imath(l/\dot{m})[l/m \approx_z Bz]\rangle\\
&\qquad\qquad \in \imath(s/twrc)[s/twr\dot{c} \approx^{((o)(o))}_{xyzu} C(xyzu)]\\
&\quad .\equiv.\ (\exists pa)[p/a \approx_z Az\ .\&.\ (\exists lm)(l/m \approx_z Bz\ .\&.\ C(palm)).
\end{aligned}
$$

The pattern is exactly akin to that of classes. The expected analogs for the schema of abstraction and the other principle schemata for the theory of classes are readily forthcoming.

Cantor's Paradox of the Greatest Cardinal

7.4 To illustrate the substitutional proxy for relations$_e$ we shall prove Cantor's famous power-class theorem. To do so we need to be able proxy the notion of the cardinal similarity of classes. This is possible in the substitutional theory. Russell wrote (*SCR*, p. 178):

> We can . . . define what we mean by saying that two cardinals of different specified types are equal, or that one is greater than the other. For a given class of terms of a certain type, we can define a one-one relation of the terms of the said class to the terms of a class of the same or any other specified type, and thus we can define what we mean by saying that classes of different type are similar.

Let us begin with some handy definitions. Recall the usual definitions of the domain, range, and field of a relation.

$$D(\hat{u}\hat{v}Auv) = \text{df } \hat{z}\ (\exists w)(\langle zw\rangle \in \hat{u}\hat{v}Auv)$$
$$R(\hat{u}\hat{v}Auv) = \text{df } \hat{z}\ (\exists w)(\langle wz\rangle \in \hat{u}\hat{v}Auv).$$
$$\text{Fld}(\hat{u}\hat{v}Auv) = \text{df Dom}(\hat{u}\hat{v}Auv) \cup \text{Rng}(\hat{u}\hat{v}Auv).$$

To capture these classes, we shall first put:

$$\text{df(Dom)}^{(\text{oo})}$$
$$\text{Dom}^{(\text{oo})}(qce)(z) = \text{df}$$
$$(\exists w)(\langle zw\rangle \in \imath(r/ab)[r/ab \approx^{(\text{oo})} q/ce])$$
$$\text{df(Rng)}^{(\text{oo})}$$
$$\text{Rng}^{(\text{oo})}(qce)(z) = \text{df}$$
$$(\exists w)(\langle wz\rangle \in \imath(r/ab)[r/ab \approx^{(\text{oo})} q/ce])$$
$$\text{df(Fld)}^{(\text{oo})}$$
$$\text{Fld}^{(\text{oo})}(qce)(z) = \text{df Dom}^{(\text{oo})}(qce)(z)\ .\vee.\ \text{Rng}^{(\text{oo})}(qce)(z).$$

For higher types we have:

$$\text{df(Dom)}^{((\text{o})(\text{o}))}$$
$$\text{Dom}^{((\text{o})(\text{o}))}(qlmpa)(fx) = \text{df}$$
$$(\exists ku)(\langle \imath(d/e)[d/e \approx f/x], \imath(d/e)[d/e \approx k/u]\rangle \in \imath(s/twrc)[s/twrc \approx^{((\text{o})(\text{o}))} q/lmpa])$$
$$\text{df(Rng)}^{((\text{o})(\text{o}))}$$
$$\text{Rng}^{((\text{o})(\text{o}))}(qlmpa)(fx) = \text{df}$$
$$(\exists ku)(\langle \imath(d/e)[d/e \approx k/u], \imath(d/e)[d/e \approx f/x]\rangle \in \imath(s/twrc)[s/twrc \approx^{((\text{o})(\text{o}))} q/lmpa])$$
$$\text{dt(Fld)}^{((\text{o})(\text{o}))}$$
$$\text{Dom}^{((\text{o})(\text{o}))}(qlmpa)(fx)\ .\vee.\ \text{Rng}^{((\text{o})(\text{o}))}(qlmpa)(fx).$$

But we also need definitions for unequal-leveled relations$_e$. For example,

$$\text{df(Dom)}^{(\text{o}(\text{o}))}$$
$$\text{Dom}^{(\text{o}(\text{o}))}(qclm)(z) = \text{df}$$
$$(\exists ku)(\langle z, \imath(d/e)[d/e \approx k/u]\rangle \in \imath(r/atw)[r/atw \approx^{(\text{o}(\text{o}))} q/clm])$$
$$\text{df(Rng)}^{(\text{o}(\text{o}))}$$
$$\text{Rng}^{(\text{o}(\text{o}))}(qclm)(fx) = \text{df}$$
$$(\exists z)(\langle z, \imath(d/e)[d/e \approx f/x]\rangle \in \imath(r/atw)[r/atw \approx^{(\text{o}(\text{o}))} q/clm]).$$

And so forth. Inhomogeneous relations have no fields, for the field is a class—the union of the domain and the range—and in this case such a "class" would violate types.

To capture the characteristic feature of a function we shall need to proxy the familiar:

$$\text{funct}(f) = \text{df } (xyz)(xfy \ \&\ xfz\ .\supset.\ y = z).$$
$$\text{1-1funct}(f) = \text{df funct}(f) \ \&\ (xyz)(xfy \ \&\ zfy\ .\supset.\ x = z).$$

Substitutionally, this is:

$$\text{df(funct)}^{(\text{oo})}$$
$$\text{funct}^{(\text{oo})}(\imath(q/ce)[q/ce \approx^{(\text{oo})}_{uv} A(uv)]) = \text{df}$$

$$(xyz)(\langle xy\rangle \in \iota(q/ce)[q/ce \approx^{(oo)}_{uv} A(uv)] \text{ .\&. } \langle xz\rangle \in \iota(q/ce)[q/ce \approx^{(oo)}_{uv} A(uv)] \text{ :}\supset\text{: } y = z)$$

df(1-1funct)$^{(oo)}$

$$\text{1-1funct}^{(oo)}(\iota(q/ce)[q/ce \approx^{(oo)}_{uv} A(uv)]) = \text{df funct}^{(oo)}(\iota(q/ce)[q/ce \approx^{(oo)}_{uv} A(uv)]) \ \& \ (xyz)(\langle xy\rangle \in \iota(q/ce)[q/ce \approx^{(oo)}_{uv} A(uv)] \text{ .\&. } \langle zy\rangle \in \iota(q/ce)[q/ce \approx^{(oo)}_{uv} A(uv)] \text{ :}\supset\text{: } x = z).$$

The same pattern repeats for the type of dyadic relations$_e$ of classes of individuals.

df(funct)$^{((o)(o))}$

$$\text{funct}^{((o)(o))}(\iota(q/lmpa)[q/lmpa \approx^{((o)(o))}_{xyuv} A(xyuv)]) = \text{df } (kn)(fw)(hc)(\langle \iota(d/e)[d/e \approx k/n], \iota(d/e)[d/e \approx f/w]\rangle \in \iota(q/lmpa)[q/lmpa \approx^{((o)(o))}_{xyuv} A(xyuv)] \text{ .\&. } \langle \iota(d/e)[d/e \approx k/n], \iota(d/e)[d/e \approx h/c]\rangle \in \iota(q/lmpa)[q/lmpa \approx^{((o)(o))}_{xyuv} A(xyuv)] \text{ :}\supset\text{: } \iota(d/e)[d/e \approx f/w] = \iota(d/e)[d/e \approx h/c])$$

df(1-1funct)$^{((o)(o))}$

$$\text{1-1funct}^{((o)(o))}(\iota(q/lmpa)[q/lmpa \approx^{((o)(o))}_{xyuv} A(xyuv)]) = \text{df funct}^{((o)(o))}(\iota(q/lmpa)[q/lmpa \approx^{((o)(o))}_{xyuv} A(xyuv)]) \ \& \ (kn)(fw)(hc)(\langle \iota(d/e)[d/e \approx k/n], \iota(d/e)[d/e \approx f/w]\rangle \in \iota(q/lmpa)[q/lmpa \approx^{((o)(o))}_{xyuv} A(xyuv)] \text{ .\&. } \langle \iota(d/e)[d/e \approx h/c], \iota(d/e)[d/e \approx f/w]\rangle \in \iota(q/lmpa)[q/lmpa \approx^{((o)(o))}_{xyuv} A(xyuv)] \text{ :}\supset\text{: } \iota(d/e)[d/e \approx k/n] = \iota(d/e)[d/e \approx h/c]).$$

We also have separate definitions for unequal-leveled relations$_e$ of various types and adicity. For instance there will be:

df(funct)$^{(o(o))}$

$$\text{funct}^{(o(o))}(\iota(q/clm)[q/clm \approx^{(o(o))}_{zrc} A(zrc)]) = \text{df } (y)(kn)(fw)(\langle y, \iota(d/e)[d/e \approx k/n]\rangle \in \iota(q/clm)[q/clm \approx^{(o(o))}_{zrc} A(zrc)] \text{ .\&. } \langle y, \iota(d/e)[d/e \approx f/w]\rangle \in \iota(q/clm)[q/clm \approx^{(o(o))}_{zrc} A(zrc)] \text{ :}\supset\text{: } \iota(d/e)[d/e \approx k/n] = \iota(d/e)[d/e \approx f/w])$$

df(1-1funct)$^{(o(o))}$

$$\text{1-1funct}^{(o(o))}(\iota(q/clm)[q/clm \approx^{(o(o))}_{zrc} A(zrc)]) = \text{df funct}^{(o(o))}(\iota(q/clm)[q/clm \approx^{(o(o))}_{zrc} A(zrc)]) \ \& \ (y)(e)(kn)(\langle y, \iota(d/e)[d/e \approx k/n]\rangle \in \iota(q/clm)[q/clm \approx^{(o(o))}_{zrc} A(zrc)] \text{ .\&. } \langle e, \iota(d/e)[d/e \approx k/n]\rangle \in \iota(q/clm)[q/clm \approx^{(o(o))}_{zrc} A(zrc)] \text{ :}\supset\text{: } y = e).$$

The pattern should be clear.

7.5 Now the definition of "α Sim β" (i.e., "α is similar to β") can be rendered as follows:

$$\hat{z}Az \text{ Sim } \hat{z}Bz = \text{df } (\exists f)(\text{1-1funct(f) .\&. Dom}(f) = \hat{z}Az \text{ .\&. Rng}(f) = \hat{z}Bz).$$

In order to proxy this substitutionally, we define:

$$\begin{array}{l} \mathrm{df(Sim)}^{((o)(o))} \\ \imath(p/a)[p/a \approx_z Az]\ \mathrm{Sim}^{((o)(o))}\ \imath(p/a)[p/a \approx_z Bz)] = \mathrm{df} \\ \quad (\exists qce)(\ \text{1-1funct}^{(oo)}(\imath(q^\circ/c^\circ e^\circ)[q^\circ/c^\circ e^\circ \approx^{(oo)} q/ce]) \ .\&. \\ \qquad \imath(p/a)[p/a \approx_z\ \mathrm{Dom}^{(oo)}(qce)(z)] = \imath(p/a)[p/a \approx_z Az]\ .\&. \\ \qquad\quad \imath(p/a)[p/a \approx_z\ \mathrm{Rng}^{(oo)}(qce)(z)] = \imath(p/a)[p/a \approx_z Bz]). \end{array}$$

As expected $\mathrm{Sim}^{((o)(o))}$ is reflexive, symmetric, and transitive. Of course, we can define "similarity" for classes of higher type. Since we have unequal-leveled relations$_e$, we can also define a notion of similarity for classes of different types:

$$\begin{array}{l} \mathrm{df(Sim)}^{((o)((o)))} \\ \imath(p/a)[p/a \approx_z A(z)]\ \mathrm{Sim}^{((o)((o)))}\imath(s/tw)[s/tw \approx_{rc} B(rc)] = \mathrm{df} \\ \quad (\exists qelm)(\ \text{1-1funct}^{(o(o))}(\imath(j/kuv)[j/kuv \approx^{(o(o))} q/elm]) \ .\&. \\ \qquad \imath(p/a)[p/a \approx_z\ \mathrm{Dom}^{(o(o))}(qelm)(z)] = \imath(p/a)[p/a \approx_z Az]\ .\&. \\ \qquad\quad \imath(s/tw)[s/tw \approx_{rc} \mathrm{Rng}^{(o(o))}(qelm)(rc)] = \imath(s/tw)[s/tw \approx_{rc} B(rc)]). \end{array}$$

The process continues for higher types. Again, $\mathrm{Sim}^{((o)((o)))}$ is reflexive, symmetric, and transitive.

The availability of unequal-leveled relations enables us to compare the cardinalities of classes of different types.[1] To illustrate, we now prove Cantor's theorem that no class is cardinally similar to its power-class (i.e., the class of all its subclasses):

$$(\forall\alpha)(\sim(\exists f)(\text{1-1funct}(f)\ .\&.\ \mathrm{Dom}(f) = \alpha\ .\&.\ \mathrm{Rng}(f) = P(\alpha))$$

For convenience we begin with the following definition:

$$\begin{array}{l} \mathrm{df}(P) \\ P(p^\circ a^\circ)(rc) = \mathrm{df}\ \imath(p/a)[p/a \approx r/c] \subseteq \imath(p/a)[p/a \approx p^\circ/a^\circ]. \end{array}$$

This will help in recognizing the proxy for the power-class when written in substitutional notation. Cantor's power-class theorem is now expressed as follows:

$$\begin{array}{l} \text{(Cantor)} \\ \quad \vdash\ (p^\circ a^\circ) \sim (\exists qelm)(\mathrm{funct}^{(o(o))}(\imath(j/kuv)[j/kuv \approx^{(o(o))} q/elm]) \\ \qquad .\&.\ \imath(p/a)[p/a \approx_z \mathrm{Dom}^{(o(o))}(qelm)(z)] = \imath(p/a)[p/a \approx p^\circ/a^\circ]\ .\&. \\ \imath(s/tw)[s/tw \approx_{rc} \mathrm{Rng}^{(o(o))}(qelm)(rc)] = \imath(s/tw)[s/tw \approx_{rc} P(p^\circ a^\circ)(rc)]). \end{array}$$

Proof:

$$\begin{array}{l} \text{df(Cantor)} \\ Cz = \mathrm{df}\ (\exists fx)\,(z \in \imath(p/a)[p/a \approx p^\circ/a^\circ]\ \ {:}\&{:} \\ \qquad\qquad \langle z, \imath(p/a)[p/a \approx f/x]\rangle \in \imath(j/kuv)[j/kuv \approx^{(o(o))} q/elm] \\ \qquad\qquad\quad .\&. \sim (z \in \imath(p/a)[p/a \approx f/x])) \end{array}$$

1. As we noted above, this is impossible if the Weiner-Kuratowski method is added to type theory. The class of lower type would have to be replaced by the class of all its singletons for comparability with the class of next higher type.

1. $\rightarrow \mathrm{funct}^{(o(o))}(\imath(j/kuv)[j/kuv \approx^{(o(o))} q/elm])$
$.\&.\ \imath(p/a)[p/a \approx_z \mathrm{Dom}^{(o(o))}(qelm)(z)] = \imath(p/a)[p/a \approx p^\circ/a^\circ]$
$.\&.\ \imath(s/tw)[s/tw \approx_{rc} \mathrm{Rng}^{(o(o))}(qelm)(rc)] =$
$\imath(s/tw)[s/tw \approx_{rc} P(p^\circ a^\circ)(rc)])$.
2. $(\exists gc)(g/c \approx_z Cz)) \quad *(\mathrm{CP}_\in)^1$.
3. $\imath(p/a)[p/a \approx_z Cz] \subseteq \imath(p/a)[p/a \approx p^\circ/a^\circ] \quad \mathrm{df}(\subseteq)^1, \mathrm{df(C)}, *(\mathrm{Abs})^1$.
4. $(\exists gc)(g/c \approx_z Cz :\&: P(p^\circ a^\circ)(gc)) \quad$ 3, df(P), $(\mathrm{T}_{6.19})^1$.
5. $\imath(g/c)[g/c \approx_z Cz] \in \imath(s/tw)[s/tw \approx_{rc} P(p^\circ a^\circ)(rc)]$
4, $*(\mathrm{Abs})^2$.
6. $\imath(g/c)[g/c \approx_z Cz] \in \imath(s/tw)[s/tw \approx_{rc} \mathrm{Rng}(qelm)(rc)]$
1, 5, $*(\mathrm{Ext})^2$, $*(\mathrm{T}_{6.20})^1$.
7. $(\exists gc)(g/c \approx_z Cz :\&: \mathrm{Rng}^{(o(o))}(qelm)(gc)) \quad$ 6, $*(\mathrm{Abs})^2$.
8. $(\exists gc)(g/c \approx_z Cz :\&:$
$(\exists v)(\langle v, \imath(p/a)[p/a \approx g/c]\rangle \in \imath(j/kuv)[j/kuv \approx^{(o(o))} q/elm]))$
7, $\mathrm{df(Rng)}^{(o(o))}$.
9. $\langle v, \imath(p/a)[p/a \approx_z Cz]\rangle \in \imath(j/kuv)[j/kuv \approx^{(o(o))} q/elm]$
8, $*(\mathrm{T}_{6.19})^1$, EI.
10. $\rightarrow v \in \imath(p/a)[p/a \approx_z Cz]$
11. $Cv \quad$ 10, $*(\mathrm{Abs})^1$.
12. $(\exists fx)(v \in \imath(p/a)[p/a \approx p^\circ/a^\circ] :\&:$
$\langle v, \imath(p/a)[p/a \approx f/x]\rangle \in \imath(j/kuv)[j/kuv \approx^{(o(o))} q/elm]$
$.\&. \sim (v \in \imath(p/a)[p/a \approx f/x]) \quad$ 11, df(Cantor)
13. $v \in \imath(p/a)[p/a \approx p^\circ/a^\circ] :\&:$
$\langle v, \imath(p/a)[p/a \approx f/x]\rangle \in \imath(j/kuv)[j/kuv \approx^{(o(o))} q/elm]$
$.\&. \sim (v \in \imath(p/a)[p/a \approx f/x]) \quad$ 12, EI.
14. $\imath(p/a)[p/a \approx f/x] = \imath(p/a)[p/a .\approx_z. Cz]$
9, 13, $\mathrm{df(funct)}^{(o(o))}$.
15. $v \in (p/a)[p/a \approx f/x] \quad$ 10, 14, $*(\mathrm{Ext})^1$.
16. $\underline{\quad} F \quad$ 13, 15, conj, (12ei).
17. $10 \supset F \quad$ cp.
18. $\sim (v \in \imath(p/a)[p/a \approx_z Cz]) \quad$ 17, reductio.
19. $\sim Cv \quad$ 18, $*(\mathrm{Abs})^1$.
20. $(fx)(v \in \imath(p/a)[p/a \in p^\circ/a^\circ] :\supset:$
$\langle v, \imath(p/a)[p/a \approx f/x]\rangle \in \imath(j/kuv)[j/kuv \approx^{(o(o))} q/elm] .\supset.$
$v \in \imath(p/a)[p/a \approx f/x])$
19, df(Cantor).
21. $v \in \imath(p/a)(p/a \in p^\circ/a^\circ) :\supset:$
$\langle v, \imath(p/a)[p/a \approx_z Cz]\rangle \in \imath(j/kuv)[j/kuv \approx^{(o(o))} q/elm] .\supset.$
$v \in \imath(p/a)[p/a \approx_z Cz] \quad$ 20, $*(\mathrm{T}_{6.20})^1$.
22. $(\exists gc)(\langle v, \imath(p/a)[p/a \approx g/c]\rangle \in \imath(j/kuv)[j/kuv \approx^{(o)(o))} q/elm])$
9, $*(\mathrm{T}_{6.20}\mathrm{a})^1$.
23. $\mathrm{Dom}^{(o(o))}(qelm)(v) \quad$ 22, $\mathrm{df(Dom)}^{(o(o))})$.
24. $v \in \imath(p/a)[p/a \approx_z \mathrm{Dom}^{(o(o))}(qelm)(z)] \quad$ 23, $*(\mathrm{Abs})^2$.
25. $v \in \imath(p/a)[p/a \approx p^\circ/a^\circ] \quad$ 1, 24, $*(\mathrm{Ext})^1$.
26. $\langle v, \imath(p/a)[p/a \approx_z Cz]\rangle \in \imath(j/kuv)[j/kuv \approx^{(o(o))} q/elm] .\supset.$
$v \in \imath(p/a)[p/a \approx_z Cz] \quad$ 21, 25, mp.

27. $v \in \imath(p/a)[p/a \approx_z Cz]$ 9, 26, mp.
28. $\underline{\quad} F$ 18, 27, conj, (*qei*).
29. $1 \supset F$ cp.
30. $\sim (\exists qelm)(\text{funct}^{(o(o))}(\imath(j/kuv)[j/kuv \approx^{(o(o))} q/elm])$
 $.\&.\ \imath(p/a)[p/a \approx_z \text{Dom}^{(o(o))}(qelm)(z)] = \imath(p/a)[p/a \approx p^\circ/a^\circ]$
 $.\&.\ \imath(s/tw)[s/tw \approx_{rc} \text{Rng}^{(o(o))}(qelm)(rc)] =$
 $\imath(s/tw)[s/tw \approx_{rc} P(p^\circ a^\circ)(rc)]))$ 29, reductio ug, log.
31. $(p^\circ a^\circ)30$ ug.

This completes the proof.

Now the usual definition for what it is for one class to be larger in cardinality than another is as follows:

$$\alpha \prec \beta = \text{df } (\exists\delta)(\delta \subset \beta\ .\&.\ \alpha \text{ Sim } \delta\ .\&.\ \sim(\alpha \text{ Sim } \beta)).$$

Substitutionally (and for lowest homogeneous type) this becomes:

$$\begin{aligned}&\text{df}(\prec)^{((o)(o))}\\ &\imath(p/a)[p/a \approx_z Az] \prec \imath(p/a)[p/a \approx_z Bz] = \text{df}\\ &(\exists lm)(\imath(p/a)[p/a \approx l/m] \subset \imath(p/a)[p/a \approx_z Bz]\ .\&.\\ &\quad \imath(p/a)[p/a \approx_z Az] \text{ Sim}^{((o)(o))}\imath(p/a)[p/a \approx l/m]\ .\&.\\ &\quad\quad \sim(\imath(p/a)[p/a \approx_z Az] \text{ Sim}^{((o)(o))}\imath(p/a)[p/a \approx_z Bz])).\end{aligned}$$

But we can also have definitions for unequal-leveled cases. For instance, we have:

$$\begin{aligned}&\text{df}(\prec)^{((o)((o)))}\\ &\imath(p/a)[p/a \approx_z Az] \prec \imath(s/tw)[s/tw \approx_{rc} B(rc)] = \text{df}\\ &\quad(\exists hde)(\imath(s/tw)[s/tw \approx h/de] \subset \imath(s/tw)[s/tw \approx_{rc} B(rc)]\ .\&.\\ &\quad\quad \imath(p/a)[p/a \approx_z Az] \text{ Sim}^{((o)((o)))}\imath(s/tw)[s/tw \approx h/de]\ .\&.\\ &\quad\quad\quad \sim(\imath(p/a)[p/a \approx_z Az] \text{ Sim}^{((o)(o)))}\imath(s/tw)[s/tw \approx_{rc} B(rc)])).\end{aligned}$$

Of course, any class is similar to the class of its singletons. That is,

$$\vdash (\forall\alpha)(\alpha \text{ Sim}^{((o)(o))}\ \hat{y}(\exists v)(v \in \alpha \ \&\ y = \hat{z}(z = v)))$$

Abbreviating with

$$\begin{aligned}I(lm)(rc) = \text{df } &(\exists v)(v \in \imath(p/a)[p/a \approx l/m]\ .\&.\\ &\imath(p/a)[p/a \approx r/c] = \imath(p/a)[p/a \approx_z z = v])\end{aligned}$$

we can express this as:

$$\vdash (lm)(\imath(p/a)[p/a \approx l/m] \text{ Sim}^{(o(o))}\imath(s/tw)[s/tw \approx_{rc} I(lm)(rc)]).$$

This is easily proved. Simply consider the fact that by $*(\text{CP}_\in)$ we know that:

$$\begin{aligned}(\exists qclm)(q/clm :\approx^{(o(o))}_{xrc}:\ &x \in \imath(p/a)[p/a \approx l^\circ/m^\circ]\ .\&.\\ &\imath(p/a)[p/a \approx r/c] = \imath(p/a)[p/a \approx_z z = x]).\end{aligned}$$

It is easy to show that this relation$_e$ ensures the similarity of the classes in question.

Now by universal instantiation of our theorem (Cantor) we can easily prove the following:

$$\vdash \sim (\imath(p/a)[p/a \approx_z Vz] \text{ Sim}^{((\text{o})((\text{o})))} \imath(s/tw)[s/tw \,.\!\approx_{rc}.\, \imath(d/e)[d/e \approx r/c] \subseteq \imath(p/a)[p/a \approx_z Vz]]).$$

This says that the universal class of entities is not similar to the power-class of the universal class of entities. But we just showed that every class is similar with the class of its singletons. Hence, in particular, we also have:

$$\imath(p/a)[p/a \approx_z Vz] \text{ Sim}^{((\text{o})((\text{o})))} \imath(s/tw)[s/tw \approx_{rc} (\exists x)(\imath(p/a)[p/a \approx r/c] = \imath(p/a)[p/a \approx_z z = x]).$$

The universal class of entities is similar to the class of all singleton classes (of entities). Accordingly, the power class of the universal class of entities is larger in cardinality than the universal class of entities.

7.6 What then of Cantor's paradox of the greatest cardinal? It might seem, *prime facie*, that the cardinal number of the class of the universal class of entities has the greatest cardinal number. It must be so, if classes are themselves genuine entities and if (as Russell was resolved to maintain) the laws of logic are universal and range over all entities whatsoever. Yet Cantor's power-class theorem shows that the power-class of the universal class of entities is larger in cardinality! The solution of the paradox is that classes are not entities. Russell put matters as follows (*SCR*, p. 179):

> Consider now the theorem which gives rise to the difficulty concerning the greatest cardinal. This theorem asserts that whatever cardinal α may be 2^α is greater than α. It follows that there can be no greatest cardinal, although the cardinal number of *all* entities might seem, *prime facie*, to be as great as could be reasonably expected. We can see now, however, that in higher types we *shall* get greater cardinals than the cardinal number of all entities, and that $2^\alpha \succ \alpha$ is true without exception. . . . The solution of the problem concerning the greatest cardinal lies simply in the fact that a class is not an entity, so that classes are not (as was supposed) a sub-class of entities.

In the substitutional theory Cantor's paradox disappears. Classes are not entities. If they were, they would be members of the universal class of entities. Mind you, the paradox is not *solved* by splitting entities (classes) into types! It is solved by the fact that the substitutional theory is a "no-classes" theory. Classes are not assumed as entities. Nonetheless, the substitutional theory recovers the result that the power-class of any class is greater in cardinality. This result is quite important. We can prove that there are $\aleph_0$ many entities.[2] But Russell tells us that he can find no good argument that there are more than $\aleph_0$ entities (*SCR*, p. 182). Since class symbols are a *façon de parler*, so also is Cantor's hierarchy of greater and greater infinite cardinals. All the same, the

2. See Appendix A.

substitutional ("no-classes") theory allows a proxy for Cantor's power-class theorem.[3]

The Burali-Forti Paradox

7.7 The Burali-Forti paradox is a paradox of ordinal numbers. To understand it we must first discuss the difference between ordinal numbers and cardinal numbers. Let us begin without consideration of types and then proceed to Russell's solution of the paradox.

A class A is a member of a finite cardinal, say 3, in virtue of its being (cardinally) similar to a three-membered class. The arrangement or order of these three members is irrelevant. An infinite class B is a member of the infinite cardinal $\aleph_0$ (aleph nought) when it is similar to N, the class of natural numbers. That is, its members can be put in a one-one correspondence with the finite cardinals. Again, the order of the members makes no difference to cardinal similarity.

Order is generated by relations. Consider the following definitions concerning order.

$$
\begin{aligned}
&R \text{ partially orders } \alpha = \text{df} \\
&\qquad (x)(x \in \alpha \ .\supset.\ \sim xRx)\ .\&. \\
&\qquad (xy)(z \in \alpha \ \&\ y \in \alpha\ :\supset:\ xRy\ .\supset.\ \sim yRx)\ .\&. \\
&\qquad (xyz)(x \in \alpha \ \&\ y \in \alpha \ \&\ z \in \alpha\ :\supset:\ xRy \ \&\ yRz\ .\supset.\ xRz) \\
&R \text{ linearly (serially) orders } \alpha = \text{df } R \text{ partially orders } \alpha\ .\&. \\
&\qquad\qquad (x, y)(x \in \alpha \ \&\ y \in \alpha\ :\supset:\ xRy\ .\vee.\ x = y\ .\vee.\ yRx) \\
&R \text{ well-orders } \alpha = \text{df } R \text{ linearly orders } \alpha\ .\&. \\
&\qquad (B)(B \subseteq \alpha\ .\&.\ B \neq \Lambda\ :\supset:\ (\exists x)(x \in B \ \&\ (y)(y \in B\ .\&.\ y \neq x \\
&\qquad :\supset:\ xRy)
\end{aligned}
$$

The ordinal similarity of relations$_e$ is defined as follows:

$$
\begin{aligned}
R \text{ smor } S = \text{df } (\exists f)(\text{1-1funct}(f) \ \&\ \text{Dom}(f) = \text{Fld}(R) \ \& \\
\text{Rng}(f) = \text{Fld}(S) \ \&\ (xy)(xRy \equiv f(x)Sf(y))).
\end{aligned}
$$

In 1902 Russell defined the *relation number* of a given relation$_e$ R to be the class of relations$_e$ ordinally similar to R. Accordingly, the *serial relation number* of a given serial relation$_e$ R is the class of all serial relations$_e$ ordinally similar to R. Russell's serial relation numbers are the analog of what Cantor called "order types." Ordinal numbers are a subclass of these; they are defined as classes of similar well-ordering relations$_e$. This is a quite natural extension of the definition of cardinal numbers. The natural numbers $\mathbb{N}$ are arranged in a progression by the well-ordering relation$_e$ $<$, i.e., *less than*. Thus, Russell defines the ordinal number ω (Omega) as the class of all well-ordering relations$_e$ ordinally similar to $<$.

3. We shall return to this in Chapter 9 when we discuss the paradoxes of propositions.

Now the series generated by R is said to be equally "long" as the series generated by S when R and S are ordinally similar. If the series generated by R and the series generated by S are both finite, then R and S have the same serial number if and only if their fields have the same cardinality. Thus a finite ordinal number is determinate when we know the cardinal number of terms in the field of the well-ordering relation having the ordinal number in question. For example, having defined the notion of an empty (dyadic) relation$_e$,

$$\dot{\Lambda} = \text{df } \hat{x}\hat{y}(\sim(x = x \ .\&.\ y = y))$$

we get the finite ordinal number 0_r,

$$*56.03 \quad 0_r = \text{df } \hat{R}(R = \dot{\Lambda}).$$

For the finite ordinal number 2_r, first note that *Principia* has,

$$*35.04 \quad \alpha \uparrow \beta = \text{df } \hat{z}\hat{u}(z \in \alpha \ .\&.\ u \in \beta)$$
$$*55.01 \quad x \downarrow y = \text{df } \iota x \uparrow \iota y.$$

Then 2_r is to be:

$$*56.02 \quad 2_r = \text{df } \hat{R}(x, y)(x \neq y \ .\&.\ R = x \downarrow y).$$

These are relation$_e$ numbers, but they are also well-ordering relation$_e$ numbers whose fields are finite in cardinality. Hence, they are finite ordinals. (If their fields are composed of entities, rather than relations$_e$ or classes, Russell calls them "entity-ordinals.") The finite entity-ordinal number *zero* is thus the class of all well-ordering relations whose fields have no individuals as terms; the finite entity-ordinal number *two*, is the class of all well-ordering relations whose fields have exactly two entities as terms. In general then, we have: Where n is a finite cardinal, the class of all well-ordering relations$_e$ whose fields have n terms will be "the ordinal number n" (*IMP*, p. 57). One might wonder about the finite ordinal number *one*. There is none. There is a relation number 1_s,

$$*153.01 \quad 1_s = \text{df } \hat{R}((\exists x)(R = x \downarrow x)),$$

but one cannot well-order a field with one term, and $x \downarrow x$ is not a well-ordering relation. Russell gets on without an entity-ordinal number *one* (see *Principia* $*181$).

Compare the construction of ordinals adopted by von Neumann. Remember we are not yet worried about types. For finite ordinals just take the field of a representative relation from each order type, that is, from each of Russell's finite ordinals. The first finite von Neumann ordinal is then Λ. For the second we need a one-membered class and so $\{\Lambda\}$ is picked; for the third a two-membered class is needed and for this $\{\Lambda, \{\Lambda\}\}$ is picked. The series is generated by the operation $z \cup \{z\}$. The ordering relation is just membership itself and the series is well-ordered. The generative process for the series relies on the idea that for any given segment of the series, the ordinal number of that segment can be added at the end of the segment so as to give a new series whose ordinal number exceeds by 1 that of the given segment.

Consider next the limit of the series of finite von Neumann ordinals. That is, consider the class of all such finite ordinals:

$$\hat{z}((y)(\Lambda \in y \;\&\; (z)(z \in y \;.\!\supset.\; z \cup \{z\} \in y) \;.\!\supset.\; z \in y)).$$

This shall now be von Neumann's class ω. (Neglecting types, it would have been a representative from Russell's ω.) This is the first infinite ordinal. But we may continue the process by putting ω itself at the end of the series ω. In this case, of course, there is no last term of the series. This forms $\omega + 1$, that is $\omega \cup \{\omega\}$. And we can go on to $\omega + 2, \ldots$, and so forth. If we form the class of all members of this series we have 2ω. This too is an ordinal number; and we continue on.

The Burali-Forti paradox comes if we consider the class *Ord* of all von Neumann ordinal numbers. Since the members of this class form a well-ordered series, the class is itself an ordinal number and would seem to be the greatest ordinal. This not withstanding, there is a greater ordinal than the "greatest"; we need only take *ord* ∪ {*ord*}.

The theory of types blocks the Burali-Forti paradox. First of all, it prohibits the very generating process essential for the von Neumann ordinals. The von Neumann ordinals violate types. Russell's definition of ordinal number, however, can operate within type constraints. As before, an ordinal number is again a class of all similar well-ordering relations_e. A relation_e of classes (of individuals) can be ordinally similar to a relation_e of individuals, but these two relations_e cannot both belong to the same class of relations_e. Hence the relations_e in a given ordinal number must all be of the same type. Indeed, one must distinguish *entity-ordinals* (when the relations_e in question have only individuals in their fields) from others. There will be different *class-ordinals* (where the relations_e have classes of some type as relata); and one can go on to *ordinal-ordinals* (where the relations_e have as relata relations_e of some type), and so on (*SCR*, p. 180f.).

Now consider again the finite entity-ordinals. These can be arranged into a series by some well-ordering relation_e (say R). Observe, however, that R does not have entities as its relata. The class of all well-ordering relations_e similar to R is not itself an entity-ordinal. Hence the generating process, illustrated in the von Neumann construction, is no longer possible. Given a segment of a series of entity-ordinals, the ordinal number of the segment cannot be added at the end; it cannot be a member of a class of entity-ordinals. This is the essence of the solution of the Burali-Forti paradox. A class of all ordinal numbers violates types.

7.8 Be this as it may, Russell wants to preserve as much of Cantor's work as possible. He must find a new generative process for ordinals, and entity-ordinals in particular. For instance, he must find an infinite entity-ordinal replacing ω which can be placed at the end of the series of finite entity ordinals and classed together with them to form a new infinite entity-ordinal $\omega + 1$. He does this as follows.[4]

When an infinite class is serially ordered by a relation$_e$ the length of the series is not fixed by the cardinal number of the class. Consider, for example, the class $\mathbb{N}$ of natural numbers. It has cardinality $\aleph_0$. Now $\mathbb{N}$ is serially ordered by the relation $<$. The class of all relations$_e$ order isomorphic with $<$ is the ordinal number ω. But there are other serial orderings not similar to $<$. Consider

$$1, 3, 4, 5, \ldots, 2$$

ordered by a serial relation R. Here the serial relation has ordered the natural numbers with 2 "at the end" (as it were). The relation$_e$ which adds one entity at the end of a relation$_e$ P is defined in *Principia* as follows:

$$*161.10 \quad P \mathbin{+\!\!\!\rightarrow} x = \mathrm{df}\ P \cup (C\text{‘}P \uparrow \iota x).$$

The notation, $C\text{‘}P$, is for the field ("campus") of P. That is:

$$*33.03 \quad C\text{‘}P = \mathrm{df}\ \hat{z}((\exists y) z\, P\, y\ .\!\vee\!.\ (\exists y) y\, P\, z).$$

The above series is thus generated by the relation$_e$,

$$< -(\iota 2 \uparrow C\text{‘} <) \mathbin{+\!\!\!\rightarrow} 2.$$

Otherwise put, the relation$_e$ is (*IMP*, p. 91):

$$\hat{x}\hat{y}\ (x \in \mathbb{N}\ \&\ y \in \mathbb{N}\ :\!\&\!:\ (y = 2\ \&\ x \neq 2)\ \vee\ (x \neq 2\ \&\ y \neq 2\ \&\ x < y)).$$

Now this new relation is not order isomorphic with $<$ even though the fields of the two relations have the same cardinality. An infinite ordinal number m is said to be "greater" than an infinite ordinal number n when any well-ordering relations$_e$ T, S such that T has m and S has n are such that S is ordinally similar to a part of T, and T is not ordinally similar to any part of S. (A relation R is "part" of a relation T, when $R = T{-}P$, for some non-empty relation P that is not equal to T). Russell defines $\omega + 1$ as the class of all well-ordering relations$_e$ similar to

$$< -(\iota 2 \uparrow C\text{‘} <) \mathbin{+\!\!\!\rightarrow} 2.$$

The name "$\omega + 1$" indicates that we are taking the field of a relation$_e$ in ω and rearranging by means of a new relation so that we get a series which puts one of the members of the field at the end. The name "$1 + \omega$", on the other hand, indicates that we are taking the field of a relation$_e$ in ω and unioning it with a one-membered class. Since $\mathbb{N}_0 \cup \{a\}$ is cardinally similar to $\mathbb{N}_0$, the ordering is similar to the original. That is, the ordinal number $1 + \omega$ is just ω.

By a similar process of "thinning out" we can get a well-ordering relation$_e$, which generates the infinite series:

$$1, 3, 5, 6, 7, \ldots, 2, 4.$$

4. There is an excellent introductory discussion of this in Russell's *Introduction to Mathematical Philosophy* [IMP]. We shall follow this discussion here.

This relation$_e$ cannot be similar to $<$ or to the relation$_e$, $< -(\iota 2 \uparrow C` <) \nrightarrow 2$. The class of all well-ordering relations$_e$ similar to it can be shown to be the next in the new series of entity ordinals—namely, $\omega + 2$. The process continues. If we arrange all the odd natural numbers then all the even natural numbers

$$1, 3, 5, 7, \ldots, 2, 4, 6, 8 \ldots$$

we have ordered the numbers so that we have a couple of series. The general definition an addition of two relations$_e$ which puts all the members of the field of the first before those of the second and yet preserves the ordering generated by each is given in *Principia* as follows:

$$*160.01 \quad P \mathbin{\dot{+}} Q = \text{df } P \dot{\cup} Q, \dot{\cup}(C`P \uparrow C`Q).$$

If the fields of the two relations$_e$ have no common elements, then this definition will suffice for the sum of the two relations$_e$ (and for the sum of two such well-ordering relations$_e$ as well). To define the sum of two ordinal numbers (or more generally the sum of two relation$_e$ numbers), Russell puts:

$$m \dot{+} n = \text{df}$$
$$\hat{R}((\exists P, Q)(m = N_o r`P \;.\&.\; n = N_o r`Q \;.\&.\; R \text{ smor } P + Q).$$

That is, it is the class of all relations$_e$ that are ordinally similar to the relation$_e$ $P + Q$, where m is the class of relations$_e$ ordinally similar to P (and the same type as P) and n is the class of relations$_e$ ordinally similar to Q (and the same type as Q). The relation $P \mathbin{\dot{+}} Q$ won't suffice because the fields of the relations might overlap. Instead, Russell puts:

$$*150.01 \quad S ; Q = \text{df } S|Q|\breve{S}$$
$$*180.01 \quad P + Q = \text{df}$$
$$\{\downarrow (\Lambda \cap C`Q);\iota;P\} \mathbin{\dot{+}} \{(\Lambda \cap C`Q) \downarrow ;\iota;Q\}.$$

This definition is used so that the sum of two relations$_e$ is defined even when their fields might overlap.

As we see, the relation$_e$ that puts all the odd cardinals before the even ones will be a member of the ordinal number 2ω (or alternatively $2 \cdot \omega$). This is not the same as $\omega \cdot 2$, for this is a series of couples and not a couple of series. Its length would be ω. By continuing the process of thinning, we can proceed to $3\omega, 4\omega, \ldots, \omega^2, \omega^3, \ldots, \omega^\omega$, and so on, without end (*IMP*, p. 92).

Now the cardinal number of the class of all such entity-ordinals can be shown to be greater than $\aleph_0$. Russell calls this cardinal $\aleph_1$. The ordinal number of the series of all ordinals that can be made out of $\aleph_0$ terms taken in order of magnitude, is called ω_1. That is, ω_1 is the smallest ordinal whose field has more than $\aleph_0$ terms; it is the class of all well-ordering relations$_e$ P such that the relations$_e$ less than P are those well-ordering relations which have either $\aleph_0$ terms in their fields or have finite fields (*PM*, vol. 3, p. 170):

$$*265.01 \quad \omega_1 = \text{df } \hat{P}(\overrightarrow{\text{less}}\,`P = (\aleph_0)_r \cup \Omega\text{fin}).$$

By a similar process we can advance to ω_2:

$$*265.03 \quad \omega_2 = \text{df } \hat{P}(\overrightarrow{\text{less}}\text{ `}P = (\aleph_1)_r \cup (\aleph_o)_r \cup \Omega\text{fin}),$$

and the process continues indefinitely to the generation of new cardinals and new ordinals. We can get the series $\omega_1, \omega_2, \ldots, \omega_n$, where n is any finite cardinal. In his April/May 1906 article "On The Substitutional Theory of Classes and Relations," Russell says he has not succeeded in showing we can extend the series to the case where n is infinite. (*SCR*, p. 187). In the September 1906 "Les paradoxes de la logique" he again addressed the question (*InS*, p. 203):

> In this way we find that, although an existence-theorem can be proved for $\omega, \omega_1, \omega_2, \ldots, \omega_n, \ldots$, it is proved at each stage for a higher type. Consequently, there is no proof (at least so far as appears by this method) of the existence of ω_ω, and there is a proof that there is no such thing, in any sense, as the whole series of ordinals of all types.

It must not be thought, however, that failure to get beyond ω_n is due to some feature of the substitutional theory. In *Principia* (where substitution is abandoned) we get the same result (*PM*, vol. 3, p. 170). The problem is independent of substitution and is generated by the theory of types itself. Each in the series $\omega_1, \omega_2, \ldots, \omega_n, \ldots$, is a different type and the theory does not allow transfinite types. Thus Cantor's ω_ω seems to be indefinable, a casualty of the theory of types.[5] Amusingly, when Russell originally communicated to Whitehead the difficulties in reconstructing $\omega_1, \omega_2, \ldots$, and so on, within types he got the following response in a note of 4 December 1905:

> Dear Bertie,
>
> I congratulate you on the progress with you have made—It is excellent—I sincerely hope that ω_1 is indefinable. The aleph-series of ordinals has always bored me extremely, and I shall be glad to assist at its funeral.

Russell was able to construct most of them, but not ω_ω. The loss here must be weighed against the gain of a solution to the Burali-Forti paradox. There can be no class of all ordinals.

5. See also Grattan-Guinness (1977b, p. 106).

PART III

Ramification

CHAPTER EIGHT

Les Paradoxes de la Logique

8.0 As we saw in Part II, the substitutional theory parallels the so-called "simple theory of types" of attributes ("propositional functions in an ontological sense"). As in the simple theory, cardinal numbers repeat on each type level. But the reduplication of arithmetic within the types—a zero, one, and so on, on each type level—is now wholly unobjectionable. It is an entirely natural consequence of the proxy—that is, the non-assumption that every open wff comprehends an attribute. Indeed, substitution avoids the simple theory's problem of justifying a hierarchy of attributes. There is no hierarchy of entities at all! Neither is there any need for an axiom of infinity.[1] With the assumption of proposition as individuals, a theorem that there is an infinity of individuals is readily forthcoming. This pristine substitutional theory, however, fell to certain "paradoxes of propositions."

We noted at the opening of Chapter 4 that Russell's "On the Substitutional Theory of Classes and Relations" was received by the London Mathematical Society on 24 April 1906 and read before it on 10 May. This was a continuation of the substitutional theory developed in the 22 December 1905 manuscript "On Substitution,"and hinted at in "On Some Difficulties in the Theory of Transfinite Numbers and Order Types" (received by the Society on 24 November 1905). In May he read "On Some Difficulties" to the Society and on 5 February had added the following footnote:

1. It should be noted that Russell never formally added an "axiom" of infinity to *Principia*. When infinity was needed for a theorem, it was taken as the antecedent of a conditional whose consequent is the theorem in question (*PM*, p. 183). The statement of infinity was that no natural number (of the type of a class of classes of individuals) is empty. This ensures an infinite class in every higher type.

> From further investigation I now feel hardly any doubt that the no-classes theory affords the complete solution of all the difficulties stated in the first three sections of this paper.

The same enthusiasm is present in "On the Substitutional Theory of Classes and Relations." Its concluding paragraph has (*SCR*, p. 189):

> It [the substitutional theory] affords what at least seems to be a complete solution of all the hoary difficulties about the one and the many; for while allowing that there are many entities, it adheres with drastic pedantry to the old maxim that 'whatever is, is one'" (*SCR*, p. 189).

This maxim had been the fundamental doctrine of *Principles* and the substitutional theory should therefore be regarded as realizing the original logicist program set forth in the work.[2]

Grattan-Guinness reports that the paper was submitted to two referees: one kept the paper for four months; the other was the logician, A. B. Kempke, who received it in October and quickly recommended it (Grattan-Guinness 1977b, p. 94). In October the Society finally voted for publication, but Russell was no longer satisfied with it and decided to withdraw. Griffin reports that it had even been typeset (Griffin 1992, p. 301). Why the decision to withdraw? Had Russell's faith in the substitutional theory now been shaken? Some think so and mark the paper's demise as evidence that the theory "fell soon after it arose."[3] In a 10 October letter to Jourdain, however, we see that this is far from the case.

> I am glad you feel attracted by the no-classes theory. I am engaged at present in purging it of metaphysical elements as far as possible, with a view to getting the bare residuum on which its success depends.
>
> I decided not to publish the paper I read at the London Mathematical [Society] in May; there was much in it that wanted correction, and I preferred to wait till I got things into more final shape. (Quoted from Grattan-Guinness, 1977b, p. 93.)

The phrase "wanted correction" is vague and leaves open what parts of the theory Russell saw in need of modification. But there is a hint at the end of "On the Substitutional Theory of Classes and Relations." Russell remarked (*SCR*, p. 188):

> The only serious danger, so far as appears, is lest some contradiction should be found to result from the assumption that *propositions* are entities, but I have not found any such contradiction, and it is very hard to believe that there are no such things as propositions.

The passage proved to be portentous. It is now clear that the wanted corrections turned on the problem of solving paradoxes of propositions.

2. Cocchiarella (1980) has seen this as well.
3. See Grattan-Guinness (1977b), p. 94.

Three Paradoxes of Propositions

8.1 There are three main paradoxes that are of importance to us. Two of the three were known to Russell well before he advanced his substitutional theory in 1905. The first is the propositional *liar* (or Epimendes paradox). It was discussed in Russell's June 1905 manuscript "On Fundamentals." Curiously, this paradox has been thought to have played a pivotal role in the historical development of ramification. But we shall see that in the context of substitution, its role is quite minor. Russell rightfully showed little concern in 1905.

In "On Fundamentals" Russell formulated the propositional *liar* as follows. He first puts (*OF*, p. 63):

$$\phi p = \text{df I now assert } p.$$

He then has,

$$(\exists p)(\iota q\phi q = p \ \& \sim p),$$

where it is assumed that the proposition asserted is unique—that is,

$$(\exists p)(\phi p \ \& \ (q)(\phi q \ .\supset. \ q = p)).$$

In "Les paradoxes de la logique,"[4] however, Russell improved his formulation of the *liar* by avoiding the definite description. He puts

$$(\exists p)(\text{ I now assert } p \ \& \ \sim p).$$

He (tacitly) assumes that everything else (if anything) now asserted is true.[5] Putting, "$A^{it}p$" for "I assert p at time t", we have:

$$A^{it}\{(\exists p)(A^{it}p \ \& \sim p)\}.$$
$$(q)(A^{it}q \ .\&. \ q \neq \{(\exists p)(A^{it}p \ \& \sim p)\} \ :\supset: \ q).$$

The *liar* goes through. An alternative formulation would use "believes" rather than "asserts," putting "$\text{Bel}^{at}p$" for "a believes p at time t."

In light of theorems such as ($T_{4.20}$) of the substitutional calculus set out in our Chapter 4, we see that the propositional *liar* cannot be generated in the pure calculus of substitution. That calculus has no predicates "x asserts y at time t" or "x believes y at time t." Accordingly, Russell might have found an avenue for solving the *liar* outside of the general theory of propositions. One might regard the *liar* as not properly a paradox of propositions. That is, "believes," "knows," "asserts," and the like, could be regarded as *intentional* contexts which differ from *intensional* contexts. "Believes" might not stand for a dyadic relation between a person and a proposition after all.

4. The English title is "On 'Insolubilia' and Their Solution by Symbolic Logic."

5. This is a more plausible background assumption than that I can assert one and only one proposition at a given time. This may be why Russell chose to avoid a definite description in his formulation of the propositional *liar*.

The next two paradoxes are intensional in nature and directly formulable within the pure calculus of substitution. The first is a version of the paradox of propositions Russell set out in Appendix B of *Principles*. We call it the "Appendix B" paradox. The second is a little-known paradox that lay buried in manuscripts until I rediscovered it in 1987. I call the "p_o/a_o" paradox.[6]

Russell knew of all three paradoxes of propositions by May of 1906. We lately observed him working on the *liar* in his June 1905 "On Fundamentals," and of course, the "Appendix B" is from 1903. Unlike the *liar*, the other two involve the same sort of diagonal construction involved in Cantor's power-class theorem. Unaware of the p_o/a_o, Cocchiarella (1980) first discovered that the substitutional technique conflicts with Cantor's power-class theorem. In writing on Cantor's paradox of the greatest cardinal, Russell had said that the solution provided by the substitutional theory is that classes are not single entities. In Russell's words (*SCR*, p. 180):

> Any two entities p, a define a class p/a; but a pair of entities is not one entity, and the number of pairs of entities may therefore, without the slightest contradiction, be greater than the number of entities.

Cocchiarella objects (1980, p. 90):

> This would be so, however, only if the number of entities were finite; and having "proved" that the number of entities is infinite, Russell could have gone on to prove that the number of pairs of entities is the same as the number of entities.

The point is really about propositional functions (proxied by use of matrices) and not about classes directly. But the point remains. We can generate a contradiction in the substitutional theory by forming a diagonal argument akin to that formed in Cantor's proof of his power-class theorem. For every propositional function (proxied by a use of matrix "r/c") there is a unique proposition correlated. This is so, given there are infinitely many entities, because the number of pairs of entities is the same as the number of entities. Consider then the propositional function exemplified by all and only those propositions not exemplifying its correlated function. This propositional function will be proxied by use of a matrix "p/a." In turn it has a uniquely correlated proposition. But this proposition will exemplify its correlated function if and only if it does not exemplify it.

A formal derivation of the paradox of propositions set out in Appendix B of *Principles* follows:

1. $(\exists p, a)(\alpha \text{ in } p\ .\&.\ (z)(p/a;z!\{(\exists r, c)(z = \{(y)(r/c;y\ .\supset.\ y)\}\ \&$
   $\sim (r/c;z))\}))$ $\quad * (\text{CP}_{\text{sub}})^1$
2. $(z)(p/a;z!\{(\exists r, c)(z = \{(y)(r/c;y\ .\supset.\ y)\}\ \& \sim (r/c;z))\})$ $\quad$ 1, EI, simp.
3. Abbreviate putting "s" for "$\{(y)(p/a;y\ .\supset.\ y)\}$".
4. $p/a;s!\{(\exists r, c)(s = \{(y)(r/c;y\ .\supset.\ y)\}\ \& \sim (r/c;s))\}$ $\quad$ 2, ui.

6. See Landini (1989).

5. → $(\exists r, c)(s = \{(y)(r/c;y \;.\!\supset\!.\; y)\} \mathbin{\&} \sim (r/c;s))$
6. $s = \{(y)(r/c;y \;.\!\supset\!.\; y)\} \mathbin{\&} \sim (r/c;s)$ 5, EI.
7. $s = \{(y)(r/c;y \;.\!\supset\!.\; y)\}$ 6, simp.
8. $\{(y)(p/a;y \;.\!\supset\!.\; y)\} = \{(y)(r/c;y \;.\!\supset\!.\; y)\}$ 7, simp, df(s).
9. $(y)(\{p/a;y \;.\!\supset\!.\; y\} = \{r/c;y \;.\!\supset\!.\; y)\})$ 8, $*(A_{20})$.
10. $\{p/a;y \;.\!\supset\!.\; y\} = \{r/c;y \;.\!\supset\!.\; y\}$ 9, ui.
11. $\{(\exists q)(p/a;y!q \;:\!\&\!:\; (t)(p/a;y!t \;.\!\supset\!.\; t = q) \;.\&.\; q \supset y)\} =$
    $\{(\exists q)(r/c;y!q \;:\!\&\!:\; (t)(r/c;y!t \;.\!\supset\!.\; t = q) \;.\&.\; q \supset y)\}$
    10, df(desc)^1.
12. $\{\sim (q) \sim (p/a;y!q \;:\!\&\!:\; (t)(p/a;y!t \;.\!\supset\!.\; t = q) \;.\&.\; q \supset y)\} =$
    $\{\sim (q) \sim (r/c;y!q \;:\!\&\!:\; (t)(r/c;y!t \;.\!\supset\!.\; t = q) \;.\&.\; q \supset y)\}$
    11, df($\exists$).
13. $\{(q) \sim (p/a;y!q \;:\!\&\!:\; (t)(p/a;y!t \;.\!\supset\!.\; t = q) \;.\&.\; q \supset y)\} =$
    $\{(q) \sim (r/c;y!q \;:\!\&\!:\; (t)(r/c;y!t \;.\!\supset\!.\; t = q) \;.\&.\; q \supset y)\}$
    12, $(T_{4.14})$.
14. $(q)(\{\sim (p/a;y!q \;:\!\&\!:\; (t)(p/a;y!t \;.\!\supset\!.\; t = q) \;.\&.\; q \supset y)\} =$
    $\{\sim (r/c;y!q \;:\!\&\!:\; (t)(r/c;y!t \;.\!\supset\!.\; t = q) \;.\&.\; q \supset y)\}$
    13, $*(A_{20})$.
15. $\{\sim (p/a;y!q \;:\!\&\!:\; (t)(p/a;y!t \;.\!\supset\!.\; t = q) \;.\&.\; q \supset y)\} =$
    $\{\sim (r/c;y!q \;:\!\&\!:\; (t)(r/c;y!t \;.\!\supset\!.\; t = q) \;.\&.\; q \supset y)\}$
    14, ui.
16. $\{p/a;y!q \;:\!\&\!:\; (t)(p/a;y!t \;.\!\supset\!.\; t = q) \;.\&.\; q \supset y\} =$
    $\{r/c;y!q \;:\!\&\!:\; (t)(r/c;y!t \;.\!\supset\!.\; t = q) \;.\&.\; q \supset y\}$
    15, $(T_{4.14})$.
17. $\{p/a;y!q\} = \{r/c;y!q\}$ 16, df(&), $(T_{4.14})$, $(T_{4.17})$a.
18. $p = r \;.\&.\; a = c$ 17, $(T_{4.19})$b.
19. $\sim (p/a;s)$ 6, 18, simp, $*(=)$.
20. $(\exists h)(p/a;s!h \;:\!\&\!:\; (t)(p/a;s!t \;.\!\supset\!.\; t = h) \;.\&.\; \sim h)$ 19, df(desc)^1.
21. $p/a;s!h \;:\!\&\!:\; (t)(p/a;s!t \;.\!\supset\!.\; t = h) \;.\&.\; \sim h$ 20, EI.
22. $h = \{(\exists r, c)(s = \{(y)(r/c;y \;.\!\supset\!.\; y)\} \mathbin{\&} \sim (r/c;s))\}$ 4, 21, $(T_{4.3})$.
23. $\sim (\exists r, c)(s = \{(y)(r/c;y \;.\!\supset\!.\; y)\} \mathbin{\&} \sim (r/c;s))$ 21, 22, $*(=)$, simp.
24. __ F 5, 23, conj, (21ei), (6ei).
25. $5 \supset F$ cp.
26. $\sim (\exists r, c)(s = \{(y)(r/c;y \;.\!\supset\!.\; y)\} \mathbin{\&} \sim (r/c;s))$ 25, reductio.
27. $(r, c)(s = \{(y)(r/c;y \;.\!\supset\!.\; y)\} \;.\!\supset\!.\; r/c;s)$ 26, log.
28. $s = \{(y)(p/a;y \;.\!\supset\!.\; y\} \;.\!\supset\!.\; p/a;s$ 27, ui.
29. $s = \{(y)(p/a;y \;.\!\supset\!.\; y\} \;.\!\supset\!.\; (\exists^1 h)(p/a;s!h \;.\&.\; h)$ 28, $*$(scope)^1.
30. $s = \{(y)(p/a;y \;.\!\supset\!.\; y\}$ $(T_{4.8})$a, df(s).
31. $(\exists^1 h)(p/a;s!h \;.\&.\; h)$ 29, 30, mp.
32. $p/a;s!h \;:\!\&\!:\; (t)(p/a;s!t \;.\!\supset\!.\; t = h) \;.\&.\; h$ 31, EI, df($\exists^1$).
33. $(t)(p/a;s!t \;.\!\supset\!.\; t = h) \;.\&.\; h$ 32, simp.
34. $h = \{(\exists r, c)(s = \{(y)(r/c;y \;.\!\supset\!.\; y)\} \mathbin{\&} \sim (r/c;s))\}$ 4, 32, $(T_{4.3})$.
35. $(t)(p/a;s!t \;.\!\supset\!.\; t = h) \;.\&.\; (\exists r, c)(s = \{(y)(r/c;y \;.\!\supset\!.\; y)\} \mathbin{\&} \sim (r/c;s))$
    33, 34, $*(=)$.
36. $(\exists r, c)(s = \{(y)(r/c;y \;.\!\supset\!.\; y)\} \mathbin{\&} \sim (r/c;s))$ 35, simp.
37. F 26, 36, conj, (32ei), (2ei).

Our last paradox is what I have called the "p_o/a_o" paradox. Like the *PoM* Appendix B this paradox is based on the diagonal method employed in Cantor's power-class theorem. In the April–May 1906 manuscript, "On Substitution$_2$," Russell wrote (*OS*, p. 7):

$$p_o \ .=:\ (\exists p, a)\colon a_o = (p\tfrac{b}{a}!q)\colon p\tfrac{a_o}{a}!r \ .\supset_r.\ \sim r$$

$$p_o\tfrac{x}{a_o} \ .=.\ (\exists p, a)\colon x = (p\tfrac{b}{a}!q)\colon p\tfrac{x}{a}!r \ .\supset_r.\ \sim r$$

$$p_o\frac{p_o\frac{b}{a_o}!q}{a_o} \ .=:\ (\exists p, a)\colon p_o\tfrac{b}{a_o}!q = p\tfrac{b}{a}!q\colon p\frac{p_o\frac{b}{a_o}!q}{a}!r \ .\supset_r.\ \sim r$$

$$.=:\ (p, a)\colon p_o\tfrac{b}{a_o}!q \ .=.\ p\tfrac{b}{a}!q\colon \supset.\ \sim\left(p\frac{p_o\frac{b}{a_o}!q}{a}\right).$$

This formulation uses complex prototypes. But formally we set out the paradox as follows:

1. $(\exists p, a)(a$ in $p\ .\&.\ (z)(p/a;z!\{(\exists r, c)(z = \{r/c;b!q\}\ \&\sim (r/c;z))\}))$
   $*(\mathrm{CP_{sub}})^{1}$
2. $(z)(p/a;z!\{(\exists r, c)(z = \{r/c;b!q\}\ \&\sim (r/c;z))\})$ 1, EI, simp.
3. Abbreviate putting "s" for "$\{p/a;b!q\}$".
4. $p/a;s!\{(\exists r, c)(s = \{(y)(r/c;y\ .\supset.\ y)\}\ \&\sim (r/c;s))\}$ 2, ui.
5. $\rightarrow (\exists r, c)(s = \{r/c;b!q\}\ \&\ \sim (r/c;s))$
6. $s = \{r/c;b!q\}\ \&\sim (r/c;s)$ 5, EI.
7. $s = \{r/c;b!q\}$ 6, simp.
8. $\{p/a;b!q\} = \{r/c;b!q\}$ 7, simp, df(s).
9. $p = r\ .\&.\ a = c$ 8, $(\mathrm{T}_{4.19})$b.
10. $\sim (p/a;s)$ 6, 9, simp, $*(=)$.
11. $(\exists h)(p/a;s!h\ :\&:\ (t)(p/a;s!t\ :\supset:\ t = h)\ .\&.\sim h)$ 10, df(desc)^{1}.
12. $p/a;s!h\ :\&:\ (t)(p/a;s!t\ .\supset.\ t = h)\ .\&.\sim h$ 11, EI.
13. $h = \{(\exists r, c)(s = \{r/c;b!q\}\ \&\sim (r/c;s))\}$ 4, 12, $(\mathrm{T}_{4.3})$.
14. $\sim (\exists r, c)(s = \{r/c;b!q\}\ \&\sim (r/c;s))$ 12, 13, $*(=)$, simp.
15. ___ F 5, 14, conj, (12ei), (6ei).
16. $5 \supset F$ cp.
17. $\sim (\exists r, c)(s = \{r/c;b!q\}\ \&\sim (r/c;s))$ 16, reductio.
18. $(r, c)(s = \{r/c;b!q\}\ .\supset.\ r/c;s)$ 17, log.
19. $s = \{p/a;b!q\}\ .\supset.\ p/a;s$ 18, ui.
20. $s = \{p/a;b!q\}\ .\supset.\ (\exists^{1} h)(p/a;s!h\ .\&.\ h)$ 19, $*(\mathrm{scope})^{1}$.
21. $s = \{p/a;b!q\}$ $(\mathrm{T}_{4.8})$a, df(s).
22. $(\exists^{1} h)(p/a;s!h\ .\&.\ h)$ 20, 21, mp.
23. $p/a;s!h\ :\&:\ (t)(p/a;s!t)\ .\supset.\ t = h)\ .\&.\ h$ 22, EI, df$(\exists^{1})$.
24. $(t)(p/a;s!t\ .\supset.\ t = h)\ .\&.\ h$ 23, simp.
25. $h = \{(\exists r, c)(s = \{r/c;b!q\}\ \&\sim (r/c;s))\}$ 4, 23, $(\mathrm{T}_{4.3})$.
26. $(t)(p/a;s!t\ .\supset.\ t = h)\ .\&.\ (\exists r, c)(s = \{r/c;b!q\}\ \&\sim (r/c;s))$
    24, 25, $*(=)$.
27. $(\exists r, c)(s = \{r/c;b!q\}\ \&\sim (r/c;s))$ 26, simp.
28. F 17, 27, conj, (23ei), (2ei).

From one perspective this seems but a distilled version of the Appendix B. But there is a very important difference—viz., the p_o/a_o does not involve identity with a general proposition. Observe as well that "semantic" or "intentional" predicates are involved in neither the Appendix B nor the "p_o/a_o" paradox. The paradoxes are, if you will, "syntactic." They simply exploit the ideas used in a Cantorian diagonal construction.

Indeed, once we see the connection with Cantor's theorem, we realize that the p_o/a_o has many variants. The following instance of $*(\mathrm{CP})^3$ posits the function Russell used in the p_o/a_o:

$$(\exists jkuv)(xyz)(j/kuv;xyz \;.{\equiv}.\; x = \{y/z;b!q\}),$$

where b and q are some chosen entities. Instantiating, the p_o/a_o paradox could be generated from the following instance of $*(\mathrm{CP_{sub}})^1$:

$$(\exists pa)(a \text{ in } p \;.\&.\; (x)(p/a;x!\{(\exists rc)(j/kuv;xrc \;.\&.\; {\sim}\,(r/c;))\})).$$

But other functions are possible. For example, there is:

$$(\exists jkuv)(xyz)(j/kuv;xyz \;.{\equiv}.\; x = \{y \supset z\}).$$

We can generate a simpler formulation of the paradox with this function. From $*(\mathrm{CP_{sub}})^1$, we get

$$(\exists pa)(a \text{ in } p \;.\&.\; (x)(p/a;x!\{(\exists rc)(x = \{r \supset c\} \;.\&.\; {\sim}\,(r/c;x))\})).$$

Instantiating and taking $\{p \supset a\}$, we arrive at the contradiction,

$$p/a;\{p \supset q\} \;.{\equiv}.\; {\sim}\,(p/a;\{p \supset a\}).$$

Even simpler from the standpoint of the derivation of the contradiction, is the following instance of $*(\mathrm{CP_{sub}})^1$:

$$(\exists pa)(a \text{ in } p \;.\&.\; (x)(p/a;x!\{(rcq)(x = \{r \supset c\} \;.\&.\; r/c;x!q \;.{\supset}.\; {\sim}\, q)\})).$$

Notice as well that with propositions as individuals, there will also be functions correlating pairs of entities with triples. For instance, the following is a theorem:

$$(\exists qabcew)(xyzuv)(q/abcew;xyzuv \;.{\equiv}.\; \{x \supset y\} = \{z \supset u \,.{\supset}.\, v\}).$$

Hence, from $*(\mathrm{CP_{sub}})^2$ we have the theorem

$$(\exists stw)(rc)(s/tw;rc!\{(\exists hde)(\{r \supset c\} = \{h \supset d \;.{\supset}.\; e\} \;\&\; {\sim}\,(h/de;rc))\}).$$

Instantiating, we arrive at the contradiction:

$$s/t,w;\{s \supset t\},w \;.{\equiv}.\; {\sim}\,(s/t,w;\{s \supset t\},w).$$

The paradox repeats at higher types.

The source of the paradox is not the $*(\mathrm{CP_{sub}})^n$ schemata, for these are theorem schemata. The source can be traced directly to $*(r)$. The oddity and suprise is that $*(r)$ certainly *seems* to be a logical truth concerning propositions!

In Chapter 9, we shall find that Russell's troubles with the p_o/a_o paradox eventuate in the abandonment of substitution. But we must be careful to properly understand the successes and failures of Russell's efforts to solve the paradoxes, for his efforts have a direct bearing on the proper understanding of the origins of ramified type theory. Hylton (1980) has argued that it must have occurred to Russell to formulate something like the Appendix B paradox substitutionally. He offers this as an explanation of Russell's article "Les paradoxes de la logique," where, in Hylton's view, a substitutional theory ramified by a hierarchy of orders of propositions is offered. The restricted variables of the substitutional theory, he continues, conflict with its very *raison d'être* (i.e., the doctrine of the unrestricted variable), and thereby destroy substitution. We shall find that this cannot be correct. Russell was successful in solving both the *liar* and Appendix B paradoxes in "Les paradoxes de la logique" and that paper nowhere introduces a substitutional theory ramified by a hierarchy of orders of propositions.

Substitutional Manuscripts of April/May 1906

8.2 Russell's extant work notes of 1906 are dedicated almost solely to finding a solution of the p_o/a_o paradox (and its variants). The relevant manuscripts are:

> "Logic in which Propositions are not Entities," (April/May 1906) cf. pp. 15f.
> "On Substitution$_2$," (April/May/June 1906) cf. pp. 7 & 57 & 71
> "The Paradox of the Liar," (September 1906) cf. pp. 72f.

The manuscript "Logic in which Propositions are not Entities" and "On Substitution$_2$" predate Russell's reply to Poincaré which appeared in September of 1906. The long manuscript "The Paradox of the Liar" is dated September 1906 and seems in general to be a return to substitutional issues after "Les paradoxes." We therefore postpone discussion of it until later.

In May of 1906 Poincaré published an article entitled "Les mathématiques et la logique." It contained criticisms of the new mathematics of the infinite and paid special attention to Russell's paper "On Some Difficulties." Drawing a lesson from the Epimenides (liar) paradox, it proposed a solution—namely, adherence to a principle of avoiding "vicious circles." Russell was eager to redress the criticisms in a reply. There was now no point in going ahead with the publication of "On the Substitutional Theory of Classes and Relations," with its problems over propositions. He decided to wait to get free of the paradoxes of propositions. He would reply to Poincaré and this would afford a new opportunity to explain how symbolic logic solves the paradoxes in virtue of the substitutional theory. This interpretation is supported by a letter Russell wrote to Conturat of 15 May 1906. Russell noted:

> I have retired here [Providence House, Clovelly] alone for nearly two months, in order to get through my work a little faster than usual. I still believe that my solution to the contradictions is good, but it seems to me that it needs to be extended to propositions, that is to say that the latter, like classes and relations, cannot replace ordinary entities. To say,

> for example, that the law of excluded middle is not red, would be to utter nonsense, and not a truth. I will follow your advice in replying to M. Poincaré. For this reason, I will not reply quickly, because I would like to get into order what I have to say about the solution of the contradictions. (Quoted from Griffin 1992, p. 301).

The manuscript "On Substitution$_2$" makes references to the *vicious circle principle* (*OS*, p. 31) and the *liar* paradox as well (*ibid.*, p. 32). Likely it was composed at Clovelly.

In reading Russell's unpublished manuscripts of 1906, it is important to see that the p_0/a_0 paradox is driving the discussion. If not for this, one may easily be led astray by Russell's often cryptic and involuted dialogue with himself over fundamental ideas. I disagree, for example, with Grattan-Guinness's interpretation of some passages in the early 1906 manuscripts. Russell suggests that he must distinguish propositional from non-propositional substitutions. In Chapter 7 we saw that Grattan-Guinness asked how dyadic relations of individuals are to be distinguished from functions of monadic functions of individuals. (Matrices used to proxy both are of the form "$x/y,z$".) He went on to conclude that the theory "cannot operate successfully" unless propositional and non-propositions substitutions are distinguished (Grattan-Guinness 1977a, p. 105). Referring to a passage in "On Substitution$_2$,"[7] he writes (*ibid.*, p. 399):

> Some of Russell's work notes for the summer of 1906 survive in the Archives. They show that he had realized the need to distinguish between propositional and constant substitutions. Indeed, in one passage he not only distinguished between $\frac{q}{p}$ and $\frac{x}{a}$ but also explicitly declared $\frac{q}{a}$ and $\frac{x}{p}$ 'to be meaningless', and then he followed up with two new definitions of identity, one . . . for constants:
>
> $$x = y \ .{=}.\ (p)(p\tfrac{y}{x}!p) \quad \text{Df.}$$
>
> and the other for propositions:
>
> $$p = q \ .{=}.\ (x)(p\tfrac{x}{x}!q) \quad \text{Df.}$$

Now Grattan-Guinness's question is an important one. But our Chapters 6 and 7 demonstrated that his conclusion is unfounded. Nothing in the development of the proxy for propositional functions, classes, or relations-in-extension, requires a distinction between propositional and non-propositional substitutions. In my view, the passage of "On Substitution$_2$" from which Grattan-Guinness quotes has nothing to do the problem he raised.[8] The passage is one of Russell's many attempts to block the p_0/a_0 paradox.

7. We designate this as "*OS*" to distinguish it from our "*S*" used for the 22 December manuscript of the same name.

8. The same mistake is made by Rodriguez-Consuegra. He argues that in unpublished manuscripts Russell came to see the "danger" of "substituting a proposition p for a constituent a, and vice versa, which would lead (obviously) to nonsenses" (Rodriguez-Consuegra 1987, p. 147n).

In fact, the passage immediately follows a reference to the paradox. Russell writes (*OS*, p. 94):

> All we need is such a modification as will permit
>
> $p\frac{b}{a}!q \;.=.\; p_o\frac{b}{a_o}!q : p/a \neq p_o/a_o.$
>
> Again we might notationally distinguish propositions and entities, and have two sorts of substitution, propositional and entity. The notational distinction may, to begin with, be as hitherto, by a difference of letters. Thus *p*, *q*, *r*, *s* will be propositions. The different sorts of substitutions may be indicated in the same way: $\frac{q}{p}$ is a propositional substitution, $\frac{x}{a}$ an entity-substitution. $\frac{q}{a}$ and $\frac{x}{p}$ are to be meaningless.

Russell hopes to avoid the p_o/a_o paradox by making "$p/a;x!q$" meaningless unless "*x*" and "*a*" both stand for "entities" (i.e., nonpropositions) or both stand for propositions. On this plan, we can substitute one entity (nonproposition) for another and one proposition for another but mixed cases are to be meaningless. It was in following out this plan that Russell proposed new definitions for the identity symbol, separating cases when it is flanked by nonproposition terms and cases when it is flanked by proposition terms.

Only two pages later Russell set out the p_o/a_o fully and realized that the simple plan of distinguishing propositional and nonpropositional substitutions would not work (*OS*, p. 96). In the paradox, we have the proposition '$p_o/a_o;b!q$'. Since a_o is to be equal to '$p/a;b!q$' it must be a proposition. If *b* is not a proposition, Russell's suggestion would go into effect. But the paradox recurs if *b* is a proposition. The same point is found in "Logic in which Propositions are not Entities," where Russell had originally hoped for the simple plan. He says (*L*, p. 15):

> The above theory comes to this: that we can substitute any entity for any entity, and any proposition for any proposition, but never an entity for a proposition or a proposition for an entity.

But then he immediately notes (*ibid.*):

> But if we don't have a hierarchy of propositions it looks as if we should get into difficulties from the fact that there are more classes of propositions than propositions, and that, to all appearances, we can establish a 1-1 function from all classes of propositions to some proposition.

The original plan failed, and Russell suggests that it will have to be replaced by a hierarchy of orders of propositions. A proposition of one order cannot be meaningfully substituted for that of another.

In both manuscripts there is a reference to "functional substitutions." Grattan-Guinness viewed this as "involving the possibility of producing a hierarchy of functions rather than matrices, and so reintroduced a hybrid kind of system in which substitution and propositional functions operated simultaneously" (Grattan-Guinness 1974, p. 400). The notion of "functional substitutions" Russell introduces here must, however, be read with caution. It is not a hybrid system which embraces an outright ontology of propositional functions. Both manuscripts contain explicit warnings about this. In the former, Russell writes (*L*, p. 12):

> *N.B.* We have to be careful not to let functions creep back into being. We must have functions, but they must be quite distinct from entities.

In "On Substitution$_2$" Russell has a similar remark (*OS*, p. 79):

> All this brings us back to the old business about non-predicative functions and the old necessity for axioms as to when we have
>
> $(\exists\psi)((\phi)(x)(f(\phi,x) \equiv .\ \psi x))$.
>
> This is intolerable.

Functional quantification revives all the horrid difficulties of finding principles to determine which functions are "simple." Russell would be back at square one excluding "quadratic forms," etc. He tried hard on this approach after *Principles* and abandoned it.[9]

So what was "functional substitution" all about? In "Logic in which Propositions are not Entities" Russell gives a partial explanation. The general idea was to put "$(\phi, x)f(\phi, x)$," instead of "$(p)fp$". It then requires two variables to proxy quantification over propositions (*L*, p. 1). Russell writes (*ibid.*, p. 9):

> Thus, propositions like classes are to be replaced by two variables. But now we take entities and functions, not entities and *propositions*, as our fundamentals. It is to be understood that a proposition is not a single entity and is not to be expressed by a single letter, but always by two, ϕ and x. This excludes
>
> $(\phi x)\frac{\psi_y}{a}!(\chi z)$ if $(\phi x)\frac{y}{a}!(\psi z)$ is our Pi.

Russell's hope is to be able to go on to parallel the expressions of the substitutional theory—only now the complexity of propositions has to be shown in the notation of propositional substitution by using expressions such as "ϕa" rather than "p". Instead of the one form,

"$p/a;b!q$".

where a and b can be any entities (propositions or otherwise), Russell hopes to build structure into substitution. Thus, as primitive ideas he has:

$(\phi a)\frac{z}{a}!(\phi z)$,

for entity substitutions; and

$(\phi a)\frac{\psi}{\phi}!(\psi a)$

9. See Landini (1992).

for functional substitutions. Then one may go on to define propositional substitutions such as:

$$f(\phi a)\tfrac{\psi z}{\phi a}!f(\psi z).$$

Russell observes that the task of defining propositional substitutions from functional is formidable (*L*, p. 16). So he also examines the alternative of dropping functional substitution and taking propositional substitution as a primitive notion. Russell explains (*L*, p. 14):

> Perhaps propositional substitution is altogether unnecessary, and we can do with functional and entity substitution. *Functional* substitution can only be affected in propositions. Probably it is better to say the same of *entity*-substitution. We thus start with two primitive ideas:
>
> $(\phi a)\frac{y}{b}!\psi x \quad \text{Pi}$
>
> $(\phi a)\frac{\chi}{\psi}!(\theta x) \quad \text{Pi}$
>
> or, if it seems better, we can instead of the second, take *propositional* substitution as our primitive idea, putting
>
> $(\phi a)\tfrac{xy}{\psi b}!(\theta x) \quad \text{Pi}.$
>
> Whichever of the two we do, there won't be a hierarchy of propositions of different types, which is a comfort.

On either alternative, the point becomes clear. Russell could not rest content with a hierarchy of orders of propositions. By holding that propositions are not "entities," and building structure into substitution, he hopes to show that the restrictions on forms of substitution (which was the purpose of the hierarchy of orders of propositions) are built into the logical grammar. A hierarchy of propositions is escaped.

Russell never explictly sets out the p_o/a_o in "Logic in which Propositions are not Entities." In "On Substitution$_2$," on the other hand, it recurs again and again. The general ideas of "Logic in which Propositions are not Entities" also recur, but they are now more explicitly connected with solving the p_o/a_o and the *liar*. The plan of "functional substitution" runs into many difficulties. Russell observes that to build in structure, he will have to continually introduce new primitive ideas as more and more complicated substitutions take place. Since there is no one form for substitution, there will have to be infinitely many primitive forms; and "this is intolerable" (*OS*, p. 12).

On again, off again, "On Substitution$_2$" resorts to introducing a hierarchy of propositions. However unsatisfactory philosophically, such a hierarchy is useful in *formally* blocking the paradoxes of propositions. Late in the manuscript Russell sets out the hierarchy as follows (*ibid.*, 183):

> The simplest plan is to say: A prop. is of the 1st order when it contains no apparent var. .... 2nd order .... an apparent entity var. only. .... *n*th order .... prop order $n - 2$ variable and nothing higher.

The hierarchy of orders of propositions is based on the limitations of the ranges of the variables of quantification. Much earlier in the manuscript, Russell had already suggested that without some such hierarchy "vicious circles" come in (*OS*, p. 27). He states a version of Poincaré's "Vicious Circle Principle" in roughly the same form as it would later appear in "Les paradoxes de la logique" (*OS*, p. 31):

> Where a proposition is asserted for all values of a variable, the proposition itself must not be among the possible values of the variable.

Excluded middle is used as an illustration. The proposition

$$(p)(\sim p \vee p),$$

is of order $n + 1$ when the propositions its variable ranges over are order n. Thus the law cannot yield,

$$\sim (p)(\sim p \vee p) \vee (p)(\sim p \vee p),$$

by instantiation (*ibid.*, p. 38). The orders also block the *liar*. The order of the proposition,

$$\text{‘}(\exists p)\ (\text{I now assert } p\ .\&.\ \sim p)\text{’},$$

will now be $n + 1$ where the order of propositions in the range of the variable is n.

But to apply the hierarchy of orders to the p_o/a_o Russell must decide what the order of a proposition such as '$p/a;b!q$' is to be. He feels it would be necessary to adopt the view that in the proposition '$p/a;b!q$', the entities b and a must be of the same relative order (type) and of a lower order than $\{p/a;b!q\}$ itself (*ibid.*, p. 5).

This rules out a substitution such as expressed in "$p/a;\{p/a;b!q\}$" for here the higher-order proposition $\{p/a;b!q\}$ would be substituted for a which is of lower order (*ibid.*, p. 7). But Russell demurs, admitting that the decision seems arbitrary and out of sorts with the plan of basing the hierarchy of propositions on the admissible ranges of the apparent variable.

In any event, Russell is clearly dissatisfied with a reliance on a hierarchy of propositions (quâ entities) in "On Substitution$_2$." The hierarchy of orders is certainly not his final say at this time. With "functional substitution" at an impass, he tries things anew. One plan is to attempt to block the paradox by preventing:

$$\{p_o/a_o;b!q\} = \{p/a;b!a\}\ :\supset:\ p_o = p\ .\&.\ a_o = a$$

(See, for instance, *OS*, p 79 & 94). As Russell puts it, he wants to be able to have

$$\{p_o/a_o;b!q\} = \{p/a;b!q\} \text{ :\&: } p_o/a_o \neq p/a,$$

where the second conjunct is just a definitional abbreviation of

$$\sim (x)(p_o/a_o;x \text{ .}\equiv\text{. } p/a;x).$$

But he is uncertain how this might be accomplished; and he soon realizes that he will have to block the move in general (*OS*, p. 79):

> If we keep propositions as entities, all that is necessary is
>
> $$\sim (p, a)(p', a')(f(p,a) = f(p', a') \text{ .}\supset\text{. } p/a = p'/a')$$
>
> for all imaginable f's; $f(p, a)$ being a proposition. We secured this for $f(p, a)$.=. $p \supset a$, by $p \supset \sim a$.=. $a \supset \sim p$. We want a similar way of securing it for $p/a;b!q$, where b, q are fixed.

Russell was well aware that the paradox can be formed as follows:

$$(\exists p_o a_o)(x)(p_o/a_o;x \text{ .}\equiv\text{. } (\exists pa)(x = \{p \supset a\} \text{ .\&. } \sim (p/a;x))).$$

Here we have $\{p \supset a\}$ instead of $\{p/a;b!q\}$. He hopes that in the case of a truth-function such as 'implies' he might allow:

$$\{p \supset \sim a\} = \{a \supset \sim p\}.$$

Then he would have to find a way to deny

$$(pq)(rs)(\{p \supset q\} = \{r \supset s\} \text{ :}\supset\text{: } p = r \text{ .\&. } q = s).$$

The paradox would not go through in such cases. The trouble Russell soon discovers is that even if this works, he shall need a similar approach for other cases and in particular for $\{p/a;b!q\}$.

Russell does not seem to have contemplated dropping identity altogether where propositions are concerned. To see why, recall that $\phi(p/a;b)$ abbreviates $\phi((\iota q)(p/a;b!q))$. The contextual definition of the description expresses an identity between propositions. Still, one might think to get along with:

$$[p/a;b]\phi(p/a;b) = \text{df } (\exists q)(p/a;b!q \text{ \& } (r)(p/a;b!r \text{ .}\supset\text{. } r \equiv q) \text{ \& } \phi q).$$

Russell never tries. And he was certainly sensitive to what has become known as Quine's dictum—viz., "No entity without identity."[10] Russell's proposal of having

$$\{p \supset \sim a\} = \{a \supset \sim p\}$$

shows, however, that he did contemplate getting along with logical equivalence instead of identity (where general propositions are concerned). The proposal never went very far and Russell came quickly to conclude that the suggestion of blocking

$$\{p_o/a_o;b!q\} = \{p/a;b!q\} \text{ :}\supset\text{: } p_o = p \text{ .\&. } a_o = a$$

was hopeless.

10. See Lackey (1974).

Poincaré's Vicious Circle Principle

8.3 Eventually, Russell wrote his reply to Poincaré. Russell's article, "Les paradoxes de la logique," was published in September 1906 in French. (The English manuscript was written in June and titled "On 'Insolubilia' and Their Solution by Symbolic Logic.") As its English title attests, Russell was still undaunted by the paradoxes of propositions. Indeed, in a letter to Jourdain of 14 June, Russell wrote:

> The no-classes theory . . . shows that we can employ the *symbol* $\hat{x}\phi x$ without ever assuming that this symbol in isolation means anything. I feel more and more certain that this theory is right. In order, however, to solve the *Epimenides*, it is necessary to extend it to *general* propositions, i.e., to such as $(x).\phi x$ and $(\exists x).\phi x$. This I shall explain in my answer to Poincaré's article in the current *Revue de Métaphysique*. (Quoted from Grattan–Guinness 1977b, p. 89).

In "Les paradoxes" Russell sets out his substitutional theory and, as promised in the letter to Jourdain, it is modified to account for the propositional *liar*. The unpublished manuscripts "Logic in which Propositions are Not Entities" and "On Substitution$_2$" (the 1906 ms) discuss many plans. But neither contained the plan finally adopted in Russell's reply to Poincaré. It seems that his philosophical objections to orders of propositions had won out. In his reply to Poincaré, Russell thought he had the whole matter solved with the abandonment of general propositions.

In "Les paradoxes," Russell criticized Poincaré's view that the paradoxes result from embracing Cantor's actual infinite. He did agree, however, that "there is an element of truth" (op. cit., p. 198) in Poincaré's view that the lesson of the Epimenides (*liar*) paradox is that one must avoid vicious circles. He therefore adopts his own version of what has come to be called Poincaré's vicious circle principle in the following form (*InS*, p. 204):

> Whatever involves an apparent variable must not be among the possible values of that variable.

Now the assumption of generalized propositions does indeed involve a circle which gives rise to paradox. But is it "vicious"? Russell never thought so before. In fact, the "viciousness" of the circle can only be obtained by an added assumption. One has to add that quantified propositions, were they to exist, "presuppose" the existence of the totality of entities over which their quantifiers range. The vicious circle principle could then be put as follows:

> No totality can contain members involving or presupposing this totality.

The added assumption of "presupposition" is, however, quite unwarranted if general propositions are viewed as objective truths and falsehoods. With an ontology of general propositions, the circle is not "vicious" at all.

This would be a telling criticism if the vicious circle principle had been Russell's justification of his approach to the paradoxes of propositions. But it was not—at least not in "Les paradoxes de la logique." Russell says, rather, that the "clue" to the paradoxes is to be found in the vicious circle principle (*InS*, p. 198). He explains further (*ibid.*, p. 295):

> It is important to observe that the vicious-circle principle is not itself the solution of vicious-circle paradoxes, but merely the result which a theory must yield if it is to afford a solution of them. It is necessary that is to say, to construct a theory of expressions containing apparent variables which will yield the vicious-circle principle as an outcome. It is for this reason that we need a reconstruction of logical first principles, and cannot rest content with the mere fact that the paradoxes are due to vicious circles.

In fact, "viciousness" has no part in Russell's solution of the paradoxes of propositions at all. By abandoning an ontology of general propositions, Russell solves the paradoxes by avoiding circularity, not "vicious circularity." Assuming that when a wff ("statement") containing a quantifier symbol occurs in a subject position it is a genuine name of a single proposition p, it follows that the proposition p can be a value of the variable(s) occurring in the wff. For instance, let the wff be "$(x)(x \supset x)$". We know that

$$\vdash \ (x)(x \supset x).$$

At the same time we have the nominalization "$\{(x)(x \supset x)\}$". This was to be a genuine singular term, a proper name of a general proposition. But then by the schema of *universal instantiation*, we have

$$\vdash \ (x)(x \supset x) \ .\supset. \ \{(x)(x \supset x)\} \supset \{(x)(x \supset)\}.$$

It is just this kind of situation that allows the circle needed for the paradoxes of propositions. Russell's solution is to drop his assumption that $\{A\}$ can be a genuine singular term where A is a wff containing a quantifier symbol. There are no "general proppositions." This is compatible with the doctrine that the variables of pure logic must be unrestricted. Without general propositions there is no possibility of a circle.

In "Les paradoxes," Russell tells us that a statement containing a quantifier phrase, such as "$(x)(x = x)$", asserts the truth of all the propositions '$a = a$', '$b = b$', '$c = c$', etc. Similarly, an existential statement "$(\exists x)(x = x)$" asserts one proposition from among these, though it is ambiguous as to which (*InS*, p. 204). Odd sounding as this may be, there is really nothing new here. Russell's phraseology is just a product of not reading variables pronomially. He speaks of our use of the statement "$(x)(x = x)$" to assert (the truth of) many propositions rather than a single general proposition. But this point is better put if we embrace a pronomial reading of the variable. When the wff "$(x)(x = x)$" occurs assertorically (i.e., when it is not in a subject position), it is used to assert that everything is self-identical. There is no commitment to general propositions here.

The crucial new point is that the process of nominalizing a quantified wff to from a singular term (for a proposition) is no longer allowed. For example, "$\{(x)(x \supset x)\}$" and "$\{(x, y)(x \supset y)\}$" are ill-formed; there are to be no such terms in the new substitutional language. There are still propositions, however. Where A is any wff not containing any quantifier symbol $\{A\}$ remains a singular term. The theory of deduction for quantifier-free formulae is unaffected by the change. But the whole theory of deduction for wffs which have as subordinate clauses wffs containing quantifier symbols is lost. Since the schematic "α", "β', are for singular terms, and since $\{(x)Ax\}$ is no longer a singular term, the schemata

$$\alpha \ .\supset. \ \beta \supset \alpha$$

does not have

$$\{(x)Ax\} \ .\supset. \ q \supset \{(x)Ax\}.$$

as an instance. The theory of quantification will have to be formulated anew.

The new plan uses the fact that for any quantified wff occurring as a subordinate clause in a truth-functional context, there is a wff in *prenex normal form*[11] which is provably equivalent to it. The idea was to use prenex forms as contextual definitions. For example, the plan would allow

$$(x)Ax \supset p$$

by means of the contextual definition,

$$(x)Ax \supset p = \text{df } (\exists x)(Ax \supset p),$$

where x does not occur free in p. Although there is no singular term $\{(x)Ax\}$, the wff $(x)Ax$ may occur in a subject position in a truth-functional context in virtue of the definition. The position of "x" in "$x \supset y$" is a subject position. Russell's solution, therefore, offers a new account of statements in which quantified wffs occur as subordinate clauses. This is the "reconstruction of logical first-principles" of which he speaks. Just as the position of "x" in "Fx" may be occupied by the incomplete symbol "$\imath zGz$", the position of "x" in "$x \supset p$", may be occupied by the incomplete symbol "$(z)(z \supset z)$". As Russell put it

> To reconcile the unrestricted range of the variable with the vicious-circle principle, which might seem impossible at first sight, we have to construct a theory in which every expression which contains an apparent variable (i..e, which contains such words as *all*, *any*, *some*, *the*,) is shown to be a mere *façon de parler* (*InS*, p. 206).

11. A wff is in prenex normal form if and only if all of its quantifiers are initially placed, no two quantifiers are on the same variable, and every variable occurring in a quantifier occurs also at least once within the scope of that quantifier.

Logic without General Propositions

8.4 Unfortunately, "Les paradoxes de la logique" says little as to how this reconstruction of first principles is to proceed. Russell clearly had a plan in mind. Apologizing for the absence, he remarks that "to show in detail how this is to be done would require much mathematics, and is impossible in the present article" (*InS*, p. 206). The plan of doing without general propositions is similar[12] to ∗9 of *Principia Mathematica*—albeit different in purpose.[13] Looking there, we can surmise what Russell must have intended for "Les paradoxes."

The formal system of "Les paradoxes" adopts the primitive signs of the original system set out in our Chapter 4. But now two new signs are added: the tilde sign "∼" as a monadic predicate constant, and the sign "∃". The definition of tilde of Chapter 4 cannot be adopted in the new system because the nominalization of a wff containing quantifiers is illicit. Accordingly, we follow Russell in adopting the tilde sign as a primitive. All the same, it should be noted that this is not forced. An alternative is to put

$$f = \text{df} \ \supset / \supset; \supset !\{\supset\supset\supset\}$$
$$\sim \alpha = \text{df} \ \alpha \supset f.$$

With appropriate supporting axioms the propositional calculus can be formulated with this definition of tilde.

To characterize the terms and wffs of the system, we begin with the following: The *basic* terms are just the individual variables. The *atomic* wffs are just $\alpha \supset \beta$, $\sim \alpha$, and $\pi/\alpha;\beta!\delta$, where $\alpha, \beta, \pi, \delta$ are basic terms. (Recall that we use "α", "β", "π', and "δ" for terms. Outer brackets on wffs are again dropped for convenience.) The *quantifier-free* wffs are the smallest set K containing all atomic wffs and such that $\{A\} \supset \{B\}$, $\{A\} \supset \alpha$, $\alpha \supset \{A\}$, $\sim \{A\}$, and,

$\pi/\alpha;\beta!\{A\}$	$\pi/\alpha;\{B\}!\{A\}$	$\pi/\{C\};\{B\}!\{A\}$	$\{A\}/\{B\};\{C\}!\{D\}$
$\pi/\alpha;\{A\}!\delta$	$\pi/\{B\};\beta!\{A\}$	$\{C\}/\alpha;\{B\}!\{A\}$	
$\pi/\{A\};\beta!\delta$	$\{B\}/\alpha;\beta!\{A\}$	$\{C\}/\{B\};\beta!\{A\}$	
$\{A\}/\alpha;\beta!\delta$	$\{B\}/\{A\};\beta!\delta$	$\{C\}/\{B\};\{A\}!\delta$	
	$\{B\}/\alpha;\{A\}!\delta$		
	$\pi/\{A\};\{B\}!\delta$		

are in K whenever $\alpha, \beta, \pi, \delta$ are terms and A, B, C, and D are in K. The *Q-formulae* are just the wffs in the smallest set Q which contains $\ulcorner(\mu)A\mu\urcorner$ and $\ulcorner(\exists\mu)A\mu\urcorner$ where A is any quantifier-free wff containing the individual variable μ free, and which contains $\ulcorner(v)Bv\urcorner$ and $\ulcorner(\exists v)Bv\urcorner$ where B is any wff in Q containing the individual variable v free. With this, the terms of our language are as follows: (i) all individual variables are terms; (ii) if A is a quantifier-free

12. ∗9 of *Principia*, however, abandons all propositions.

13. In Chapter 10, we shall argue that *Principia*'s logical particles are statement connectives and not predicates at all.

wff, then $\{A\}$ is a term. The wffs of our language are just the elementary and Q-formulae. As before, nominalizing braces will often be dropped. But one must not forget that occurrences of quantified wffs in subject positions are allowed only by definition.

The axiom schemata and rules for the calculus are then adapted from *Principia*'s ∗9.[14] The starred numbers from *Principia* have been included for reference.

∗1906(A_1) $\alpha \supset \beta \;{:}{\supset}{:}\; \beta \supset \delta \;.{\supset}.\; \alpha \supset \delta$
∗1906(A_2) $\alpha \;.{\supset}.\; \sim\alpha \supset \beta$
∗1906(A_3) $\sim\alpha \supset \alpha \;.{\supset}.\; \alpha$
∗1906(A_4) $\alpha = \beta \;.{\supset}.\; A(\alpha, \alpha) \supset A(\beta|\alpha, \alpha)$,
where β is free for α in A
∗1906(A_5) ∗9.1
$A\alpha|\mu \supset (\exists\mu)A\mu$
∗1906(A_6) ∗9.11
$\sim A\alpha|\mu \supset A\beta|\mu \;.{\supset}.\; (\exists\mu)A\mu$
Modus ponens$_1$ ∗1.1
From A and $\{A\} \supset \{B\}$, infer B
Modens ponens$_2$ ∗9.12
From A and $A \supset B$, infer B
Universal generalization (UG) ∗9.13
From A, infer $(\mu)A$
Switch
From $B(\mu)(\exists v)A$, infer $B(\exists v)(\mu)A$
where all free occurrences of the variable μ in A are on one side of a logical particle and all free occurrences of the variable v in A are on the other.
Replacement rule for defined signs
Definiendum may replace and be replaced by definiens wherever it occurs.
Variable rewrite (var)
From CA, infer CB,
where CB results from replacing all bound occurrences of the individual variable μ in A by occurrences of the individual variable v and v does not occur in A.[15]
df(=) $\alpha = \beta = \text{df } \alpha/\alpha;\beta!\alpha$
df(v) $\alpha \vee \beta = \text{df } \sim\alpha \supset \beta$
df(&) $\alpha \,\&\, \beta = \text{df } \sim(\alpha \supset \sim\beta)$
df(≡) $\alpha \equiv \beta = \text{df } \alpha \supset \beta \;.\&.\; \beta \supset \alpha$
df(v)^2 $A \vee B = \text{df } \sim(A \supset B)$
df(&)^2 $A \,\&\, B = \text{df } \sim(A \supset \sim B)$
df(≡)^2 $A \equiv B = \text{df } (A \supset B) \,\&\, (B \supset A)$
df(A) $(\mu_1 \ldots \mu_n)A = \text{df } (\mu_1)\ldots(\mu_n)A$
df(E) $(\exists\mu_1 \ldots \mu_n)A = \text{df } (\exists\mu_1)\ldots(\exists\mu_n)A$

14. The propositional schemata are from Łuksiewicz (1963).
15. See *Principia* (*PM*, pp. 134, 153).

$*9.01 \sim (\mu)A\mu \ = \text{df}\ (\exists\mu) \sim A\mu$
$*9.02 \sim (\exists\mu)A\mu \ = \text{df}\ (\mu) \sim A\mu$
dfs(P)p
$*9.03 \quad (\exists\mu)A\mu \supset \alpha = \text{df}\ (\mu)(A\mu \supset \alpha)$
$*9.04 \quad \alpha \supset (\mu)A\mu = \text{df}\ (\mu)(\alpha \supset A\mu)$
$*9.05 \quad (\mu)A\mu \supset \alpha = \text{df}\ (\exists\mu)(A\mu \supset \alpha)$
$*9.06 \quad \alpha \supset (\exists\mu)A\mu = \text{df}\ (\exists\mu)(\alpha \supset A\mu)$
dfs(P)x
$*9.07 \quad (\exists\mu)A\mu \supset (\exists\nu)B\nu = \text{df}\ (\mu)(\exists\nu)(A\mu \supset B\nu)$
$*9.08 \quad (\mu)A\mu \supset (\nu)B\nu = \text{df}\ (\nu)(\exists\mu)(A\mu \supset B\nu)$
$*9.\text{ZZ} \quad (\exists\mu)A\mu \supset (\nu)B\nu = \text{df}\ (\mu)(\nu)(A\mu \supset B\nu)]$
$*9.\text{WW} \quad (\mu)A\mu \supset (\exists\nu)B\nu = \text{df}\ (\exists\mu)(\exists\nu)(A\mu \supset B\nu)$,

where μ does not occur free in B or α and ν does not occur free in A. This completes the quantificational system.

Two separate groups of definitions for pulling quantifiers out are needed. It would not suffice to allow α in dfs(P)p to take nonelementary wffs, for then certain wffs would have two distinct definiens. Modus ponens$_2$ is needed when quantified wffs flank the conditional sign.

The substitutional axioms and axiom schemata of the original substitutional theory set out in Chapter 4 remain largely intact in the system of "Les paradoxes." Of course, axiom and theorem schemata governing general propositions are lost. Without general propositions we shall need a new definition of the identity sign and the new axiom schema $*1906(\text{A}_4)$ else the system would lose the laws of identity. With $*1906(\text{A}_4)$ the derived rule $*(=)$ is valid in the system of "Les paradoxes." We have $*1906(\text{A}_7)$–$*1906(\text{A}_{18})$ from the original theory, together with df(*out*), df(*in*), df(*ex*) and df(*ind*).

But the modified theory now adds:

df(n-subtitution)
$p/a_1, \ldots, a_n;b_1, \ldots, b_n!q = \text{df}$
$(\exists x_1, \ldots, x_{n-1}, y_1, \ldots, y_{n-1}, z_1, \ldots, z_{n-1})(p/a_1;x_1!y_1 \ .\&.\ y_1/a_2;x_2!y_2$
$.\&., \ldots, .\&.\ y_{n-2}/a_{n-1};x_{n-1}!y_{n-1} \ :\&:\ y_{n-1}/a_n;b_n!z_1 \ .\&.$
$z_1/x_{n-1};b_{n-1}!z_2 \ .\&., \ldots, .\&.\ z_{n-1}/y_1;b_1!q)$
$*1906(\text{A}_{19})$
$(\exists p, a_1, \ldots, a_n)(b_1, \ldots, b_n)(\exists q)(p/a_1, \ldots, a_n;b_1, \ldots, b_n!q \ .\&.\ q \equiv A)$,
where $p, a_1, \ldots, a_n$ are not free in A, and $b_1, \ldots, b_n$ are free in A.

we shall call this "$*1906(\text{Reduc})^n$". This is a new comprehension axiom schema for "Les paradoxes." In the early system, it has an analog which is a theorem schema. The need for its addition as a fundamental axiom schema will be discussed later.

The method of proof for the system of "Les paradoxes" need not be different from that in Chapter 4, since the fundamental axiom schema of the quantification theory of Chapter 4 can be established. For any wffs A and B in prenex normal form, we have:

$(\mu)A\mu \supset A\alpha|\mu$,

where α is free for μ in A

$(\mu)(B \supset A\mu) \ .\supset. \ B \supset (\mu)A\mu$,

where μ is not free in B.

$A \ .\supset. \ B \supset A$

$A \ .\supset. \ B \supset C :\supset: A \supset B \ .\supset. A \supset C$

$\sim B \supset \sim A \ .\supset. \ A \supset B.$

That is, by employing strong mathematical induction on the length of a wff, the calculus can be shown to be adequate for quantification theory.[16] The adequacy of "Les paradoxes" as a quantification theory can only be shown by an inductive proof. Of course, Russell could not have proceeded in this way. As we saw in Chapter 1, to use (without explanation) mathematical induction might appear out of sorts with Russell's plan of demonstrating logicism. Though one cannot escape *using* logic in setting out any calculus for logic, it would not do to *use* mathematical induction—something that would be regarded as a uniquely mathematical (nonlogical principle). As in *Principia*, Russell has to be content to reveal (by examples) that by "repeated application of the process" one can see that we shall be able to get the proof in any particular case.

It should be noted that our definitions dfs(P)x differ from those presented in *Principia*'s ∗9. In fact, *Principia* only presented analogs of ∗9.07 and ∗9.08 and not also our ∗9.ZZ and ∗9.WW. Curiously, Whitehead and Russell write in their introduction (*PM*, p. 47):

> We can define a disjunction of two universal propositions: "$(x)\phi x \vee (y)\psi y$" will be defined as meaning "$(x, y)(\psi x \vee \psi y)$"[17]

Modifying matters for "$\supset$" as a primitive, this would be our ∗9.ZZ. But it never occurs. Something is amiss with *Principia*'s ∗9. What is the explanation of the omission?

If we look to *Principia*'s proof of ∗9.31 we can glean some hints. Unique among all the proofs found in ∗9, this one misapplies the definitions ∗9.03, ∗9.05, and ∗9.06. Adjusted to the primitives "$\supset$" and "$\sim$", and letting A be quantifier-free, Whitehead and Russell have:

∗9.31 $\vdash (\exists x)Ax \vee (\exists y)Ay \ .\supset. \ (\exists z)Az$

Proof:

1. $\sim Ax \supset Ay \ .\supset. \ (\exists z)Az$ ∗9.11
2. $(y)(\sim Ax \supset Ay \ .\supset. \ (\exists z)Az)$ 1, ∗9.13
3. $(\exists y)(\sim Ax \supset Ay) \supset (\exists z)Az$ 2, ∗9.03
4. $(x)((\exists y)(\sim Ax \supset Ay) \supset (\exists z)Az)$ 3, ∗9.13
5. $(\exists x)(\exists y)(\sim Ax \supset Ay) \supset (\exists z)Az$ 4, ∗9.03
6. $(x) \sim Ax \supset (\exists y)Ay \ .\supset. \ (\exists z)Az$ 5, ∗9.05, ∗9.06
7. $\sim (\exists x)Ax \supset (\exists y)Ay \ .\supset. \ (\exists z)Az$ 6, ∗9.02

16. See Landini (1998).
17. The letters "ϕ" and "ψ" are used schematically here. We use "A" and "B" as usual.

8. $(\exists x)Ax \vee (\exists y)Ay$.⊃. $(\exists z)Az$ df$(\vee)^2$.

Evidently, Whitehead and Russell thought that ∗9.ZZ and ∗9.WW are not needed because the definitions dfs(*P*)p cover such cases. But this is a slip. They apply for quantifier-free α. The definition ∗9.03 does not warrant the move from line #2 to line #3. Now one can avoid the problem at line #3 as follows:

2a. $(y)(\exists z)(\sim Ax \supset Ay$.⊃. $Az)$ 2, ∗9.06
3. $(\exists y)(\sim Ax \supset Ay) \supset (\exists z)Az$ 2a, ∗9.07

But one cannot similarly correct lines #5 and #6. Whitehead and Russell's definitions ∗9.07 and ∗9.08 require that quantifiers be moved in *pairs*. We might try skipping line #3 and derive the following

2b. $(x)(y)(\exists z)(\sim Ax \supset Ay$.⊃. $Az)$ 2a, ∗9.13
2c. $(x)(y)(\sim Ax \supset Ay$.⊃. $(\exists z)Az$ 2b, ∗9.06

But we are now stuck. No definitions allow us to proceed. Adding our ∗9.ZZ and ∗9.WW to the system helps, but it will not be enough to recover the proof. We need the inference rule (Switch), which does not occur in *Principia*.[18] The proof of *Principia*'s ∗9.31 (for quantifier-free *A*) is then straightforward:

∗9.31 ⊢ $Ax \vee Ay$.⊃. $(\exists z)Az$

Proof:
1. $\sim Ax \supset Ay$.⊃. $(\exists z)Az$ ∗9.11
2. $(\exists z)(\sim Az \supset Ay$.⊃. $Az)$ 1, ∗9.06.
3. $(y)(\exists z)(\sim Ax \supset Ay$.⊃. $Az)$ 2, ∗9.13
4. $(\exists z)(y)(\sim Ax \supset Ay$.⊃. $Az)$ 3, Switch.
5. $(\exists z)((\exists y)(\sim Ax \supset Ay)$.⊃. $Az)$ 4, ∗9.03.
6. $(x)(\exists z)((\exists y)(\sim Ax \supset Ay) \supset Az)$ 5, ∗9.13
7. $(\exists x)(\exists y)(\sim Ax \supset Ay) \supset (\exists z)Az$ 6, ∗9.07
8. $(x) \sim Ax \supset (\exists y)Ay$.⊃. $(\exists z)Az$ 7, ∗9.WW
9. $\sim (\exists x)Ax \supset (\exists y)Ay$.⊃. $(\exists z)Az$ 5, ∗9.02
10. $(\exists x)Ax \vee (\exists y)Ay$.⊃. $(\exists z)Az$ df$(\vee)^2$.

Of course, one would eventually need a proof for any wff *A*, quantifier-free or otherwise.

Some of the proofs of the theorems of Chapter 4 must be altered, because of their use of singular terms made by nominalizing wffs (e.g., the use of @ and *f* must go). But (excepting those for general propositions) they are forthcoming.

The Statement Liar

8.5 A rigorous treatment of a logistic calculus requires the formal distinction between syntax and semantics. We have seen that Russell's conception of logic

18. Such a rule does occur in Appendix A of the 1925 second edition.

is not incompatible with the distinction. Russell's conception of logic does, however, require that any *admissible* predicate may be added to the language of the pure theory. This is fundamental in what follows.

As we saw, the tilde sign is a primitive of the calculus of "Les paradoxes de la logique." In the intended interpretation the tilde stands for the property of nontruth, and for propositions nontruth is just the property of falsehood. But it is the tilde sign, not ". . . is false" that is part of the formal system. The predicates ". . . is true" and ". . . is false" are not formally a part of calculus for substitutional logic.

"Les paradoxes" does, nonetheless, discuss a hierarchy of senses of "truth" and "falsehood." This gives rise to certain questions of interpretation. Is Russell exploring the consequence of adding "truth" and "falsehood" predicates to the formal language? If so, do the object-language "truth" and "falsehood" predicates flank statement names? Or does he want object-language predicates that flank statements by means of contextual definitions akin to the dfs(P) of the last section? Or, perhaps he wants to avoid object-langage predicates altogether. It may be that he is simply rendering an informal semantics, endeavoring to explain the "truth" and "falsehood" conditions of general wffs as opposed to quantifier-free wffs.

Consider the following passage in which Russell offers a hierarchy of senses (let us call them "orders") of "truth" and "falsehood" (*InS*, p. 208):

> In the restricted sense which we have given to *proposition*, the law of excluded middle is not a proposition, since it contains an apparent variable. It is a true *statement*; but *true* here has a different meaning, namely that all the propositions which the statement ambiguously denotes are true (in the previous sense). As applied to *statements* the meaning of the word *true* varies as the number of apparent variables in the statements varies.

What Russell has in mind here is a semantic explanation of the truth conditions of quantified wffs. He may well have felt he owed his readers this. While there were general propositions, there were primitive properties of 'truth' and 'falsehood'—as the redness and whiteness of roses. But with general propositions abandoned, "truth" (as applied to a general wff) is now to be defined.

But in what sense is the "truth" predicate to be "applied" to a general wff? On one reading, the subject position of a given truth-predicate can be occupied by a quantified wff (rather than its name) via meta-linguistic contextual definitions akin to dfs(P). Put,

$$T^{\circ}A = \text{df } T\{A\},$$

where A is quantifier-free; and then put:

$$T^{n}(Q_1x_1), \ldots, (Q_nx_n)A = \text{df } (Q_1x_1)T^{n-1}(Q_2x_2), \ldots, (Q_nx_n)A,$$

where $n \geq 1$. It follows that,

$$T^{n}(Q_1x_1), \ldots, (Q_nx_n)A = \text{df } (Q_1x_1), \ldots, (Q_nx_n)\ T\{A\},$$

where A is an elementary wff. Accompanying predicates "F^n" would follow the same pattern.

The trouble with this interpretation is that the hierarchy of new "truth" predicates does nothing. We may as well have:

$$T(Q_1 x_1), \ldots, (Q_n x_n)A = \text{df } (Q_1 x_1), \ldots, (Q_n x_n)T\{A\}.$$

Moreover, Russell suggests that he wants to address the Statement Liar (not to be confused with the propositional *liar*):

"I now assert a false statement."

The following passage seems to find Russell discussing the Statement Liar (*InS*, p. 207):

> He [the Liar] cannot mean: "I am now making a statement which is false", because there is no way of speaking of statements *in general*: we can speak of statements of propositions, or statements containing one, two, three, ..., apparent variables, but not of statements in general.

The Statement Liar quantifies over statements and thus requires that "falsehood" flanks statement names rather than statements themselves. This suggests that the above interpretation cannot be correct.

Let us then construe truth and falsehood predicates as flanking statement names. On this view, the truth-definitions Russell intends would be as follows: Let "$E(\mu_1, \ldots \mu_n)$" be a meta-linguistic expression for quantifier-free wffs containing the variables $\mu_1, \ldots, \mu_n$ free. Next put:

$\ulcorner E(\mu_1, \ldots, \mu_n)\urcorner$ is true$_o$ = df $\ulcorner\{E(\mu_1, \ldots, \mu_n)\}\urcorner$ names a true proposition.
$\ulcorner E(\mu_1, \ldots, \mu_n)\urcorner$ is false$_o$ = df $\ulcorner\{E(\mu_1, \ldots, \mu_n)\}\urcorner$ names a false proposition.
A is false$_n$ = df A is not true$_n$.

The truth conditions for quantified wffs (in prenex normal form) are next given by means of the following schema:

$$\ulcorner(Q_1 \nu_1), \ldots, (Q_n \nu_n)\ E(\nu_1|\mu_1, \ldots, \nu_n|\mu_n)\urcorner \text{ is true}_n \ .\equiv.$$
$$(Q_1 \nu_1), \ldots, (Q_n \nu_n)\ E(\nu_1|\mu_1, \ldots, \nu_n|\mu_n),$$

for $n > 0$. (We allow no vacuous quantifiers.)

With this approach, we get a means of rendering the truth and falsehood conditions of the wffs of the object language. The "truth" and "falsehood" predicates are not themselves part of the object language of the pure substitutional calculus. They might, however, be added without producing the Statement Liar. On Russell's view, we cannot quantify over statements in general. His implicit reason seems to be that a "statement" must be capable of truth$_n$ or falsehood$_n$ for some fixed number n. Quantification over statements (and also the assertion of their truth$_n$ or falsehood$_n$) is a construct, and the construct fixes the number of bound variables of the statement quantified over. Russell explains (*InS*, p. 207):

> If we want say what is equivalent to "I am making a false statement containing n apparent variables", we must say something like: "There is a propositional function $\phi(x_1, x_2, \ldots, x_n)$ such that I assert that $\phi(x_1, x_2, \ldots, x_n)$ is true for any values of $x_1, x_2, \ldots, x_n$ and this is in fact false." This statement contains $n+1$ apparent variables, namely, $x_1, x_2, \ldots, x_n$ and ϕ. Hence, it does not apply to itself.

This passage is far from clear; and it has been criticized for reverting to bound predicate variables in a paper advocating a the nonassumption of attributes (propositional functions in an ontological sense).[19] What, then, was Russell's treatment of the Statement Liar?

Likely, Russell is using the expression "propositional function" to mean an open wff with n-many free variables. His point is that to affect quantification over statements, we have to write something like:

> There is some elementary wff $\ulcorner E(\mu_1, \ldots, \mu_n)\urcorner$ and some quantifiers $(Q_1\nu_1), \ldots, (Q_n\nu_n)$ and I now assert that $\ulcorner(Q_1\nu_1), \ldots, (Q_n\nu_n)$ $E(\nu_1|\mu_1, \ldots, \nu_n|\mu_n)\urcorner$ is true$_n$ and $\ulcorner(Q_1\nu_1), \ldots, (Q_n\nu_n)$ $E(\nu_1|\mu_1, \ldots, \nu_n|\mu_n)\urcorner$ is false$_n$.

This interpretation is consonant with the truth-definitions (above). Accordingly, an expression such as

> "I now assert a false statement"

is ill-formed. We have to construct quantification over statements. When done, we arrive at:

> "I now assert a false$_n$ statement."

where n is fixed. But this statement does not contain exactly n-many bound variables. The Statement Liar is solved.

In sum, "truth" and "falsehood" are not to be object language predicates of the formal language of substitution. But they can be added if one recognizes that they are not univocal in meaning. The predicate "truth$_n$" is defined so that it can only meaningfully flank a statement name with n-bound variables. Addition does yield a new and richer object language. Unlike the Tarski hierarchy, however, we do not get an infinite hierarchy of languages. We can add them all and at once. That is, for every numeral "n" we have "truth$_n$" and "falsehood$_n$".[20]

19. See, for example, Rodriguez-Consuegra (1987).

20. Hylton thinks we can add "truth$_x$" and "false$_x$", where "x" is a variable (Hylton 1990, p. 313). This is mistaken. Hylton's predicates are not characterizable by Russell's definitions.

The König, Dixon, Berry, Richard, and Grelling Paradoxes

8.6 As we see, Russell's solution of the Statement Liar does not involve a Tarski-like relativization of "truth" to a given language. There are, however, other non-propositional paradoxes which Russell solves by means of a Tarski-like relativization. Their presence in "Les paradoxes" is something of a curiosity as we shall see.

The König paradox was introduced in 1905 and involves putting two facts together. The number of possible definitions (names) in any language is denumerable. At the same time, Cantor has shown that the real numbers are nondenumerable. Now if the reals can be well-ordered, there must be a first nonnameable ("nondefinable") real number. Paradoxically, however, we seem to have just named it.

Dixon (1906) discovered this paradox independently of König. Dixon considers the expression "the least (real) number not definable without using more that *n* symbols." This names a unique number, yet for some numericals "*n*", the name involves more than *n* symbols.

Jules Richard articulated a similar paradox in 1905. Only denumerably many real numbers can be defined (named) with a finite number of words of a given language. Richard therefore considers the class *E* of all real numbers definable in a finite number of words. The members of *E* can be placed in serial order. Using Cantor's diagonal method, we can find a real number *m* which is not in this serial order. Nonetheless, the method allows us to define (name) *m* in a finite number of words. Accordingly, *m* is a member of *E*. Yet it cannot be a member of *E*, for it is not in the serial order.

Finally, there is the paradox which Russell credits to G. G. Berry of the Bodleian Library at Oxford (*InS*, p. 210, fn2). The Berry paradox considers the fact that there are only finitely many English syllables; hence only finitely many natural numbers each of which is specifiable in fewer than twenty-four syllables; and hence a least natural number not specifiable in fewer than twenty-four syllables. But it has just been specified in twenty-three syllables.[21]

Russell approaches the solution of all these paradoxes in the same way. The König, Dixon, and Berry paradoxes have to do with "nameability," "definability," "specifiability," and the like. These are not predicates of Russell's implicit object language. But any might be added together with appropriate meaning postulates. Russell's solution is to show that there is no univocal notion of "definability" (or "nameability" or "specifiability") to add. The notion is meaningful only if relativized to a particular language's set of primitive symbols.[22] Russell writes (*SCR*, p. 185):

> *Definable* is relative to some given set of fundamental notions, and if we call this set of fundamental notions *I*, 'definable in terms of *I*' is never itself definable in terms of *I*.

21. This formulation is taken from Quine (1963, p. 255). Russell takes the phrase to be "the least integer not nameable in fewer than eighteen syllables."

22. See also *InS*, p. 209.

If we add a predicate "definable in terms of I" to the terms of I, we generate a new class of terms I'. This solves the König and Dixon paradoxes. In the latter, the name "the least (real) number not definable in I without using more than n symbols" necessarily involves terms not in I. Richard's paradox is solved in the same manner. There is no univocal meaning of "definable in a finite number of words." If we rephrase with "definable$_n$ in a finite number of words" the paradox collapses. The general point is that one can add to a given language L the predicate "definable$_n$," where "n" indexes the class of terms I of L. But "definable$_n$" is never a term of n. And it cannot be. Even the finite union Σ of languages L_i each got by adding a predicate "definable$_i$" to L_i to make L_{i+1} does not contain the predicate "definable$_\Sigma$."[23]

It should be noted that Russell's solution has nothing directly to do with his solution in "Les Paradoxes" of the propositional paradoxes. That solution was simply to recover logic (and substitutional logic, in particular) without general propositions. In this respect the König, Dixon, Berry, and Richard paradoxes may seem irrelevant to the question of whether the logicist program can be salvaged by means of the substitutional theory.

Indeed, Russell's discussion of such paradoxes in substitution seems to be something of an anomaly. One might well wonder whether the König, Dixon, and Richard paradoxes are even formulable within the substitutional theory. The paradoxes require substantial reformulation. Real numbers (as with all numbers) are proxied, and neither classes nor propositional functions are entities. There are, therefore, no genuine names of such. To formulate the paradoxes, one would need to restructure them so as to concern propositions. Consider the loosely put phrase: "there is a unique proposition for each real number." If this can be given an appropriate expression in the language of substitution, then by Cantor's power-class theorem there would be more propositions than names of them. Well-ordering the propositions, a paradox might then be generated by considering the first nonnameable proposition. This proposition would seem to be nonnameable and all the same has a name.

But is there any sound argument that there are nondenumerably many propositions? Is there any diagonal argument to show this? To be sure, Cantor's power-class theorem and the nondenumerability of the class of all subclasses of the finite cardinals is proxied in the substitutional theory. So there is a proxy for "there are more classes of entities than there are entities." The proper formulation of this, however, relies upon the meaning of "similarity" in the heterogeneous case of classes of entities (individuals) and classes of classes (of some type). The formulations are consistent with there being only denumerably many entities (propositions, or otherwise). If there are, indeed, only denumer-

23. Observe again that "n" is not a variable in "definable$_n$." No language in the range of "n" contains a formula "x is definable$_n$, for some language n." This shows, incidentally, that Hylton (1990, p. 313) is mistaken in thinking that Russell could not have embraced Tarski's solution of the Statement Liar—a solution which abandons a univocal "truth" predicate for an infinite hierarchy of "true in L" predicates. In "Les paradoxes", Russell adopted precisely this strategy as a solution to the König, Dixon, Berry, and Richard Paradoxes.

ably many entities, then the ontological austerity of substitution already provides an escape from the König, Dixon, and Richard paradoxes. Russell's view that "definability" and "nameability" are not univocal in meaning would not be needed.

The Berry paradox does not succumb so easily, however. It does not depend upon a nondenumerable realm. Of course, integers are not entities. "Least integer" will be a phrase which abbreviates its formal proxy. Removing defined signs, the phrase "the least integer not nameable in fewer than twenty-four syllables" will involve greater than twenty-four syllables. But we can simply increase the number of syllables appropriately to generate the paradox. It may be that this was the reason Russell chose not to rely upon the ontological austerity of substitution as a solution of the König, Dixon, and Richard paradoxes.

Moreover, there is a paradox that Russell did not discuss which shows that if "names," "designates," and the like are univocal, then there are nondenumerably many *entities*. This is the paradox published by Kurt Grelling in 1907. In its popular formulation, the paradox begins by observing that some words designate attributes that do not belong to the words themselves. "Long" is not a long word; "German" is not German. Call such words "heterological":

x is heterological = df
x is a word which designates an attribute that does not belong to it.

What of the word "heterological"? We find that "heterological" is heterological if and only if it is not. The Grelling paradox is not formulable in the substitutional theory for the simple reason that it presupposes a relation 'designates' between expressions and an ontology of attributes. We need to find a substitutional formulation. Add a primitive predicate constant "$x\,D\,y$" (for "x designates y") to the language of substitution, with the axiom:

$$(x)(y)(x\,D\,y\ .\&.\ x\,D\,z\ :\supset:\ y = z).$$

Add as well denumerably many primitive individual constants "♀", "♀°", "♀°°", and so on. Now presumably we have:

$$(\exists pa)(x)(p/a;x \equiv (\exists rc)(xD\{r \supset c\}\ .\&.\ {\sim}\,(r/c;x))).$$

So existentially instantiate and assume,

$$♀D\{p \supset a\}.$$

We get:

$$♀D\{p \supset a\}\ :\supset:$$
$$(\exists rc)(♀D\{r \supset c\}\ .\&.\ {\sim}\,(r/c;♀))) \equiv{\sim}\,(\exists rc)(♀D\{r \supset c\}\ .\&.\ {\sim}\,(r/c;♀))).$$

Accordingly, we know that

$${\sim}\,(♀D\{p \supset a\}).$$

The lesson here is that if "designates" is univocal in meaning, the Grelling construction will show that our assumption that there is a constant ♀ as above is false. Since we added denumerably many constants to the language, the Grelling construction shows that if "designates" is univocal, then there are nondenumerably many propositions. There will be propositions that have no names in the language in question.[24] So it appears that the nondenumerability of propositions blocks the formulation of a contradiction by means of the Grelling construction. But the price is that one cannot rely upon a claim that there are only denumerability many entities to block the other paradoxes. Russell will now need the nonunivocality thesis to block the König, Dixon, and Richard paradoxes. One way or the other, the non-univocity thesis is needed.

Russell's "Mitigating Axiom"

8.7 As we have seen, Russell's plan in "Les paradoxes" is to solve all paradoxes of propositions with the abandonment of general propositions. The plan immediately blocks the *propositional liar*. It blocks the (*Principles*) Appendix B paradox because there is no singular term "$\{(z)(r/c;z \ .\supset.\ z)\}$." The expression

$$\text{“}(\exists rc)(x = \{(z)(r/c;z \ .\supset.\ z)\} \ .\&.\ {\sim}(r/c;x))\text{”}$$

is no longer well-formed. The p_o/a_o paradox (and variants) are also blocked. In this case the expression,

$$\text{“}(\exists rc)(x = \{r/c;b!q\} \ .\&.\ {\sim}(r/c;x)),\text{”}$$

is well-formed. Even without general propositions, "$\{r/c;b!q\}$" is a genuine singular term. But the paradox does not go through because the existence theorem

$$\vdash \ (\exists pa)(x)(p/a;x \ .\equiv.\ \{(\exists rc)(x = \{r/c;b!q\} \ .\&.\ {\sim}(r/c;x))\})$$

is no longer provable.

All three paradoxes of propositions are solved. But the price is high. The abandonment of general propositions destroys the existence theorems essential to the substitutional proxy of the theory of propositional functions. The theory of classes is dutifully affected. Mathematics is lost. To see this, consider again our axiom schema

$$*(r) \quad (\exists\mu)(\alpha \text{ out } \{A\mu|\nu\}) \ .\supset.\ (\mu)(\{A\alpha|\nu\}/\alpha;\mu!\{A\mu|\nu\}),$$

where α and μ are free for ν in A. With the abandonment of general propositions, the expression $\{A\mu|\nu\}$ is not well-formed unless *A contains no quantifier symbols*. Accordingly, the axiom schema will now only yield:

$*1906(\mathrm{CP_{sub}})^1$

$$\vdash \ (\exists pa)(a \ in \ p \ .\&.\ (x)(p/a;x!\{A\})),$$

24. See Myhill (1979, p. 88).

where p, a not free in A, and A is any quantifier free wff containing x free. We also have the comprehension principle:

$$*1906(\text{CP})^1, \quad (\exists pa)(x)(p/a;x \;.\!\equiv\!.\; \{A\}),$$

where p, a are not free in A, and A is any quantifier free wff containing x free. This means that we cannot now prove, say,

$$(\exists pa)(x)(p/a;x \;.\!\equiv\!.\; (z)Cxz).$$

With the definite description eliminated, we have:

$$(\exists pa)(x)(\exists q)(p/a;x!q \;.\&.\; q \equiv (z)Cxz).$$

Instantiating to p, a, m (respectively) our q here is such that

$$q \equiv (z)Cmz.$$

The clause $(z)Cmz$ occurs in a subject position here. Applying our dfs(P), we arrive at,

$$(\exists y)(z)(Cmy \supset q \;.\&.\; q \supset Cmz).$$

Without general propositions we are not assured that there is some one such q.

The same problem occurs at higher types. Without general propositions we have only,

$$*1906(\text{CP})^2 \quad \vdash (\exists stw)(rc)(s/tw;rc \;.\!\equiv\!.\; \{A\}).$$

where s, t, w are not free in A and r, c are free in A.

$$*1906(\text{CP})^3 \quad \vdash (\exists jkuv)(qpa)(k/kuv;qpa \;.\!\equiv\!.\; \{A\}),$$

where j, k, u, v are not free in A and q, p, a are free in A, and so on. This means that we no longer have:

$$*(\text{CP}_\in)^1 \quad \vdash (\exists stw)(rc)(s/tw \approx_{rc} A(rc))$$
$$((\text{CP}_\in)^2 \quad \vdash (\exists jkuv)(qpa)j/kuv \approx_{qpa} A(qpa)),$$

and so forth. The natural numbers are not formulable. For instance, we cannot arrive at the existence theorem:

$$\vdash (\exists stw)(rc)(s/tw \approx_{rc} 0(rc)).$$

Mathematical induction is woefully enfeebled as well. The only propositional functions it can now offer for induction will be ones that are proxied by using matrices whose prototypes name propositions (i.e., elementary propositions).

Observe that it is possible to proxy quantification over *general propositions* (in truth-functional contexts). We can put:

$$(p_1)Ap_1|\nu = \text{df}$$
$$(r, c)[A(x)(\exists q)(r/c;x!q \;\&\; q)|v \;.\&.\; A(\exists x)(\exists q)(r/c;x!q \;\&\; q)|\nu],$$

$$(\exists p_1)Ap_1|\nu = \text{df}$$
$$(\exists r, c)[A(x)(\exists q)(r/c;x!q \ \&\ q)|\nu \ .\vee.\ A(\exists x)(\exists q)(r/c;x!q \ \&\ q)|\nu],$$

where $A\nu$ is any truth-functional context. This proxies what would be quantification over propositions containing exactly one quantifier. Since no propositions contain quantifiers, we have the appropriate limitations built into the logical form of the proxy. The process could be continued. Moreover, the following theorem schema is valid:

$$\vdash \quad (p)Ap|\nu \ .\supset.\ (p_n)Ap_n|\nu,$$

where $A\nu$ is any truth-functional context. But this will not help with existence theorems and does not, therefore, help to restore mathematical induction.

To offset the consequences of the abandonment of general propositions, "Les paradoxes de la logique" introduces what Russell calls a "mitigating axiom" (*InS*, p. 212):

> Every statement containing x and an apparent variable is *equivalent*, for all values of x, to some statement ϕx containing no apparent variable.

Russell hoped that since mathematics involves only extensional contexts his "mitigating axiom" would solve the problem. He explains (*ibid.*):

> In most of the sort of statements we wish to make in mathematics, if a proposition p occurs otherwise than in a matrix, it may be replaced by any other equivalent proposition without altering the truth or falsehood of our statement. And if a function ϕx occurs, where the argument x is an apparent variable, ϕ may usually be replaced by any other function ψ having the same extension, i.e., true for the same values of x. And the same holds of a matrix p/a.

As before, Russell uses "ϕx" schematically in these passages. His idea is clear enough. If for any wff $A\mu$ with free variable μ we can find a wff $B\mu$ which has a free variable μ, contains no quantifiers, and is such that $(\mu)(A\mu \ .\equiv.\ \{B\mu\})$, then $*1906(\text{CP}_{\text{sub}})^1$ would give all that is needed. Theorem schema $*1906(\text{CP})^1$ yields,

$$(\exists pa)(x)(p/a;x \ .\equiv.\ \{Bx\}).$$

Then from $(x)(Ax \ .\equiv.\ \{Bx\})$ one can easily arrive at:

$$(\exists pa)(x)(p/a;x \ .\equiv.\ Ax).$$

Here Ax, which may be a general wff, occurs in a subject position as is allowed by dfs(P). For extensional contexts, this restores what would have been possible by $*(\text{CP})^1$. The same holds for higher types by introducing new axiom schemata appropriate to two variables, three, and so forth.

The upshot[25] of Russell's idea of a "mitigating axiom" can be better rendered formally by means of the following axiom schema:

$*1906(\text{Reduc})^1$
$\vdash (\exists pa)(x)(\exists q)(p/a;x!q \ .\&.\ q \equiv A)$,

where p, a are not free in A, and A is any wff (general or otherwise) containing x free. Russell will need similar schemata for higher types. For instance, he needs:

$*1906(\text{Reduc})^2$
$(\exists qpa)(rc)(\exists t)(q/pa;rc!t \ .\&.\ t \equiv A)$,

where q, p, a are not free in A, and A is any wff containing r, c free. To introduce a general schema covering all types, we put:

df(n-substitition)
$$p/a_1, \ldots, a_n;b_1, \ldots, b_n!q = \text{df}$$
$$(\exists x_1, \ldots, x_{n-1}, y_1, \ldots, y_{n-1}, z_1, \ldots, z_{n-1})(p/a_1;x_1!y_1 \ .\&.\ y_1/a_2;x_2!y_2 \ .\&., \ldots, .\&.\ y_{n-2}/a_{n-1};x_{n-1}!y_{n-1} \ :\&:\ y_{n-1}/a_n;b_n!z_1 \ .\&.\ z_1/x_{n-1};b_{n-1}!z_2 \ .\&., \ldots, .\&.\ z_{n-1}/y_1;b_1!q)$$

$*1906(\text{Reduc})^n$
$$(\exists p, a_1, \ldots, a_n)(b_1, \ldots, b_n)(\exists q)(p/a_1, \ldots, a_n;b_1, \ldots, b_n!q \ .\&.\ q \equiv A),$$
$p, a_1, \ldots, a_n$ are not free in A, and $b_1, \ldots b_n$ are free in A.

No general schema is needed in the original substitutional system set out in our Chapter 4. Any instance would always be derivable whenever needed. In "'Les paradoxes," however, the schema has to be taken as a fundamental principle.

The 1906 axiom schema of reducibility nullifies the effect of the abandonment of generalized propositions where extensional contexts are concerned. All the original theorems and theorem schema appropriate to the generation of arithmetic, the proofs of the Peano postulates, etc., are restored. The *liar* (qua paradox of propositions) and the Appendix B paradoxes cannot recur. The reason is that the formulations of these paradoxes involve the expression of an identity with a general proposition. This context is intensional, and we can find no equivalent in which the quantifiers are in an assertoric position.[26]

25. Russell's formulation of his "mitigating axiom" is problematic. While there may be an appropriate proposition, there may be no quantifier-free wff $B\mu$. Formulae have to be finite in length, but propositions may have infinitely many constituents.

26. Hylton (1980) missed this point. Formulating the Appendix B paradox of *Principles* in substitutional terms, Hylton says that "Les paradoxes de la logique" avoids the paradox by introducing a hierarchy of propositions based on the range of their apparent variables—Russell's abandonment of general propositions notwithstanding. Though mistaken about "Les paradoxes," Hylton is correct that such a hierarchy was soon to be Russell's move. I shall show that in citing the substitutional theory in the 1907 "Mathematical Logic as Based on the Theory of Types," Russell intended just this.

8.8 As we see, the system of "Les paradoxes" requires new axiom schemata in order to recover mathematics. It does not, however, require a special axiom to arrive at the infinity of the universal class. Ontologically austere as the system is, there is yet a store of nonpropositions whose subsistence is purely logical—for example, the property 'negation' and the relation 'implies.' Universals have both an individual and a predicable nature. Accordingly, $\{\supset\supset\supset\}$ is a proposition. It is a true one at that!

To see that there is an infinity of entities, one need only observe that there is an infinity of nongeneral propositions. In "Les paradoxes" Russell says that he makes an assumption of at least one individual and also assumes the following (*ibid.*, p. 202):

$$(x)(\exists y)(y\ out\ x)$$

These assumptions were part of the original substitutional theory as well. They offer no new troubles. In virtue of them, Russell argues that given an entity a there is a proposition '$a = a$'. Now by axiom there is an entity u not in '$a = a$'. Since a is in '$a = a$' and u is not, $a \neq u$, so there are at least two entities. Continuing, there must be an entity b not in '$a = u$'. This entity must be neither a nor u, for both of these are in '$a = u$'. Russell concludes that "we can in this way show that, if n is any finite number, there are more than n entities, and by taking propositions into account we can manufacture $\aleph_0$ entities" (*InS*, p. 203).

Russell's basic idea for a proof of infinity can be recovered in our formulation of the substitutional theory. A formal proof that the universal class of individuals is infinite in Frege's sense will, however, rely upon $*1906(\text{Reduc})^n$. In our Appendix A the Peano postulates are deduced within substitution. The proof of the infinity of the universal class of entities used mathematical induction. If the universal class (of individuals) were finite and nonempty, there would be a single proposition p containing every entity as its constituent. This is impossible. Both p and $\{\sim p\}$ are individuals in the universal class, so we would have p *in* $\{\sim p\}$ and $\{\sim p\}$ *in* p, and from this $p = \{\sim p\}$.

The Demise of "Les Paradoxes"

8.9 The schemata for reducibility may at first appear questionable as logical truths. Help comes if we consider that there are propositions with infinitely many constituents. But some care is in order. Consider the following instance of $*1906(\text{Reduc})^1$:

$$(\exists p, a)(z)(\exists q)(p/a;z!q\ .\&.\ q \equiv (x)(z \supset x)).$$

It might at first be thought that we can find entities p and a, by letting p be the infinite conjunction,

$$\{a \supset a\ .\&.\ a \supset b\ .\&.\ a \supset c\ .\&., \ldots, .\&., \ldots, \},$$

with $a, b, c, \ldots$ etc., as all the entities. This would not work. The entity a may well occur in some of the entities $b, c, \ldots$ etc., so the outcome of substituting an entity for a in p wouldn't render a proposition q such that

$$q \equiv (x)(z \supset x).$$

A second problem is that if $a, b, c, \ldots,$ etc. include *all* the entities, then $\{a \supset p\}$ and $\{a \supset \sim p\}$ would also be constituents of p. We would have:

$$p = \{a \supset a \text{ .\&. } a \supset b \text{ .\&. } a \supset c \text{ .\&.}, \ldots, \text{.\&. } a \supset p \text{ .\&. } a \supset \{\sim p\} \text{ .\&.}, \ldots\}$$

This is impossible, for now we have p *in* $\{\sim p\}$ and $\{\sim p\}$ *in* p, from which $p = \{\sim p\}$ follows by $*1906(A_8)$. Seeing this, it appears that $*1906(\text{Reduc})^n$ fails. General propositions seem required to recover existence theorems.

Philosophically justifying reducibility schemata is a problem for the system of "Les paradoxes," but there are far worse problems. It turns out that reducibility schemata lead the system to contradiction. The p_o/a_o paradox (and variants) proved to be a major setback. Unlike the other paradoxes, the p_o/a_o (and variants) are revived by the reducibility schemata! In the original substitutional theory $*(\text{CP}_{\text{sub}})^1$ yields:

$$(\exists pa)(x)(p/a;x!\{(\exists rc)(x = \{r/c;b!q\} \text{ .\&. } \sim (r/c;x))\}).$$

This is not an instance of the new $*1906(\text{CP}_{\text{sub}})^1$. Nonetheless, $*1906(\text{Reduc})^1$ yields a contradiction as follows:

1. $(\exists pa)(x)(p/a;x \text{ .}{\equiv}\text{. } (\exists rc)(x = \{r/c;b!q\} \text{ .\&. } \sim (r/c;x)))$ $*1906(\text{Reduc})^1$.
2. $(x)(p/a;x \text{ .}{\equiv}\text{. } (\exists rc)(x = \{r/c;b!q\} \text{ .\&. } \sim (r/c;x)))$ 1, EI.
3. $p/a;\{p/a;b!q\} \text{ .}{\equiv}\text{.}$ $(\exists rc)(\{p/a;b!q\} = \{r/c;b!q\} \text{ .\&. } \sim (r/c;\{p/a;b!q\}))$ 2, ui.
4. $(\exists^1 t)(p/a;\{p/a;b!q\}!t \text{ \& } t) \text{ .}{\equiv}\text{.}$ $(\exists rc)(\{p/a;b!q\} = \{r/c;b!q\} \text{ .\&. } \sim (r/c;\{p/a;b!q\}))$ 3, $*1906(\text{scope})^1$
5. $\rightarrow (\exists^1 t)(p/a;\{p/a;b!q\}!t \text{ \& } t)$
6. $(\exists rc)(\{p/a;b!q\} = \{r/c;b!q\} \text{ .\&. } \sim (r/c;\{p/a;b!q\}))$ 4, 5, log.
7. $\{p/a;b!q\} = \{r/c;b!q\} \text{ .\&. } \sim (r/c;\{p/a;b!q\})$ 6, EI.
8. $\sim (r/c;\{p/a;b!q\})$ 7, simp.
9. $\{p/a;b!q\} = \{r/c;b!q\}$ 7, simp.
10. $p = r \text{ .\&. } a = c$ 9, $1906(T_{4.19})$b.
11. $\sim (p/a;\{p/a;b!q\})$ 8, 10, $*1906(=)$.
12. $\sim (\exists^1 t)(p/a;\{p/a;b!q\}!t \text{ \& } t)$ 11, $*1906(\text{scope})^1$.
13. $\underline{\quad} F$ 5, 12, conj, (7ei).
14. $5 \supset F$ cp.
15. $\sim (\exists^1 t)(p/a;\{p/a;b!q\}!t \text{ \& } t)$ 14, reductio.
16. $\sim (\exists rc)(\{p/a;b!q\} = \{r/c;b!q\} \text{ .\&. } \sim (r/c;\{p/a;b!q\}))$ 15, 4, simp, mp.
17. $\{p/a;b!q\} = \{p/a;b!q\} \text{ .}{\supset}\text{. } (p/a;\{p/a;b!q\})$ 16, ui, log.
18. $\{p/a;b!q\} = \{p/a;b!q\} \text{ .}{\supset}\text{. } (\exists^1 t)(p/a;\{p/a;b!q\}t \text{ \& } t)$ 17,$*1906(\text{scope})^1$.

19. $\{p/a;b!q\} = \{p/a;b!q\}$ 1906($T_{4.8}$)a, *ui*.
20. $(\exists^1 t)(p/a;\{p/a;b!q\}!t \ \& \ t)$ 18, 19, mp.
21. F 15, 20, conj, (2ei).

The bald fact is that with the addition of ∗1906(Reduc)[1] the system is inconsistent. Yet without it the system cannot recover mathematics. Russell is trapped. "Les paradoxes" was a failure.

CHAPTER NINE

Mathematical Logic as Based on the Theory of Types

9.0 In the last chapter we set out the system of "Les paradoxes de la logique"—the subtitutional theory without general propositions. We found that the system met its end with the problems introduced by the addition of $*$1906(Reduc)[1]. With this schemata, the p_o/a_o and its variants return; without it, the system could not construct arithmetic. This interpretation is corroborated in a letter Russell wrote to Hawtrey of 22 January 1907.

> Dear Hawtrey,
>
> I forgot to send you the paradox which pilled the substitutional theory. Here it is. Put
>
> $$p_o \;.\!=\!:\; (\exists p, a) \colon a_o \;.\!=\!.\; p\tfrac{b}{q}!q \colon \sim (p\tfrac{a_o}{a})$$
>
> . . . .
>
> In trying to avoid this paradox, I modified the substitutional theory in various ways, but the paradox always reappeared in more and more complicated forms.
>
> Yours ever B. Russell

Thus it is quite clear that it was the p_o/a_o and its variants, not the *liar* or Appendix B (*PoM*) paradoxes that pilled substitution.

To pill is to despoil, but not necessarily to kill. It would be mistaken to think that "Les paradoxes" was the last of substitution. The substitutional theory was indeed abandoned. But what comes after "Les paradoxes" is of utmost importance. The really exciting point is that it is the *survival* of the substitutional theory in the 1907 "Mathematical Logic as Based on the Theory of Types" that holds the secret to understanding the historical origins and development of *Principia*'s ramified type theory. Hidden in the dusty corners of unpublished manuscripts for some 67 years, the p_0/a_0 paradox (and its variants) was the cause of ramification.

Orders of Propositions

9.1 The appearance of "Mathematical Logic" in 1908 is often regarded as the beginning of a new turn for Russell—viz., the abandonment of the unrestricted variable and the ontological assumption of a ramified hierarchy of types of attributes ("propositional functions"). But "Mathematical Logic" confounds this interpretation, explicitly endorsing the substitutional theory (*ML*, p. 77):

> In practice, a hierarchy of *functions* is more convenient than one of propositions. Functions of various orders may be obtained from propositions of various orders by the method of *substitution*. If p is a proposition, and a a constituent of p, let '$p/a;x$' denote the proposition which results from substituting x for a wherever a occurs in p. Then p/a, which we call a *matrix*, may take the place of a function. . . . Similarly, if '$p/(a, b);(x, y)$' denotes the result of first substituting x for a and then substituting y for b we may use the double matrix $p/(a, b)$ to represent a double function. In this way we can avoid apparent variables other than individuals and propositions of various orders.

Russell's point is that a hierarchy of functions (i.e., an order and order/type indexed language of predicate variables) is more convenient than the substitutional language's order indexed entity variables. The substitutional language better explains the nature of types, but the consideration of being able to introduce ordinary class symbols favors the language of predicate variables. Russell explains (*ibid.*):

> Although it is *possible* to replace functions by matrices, and although this procedure introduces a certain simplicity into the explanation of types, it is technically inconvenient. Technically, it is convenient to replace the prototype p by ϕa, and to replace $p/a;x$ by ϕx; thus where, if matrices were being employed, p and a would appear as apparent variables, we now have ϕ as our apparent variable. In order that ϕ may be legitimate as an apparent variable, it is necessary that its values should be confined to propositions of some one type.

Russell favors using the convenient language of predicate variables, but this in no way undermines the fact that he endorses the substitutional theory. Moreover, we see that it is a new substitutional theory that is being endorsed—a version modified by a hierarchy of orders ("types") of propositions.

9.2 Our task is to uncover the nature of the new substitutional theory hidden behind the "convenient" language of predicate variables employed in "Mathematical Logic." To do this, we have to understand the nature of the

1907 hierarchy of orders of propositions and the effect they have on the substitutional proxy of the hierarchy of propositional functions.

"Mathematical Logic" embraces the hierarchy of orders of propositions that "Les paradoxes" sought so desperately to avoid. As in the original theory, all statements[1] are to have nominalizing transformations. But no proposition, elementary or otherwise, is of the "order" of what Russell now calls an "individual." Unlike propositions, Russell says, individuals are "destitute of complexity" (*ML*, p. 76). The "type" (i.e., order) of a general proposition is to be determined by the order of the entities in the admissible range of its bound variables (*ML*, p. 75). Elementary propositions and those involving ontological counterparts of bound variables for individuals are "first-order"; propositions involving ontological counterparts of bound variables for first-order propositions but no higher-order bound propositional variables are "second-order", and so on. Thus, only generalized propositions of order n are in the range of the propositional variables of quantification of order $n+1$.

Unfortunately, Russell leaves only hints of how the orders affect a proposition such as '$p/a;b!q$' in "Mathematical Logic." To understand these hints, we have to first turn to discussions of the matter in his unpublished work notes.

It is quite clear from "On Substitution$_2$," "The Paradox of the Liar," and "On Types" that orders of propositions, generated by the admissible range of their apparent variables, are to be reflected in what counts as a proper substitution. Only entities of the same complexity ("order") are substitutable for one another. No proposition may be admissibly substituted for an individual; no second-order proposition may be subtituted for a first-order, and so forth. The restriction on substitution allows only a proxy for a class of individuals, a class of all first-order propositions, a class of all second-order propositions, and so forth. Each of these is a legitimate totality—each an admissible range of a type ("order") of bound variable.

The matter of the order of '$p/a;b!q$', however, is more difficult. Russell explores a number of different approaches here. One plan in "The Paradox of the Liar" was to require that a and b must be of the same order, p and q must be same order, and the whole proposition '$p/a;b!q$' is to be of the order next above p. Russell writes (*PL*, p. 72):

> Thus, $p\frac{\beta}{\alpha}!q$ is always of higher order than any of its constituents; this disposes of the fallacy which led to the abandonment of substitution before, i.e.,
>
> $$\begin{aligned} p_0 \;.=:\; (\exists p, a)\colon a_0 \;.=.\; p\tfrac{b}{a}!q : \sim p\tfrac{a_0}{a} :. \\ \supset\colon p_0/a_0;\, p_0\tfrac{b}{a_0}!q : \sim (p_0/a_0;\, p_0\tfrac{b}{a_0}!q) \end{aligned}$$
>
> for here we substitute for a_0 the proposition $p_0\frac{b}{a_0}!q$, which is necessarily of higher grade than a_0.

He continues, following out this plan, a few pages later (*PL*, 77):

1. We exclude those involving statement connectives or vacuous quantifiers.

> The rule is: In $p\frac{b}{a}!q$, p and q are of higher order than b and a, and $p\frac{b}{a}!q$ is the next order above that of p and q.

This eliminates the offending substitution causing the paradox.

There are troubles in this rule. For instance, if "$(p_1)(p_1 \supset p_1)$" is to yield "$r_1 \supset c_o$.⊃. $r_1 \supset c_o$" by *universal instantiation*, then "$\{r_1 \supset c_o\}$" should get the order index "1". The general plan for orders is that the order of a term shall be the order of its highest-order constituent. But Russell's plan in "Paradox of the Liar" requires that "$\{r_1/c_o;z_o!q_1\}$" have the index "2." This seems arbitrary and out of sorts with the general idea. Indeed, *universal instantiation* cannot yield "$r_1/c_o;z_o!q_1$.⊃. $r_1/c_o;z_o!q_1$" unless "$\{r_1/c_o;z_o!q_1\}$ has order index "1."

A different suggestion tried in "Paradox of the Liar" was to impose type (order) restrictions on the axiom (p), viz.,

$$(p, a, x)(\exists q)(p/a;x!q \text{ .\&. } (r)(p/a;x!r \text{ .⊃. } r = q))$$

Russell wrote (*PL*, p. 82):

> There seems no reason, however, against holding that $p\frac{b}{a}!q$ is significant for *all* values of p, a, x, q, but that $E!(p\frac{x}{a})$ requires p to be a proposition and a to be of lower type than p, and x to be of the same type as a.

Later in the manuscript yet another suggestion is offered. Russell is clearly dissatisfied with the requirement that the order of '$p/a;b!q$' must be higher than any of its constituents. He wrote (*PL*, p. 84):

> Why not make the presence of a.v.'s [apparent variables] again the proper test of the order of a proposition? This seems really the fundamental thing. Then, though $p\frac{x}{a}!q$ is first-order prop[osition], $p\frac{x}{a}$ is not, since it involves an apparent q. . . . When we go on to make p or x a.v.'s in $p\frac{x}{a}$, we don't raise the type. But we raise the type if we go on to $q/(p,a)$.

This is an improvement. Unfortunately, Russell seems to have thought that the hierarchy of orders is generated not just by the occurrence of apparent variables, but by their assertoric occurrence. To be sure, α is subordinate in a context such as $(\alpha \vee \beta)$. But Russell thinks that the order index of the term $\{(p)\phi \vee \beta\}$ should be assigned in accordance with the index of $\{(p)(\phi \vee \beta)\}$. This would be achieved by employing prenex forms as definitions of subordinate occurrences *à la* "Les paradoxes de la logique." Russell soon observes, however, that this imposes a problem for, say, $\{\{(p)\phi p\}/a;z!q\}$, for the prototype occurs in an intensional context (*PL*, p. 86):

> We must hold, with this view of types, that
>
> $((p).\phi p)\frac{x}{a}$
>
> is prima facie meaningless. . . . But we may *define*
>
> $\{(p).\phi p\}\frac{x}{a}$.=. $(p).\phi\ (p\frac{x}{a})$. Df.
>
> But this is not quite right. . . . We want

$$a \text{ ex } p \;.\supset_p.\; (\phi p)\tfrac{x}{a}!(\psi p) \;:\supset:\; \{(p).\phi p\}\tfrac{x}{a}!\{(p).\psi p\}.$$

We may make a definition out of this if we like.
We shall need to know that

$$(p)\phi!p\tfrac{x}{a} \;.\supset.\; \phi!\{((p)\psi p)\tfrac{x}{a}\}.$$

This is now the reducibility requirement.
It rather looks as if the complications of the substitutional point of view were too great.

That Russell wanted the assertoric occurrence of apparent variables as the generating principle of the hierarchy of orders, shows how deeply he wanted a *philosophical* explanation of the hierarchy. But this plan returns him to the same failed approach of "Les paradoxes." If quantifiers must always occur in an assertoric position, there really are no general propositions.

"Paradox of the Liar" leaves matters undecided. Russell returned to the issue in writing "On Types." He now decides that the proper plan for a propositional hierarchy should adhere to the view that the method of generation is to be from the admissible ranges of bound variables, and he disassociates this from the view that the occurrence of the quantifiers must be assertoric. Reconsidering his decision in "Paradox of the Liar," he writes (*T*, p. 5):

> When this plan was tried in the MS "Paradox of the Liar" [PL], I took the view that $p^1\tfrac{x}{a}!q^1$ was a 2-nd order proposition. I think that this was a mistake. The order of a proposition must depend only upon the a.v.'s [apparent variables] it contains. If so, we have however a difficulty. We settled that in p/a, then p must be of higher order than the a. Therefore $(p\tfrac{x}{a}!q)\tfrac{r}{p}$ will be nonsense. Perhaps it doesn't matter. A similar argument applies to $p \supset q$. Thus $(p \supset q)\tfrac{r}{p}$ will be nonsense.
>
> I think it is a mistake to say that p must be of *higher* order than the a: all that is really necessary is that it should not be of *lower* order. Then p/p will be significant, and so will $(p \supset q)\tfrac{r}{p}$ and $(p\tfrac{x}{a}!q)\tfrac{r}{p}$.

On this view, $p_m/a_n;b_{n'}!q_{m'}$ is well-formed if $n = n'$, $m = m'$ and $m \geq n$. This plan seems more natural and, as Russell recognized, has a distinct advantage. It leaves elementary propositions in the order of individuals, and this allows a theorem of the infinity of individuals (*OT*, p. 3).

We are left with an important question: What was Russell's decision in "Mathematical Logic" concerning the conditions of well-formedness for the expression $\mu_m/\alpha_n;\beta_{n'}!\delta_{m'}$? There are two hints from "Mathematical Logic." The first is that elementary propositions are considered order 1 entities. Russell writes (*ML*, p. 76):

> Propositions which contain apparent variables are generated from such as do not contain these apparent variables by a process of *generalization*. . . . By applying the process of generalization to individuals occurring in elementary propositions, we obtain new propositions. The legitimacy of this process requires only that no individuals should be propositions. That this

> is so, is secured by the meaning we give to the word *individual.* We may define an individual as something devoid of complexity; it is then obviously not a proposition, since propositions are essentially complex. Hence in applying the process of generalization to individuals we run no risk in incurring reflexive fallacies.

The second hint comes when Russell discusses the difference between types and orders in the substitutional theory of 1907. Russell says (*ML*, p. 77):

> The order of a matrix does not determine its type; in the first place because it does not determine the number of arguments for which the others are to be substituted (i.e., whether the matrix is of the form p/a or $p/(a, b)$ or $p/(a, b, c)$, etc.); in the second place because if the prototype is more than the first order, the arguments may be either propositions or individuals.

Russell (sometimes) used the word "prototype" for the proposition in which the substitution is to take place rather than the wff indicating such a proposition. But the point becomes clear in "Mathematical Logic." For the expression $\mu_m/\alpha_n;\beta_{n'};\delta_{m'}$ to be well-formed, Russell wants $m > n$ as well as $m = m'$, $n = n'$.

But we do well to ask: What sort of "reflexive fallacy" could occur in having elementary propositions included among 0-order entities (i.e., individuals)? So long as first-order propositions are excluded from individuals no proposition is in the range of its own quantifiers. No paradox can arise here. Including elementary propositions among the order 0 entities seems innocuous and has a very large benefit—viz., a theorem of infinity, on the assumption of at least one individual. What then caused Russell to exclude elementary propositions from among individuals?

It might be thought that the p_o/a_o paradox itself demands the exclusion. Suppose we can have a wff $\mu_m/\alpha_n;\beta_n!\delta_m$ with $m \geq n$. Suppose as well that elementary propositions are included among order-0 entities. Consider the following variant of the p_o/a_o paradox:

$$(\exists p_1 a_o)(x_o)(p_1/a_o;x_o!\{(\exists r_o, c_o)(x_o = \{r_o \supset c_o\}_o .\&. \sim (r_o/c_o;x_o))\}_1).$$

Now we cannot instantiate "x_o" to $\{p_1 \supset a_o\}_1$." But with elementary propositions included among individuals, one might fear that there is a t_o and w_o such that $p_1/a_o \equiv t_o/w_o$. Then instantiating "x_o" to "$\{t_o \supset w_o\}_o$" the paradox is revived. This fear, it might be suggested, provides a motivation to exclude elementary propositions from order 0. Indeed, Russell had recently been burned in "Les paradoxes" by allowing elementary propositions among individuals.

Unfortunately, this will not do as an explanation, for an exactly similar situation arises at the next higher order. In general we have:

$$(\exists p_m a_n)(x_n)(p_m/a_n;x_n!\{(\exists r_n, c_n)(x_n = \{r_n \supset c_n\}_n .\&. \sim (r_n/c_n;x_n))\}_m).$$

We know that $m = n + 1$, and we cannot instantiate "x_n" to "$\{p_m \supset a_n\}_m$." But if it so happens that there is a t_n and w_n such that $t_n/w_n \equiv p_m/a_n$, we can instantiate "x_n" to "$\{t_n \supset w_n\}_n$" and the paradox returns. Obviously, Russell cannot now put order 1 entities in order 2! "On Types" demurs over the question. Shortly after proclaiming that the plan of $m > n$ of "Paradox of the Liar" was unnecessarily stringent, Russell wrote (*OT*, 6):

> I think better say that in $p\frac{x}{a}!q$ the p and q must be of *higher* order than the x and a, and that $p\frac{x}{a}!q$ is of the *same* order as p and q.

Russell does not say why this is "better." But he knew, in any case, that this stringent plan is complicated and that its distinctions seem without philosophical ground. Surely there is nothing inadmissible about a matrix such as "$\{p_1 \supset a_o\}_1/p_1$." Indeed, with the condition $m > n$, many of the proofs of the basic logic of substitution would certainly be hard to recover. Russell therefore wonders in "On Types" whether the added complications are worth the trouble. Substitution has many merits philosophically, he admits, but its basic logic is so complicated as to be hardly worth it.

Worries aside, Russell need not adopt the inchoate $m > n$ to block the paradoxes plaguing substitution. In the sequel we set out a substitutional logic with orders of propositions which avoids this and includes elementary propositions among individuals. We shall find in Chapter 10, that the exclusion of elementary propositions from individuals has a motivation outside of the problem of the p_o/a_o paradox. The exclusion is appropriate to Russell's abandonment of an ontology of propositions and his adoption of a correspondence theory of truth—a theory which was to provide a completely new philosophical explanation for orders.

Substitutional Logic *cum* Orders of Propositions

9.3 In what follows, we modify the substitutional logic of propositions. In the last section, we witnessed Russell's fears and vacillations concerning the application of orders to substitutional wffs. We suggested reasons for the conservative decision in "Mathematical Logic" which favors the view that the prototype must be greater than the argument, and the requirement that elementary propositions are order 1. This was costly since it introduces an arbitrariness into substitutional orders. We shall set out a version of the substitutional theory *cum* orders in a way that is closer to that suggested in (part of) the 1907 "On Types." Elementary propositions are included in order 0 and the only requirement on a substitutional wff $\mu_m/\alpha_n{:}\beta_{n'}!\delta_{m'}$ is that $n = n'$. This enables the hierarchy of orders of propositions to be grafted onto the substitutional theory in a reasonably natural way. Moreover, it preserves the theorem of infinity found in "On Types" but absent in "Mathematical Logic." Since "Mathematical Logic" was written in 1907 and published in *The American Journal of Mathematics* in 1908 after a long delay, we shall call this "the 1907 substitutional theory."

9.4 Formally, orders require that what used to be called "individual" variables are now called "entity variables" for each order of entity. Let us employ order indices in the object language. Any lower-case letter of the English alphabet with a subscripted "o" is an order 0-order entity variable. Thus "x_o", "y_o", "z_o", "a_o", "b_o" . . . "p_o", "q_o", "r_o" (together with any followed by one or more superscripts "$'$" so that we cannot run out) are to be order-0 entity variables. This is the order of "individuals," and we are including non-general propositions among individuals here. Next any lower-case letter of the English alphabet with a subscripted "1" (together with one or more superscripts "$'$") is an order 1 entity variable. Thus we have "x_1", "y_1", "z_1", as well as "p_1", "q_1", "r_1". Similarly, there are order 2 entity variables, and so forth.[2]

We shall use "α_n", "α_m", "α_z", "β_n", "β_m", "β_z", etc., as meta-linguistic letters for terms of a given order. Here "n" and "m" and "z" are meta-linguistic variables for object-language order indices. They must always be numerals for finite integers greater than or equal to 0. The primitive signs for the logical particles are "$\sim$", and "$\vee$". It is important to emphasize that these are *predicate constants* and not statement connectives. In "Mathematical Logic," Russell decided to adopt "$\vee$" instead of "$\supset$" which indicated the relation 'implies.' We should therefore read $\ulcorner \alpha \vee \beta \urcorner$ as saying that α is (inclusively) disjoint from β. The sign is flanked by terms, not wffs. The remaining primitive signs are "(", ")", "/", "!", our braces "{" , "}" for nominalizing wffs, and the sign "$'$".

The *basic terms* are the entity variables of whatever order index. The *atomic* wffs are just the wffs $\mu_m \vee \nu_n$, and $\sim \mu_m$, where μ_m and ν_n are entity variables. The wffs of the language are just those in the smallest set K containing the atomic wffs and such that $\{A\}_m \vee \mu_n$, $\mu_n \vee \{A\}_m$, $\sim \{A\}_m$, $\{A\}_m \vee \{B\}_n$, $(\mu_m)C$, and

$$\begin{array}{lll}
\pi_m/\alpha_n;\beta_n!\{A\}_z & \pi_m/\alpha_n;\{B\}_n!\{A\}_z & \pi_m/\{D\}_n;\{B\}_n!\{A\}_z \\
\pi_m/\alpha_n;\{A\}_n!\delta_z & \pi_m/\{B\}_n;\beta_n!\{A\}_z & \{D\}_m\alpha_n;\{B\}_n!\{A\}_z \\
\pi_m/\{A\}_n;\beta_n!\delta_z & \{B\}_m/\alpha_n;\beta_n!\{A\}_z & \{D\}_m/\{B\}_n;\beta_n!\{A\}_z \\
\{A\}_m/\alpha_n;\beta_n!\delta_z & \{B\}_m/\{A\}_n;\beta_n!\delta_z & \{D\}_m/\{B\}_n;\{A\}_n!\delta_z \\
 & \{B\}_m/\alpha_n;\{A\}_n!\delta_z & \{A\}_m/\{B\}_n;\{D\}_n!\{E\}_z \\
 & \pi_m/\{A\}_n;\{B\}_n!\delta_z &
\end{array}$$

are in K whenever A, B, C, D and E are in K, α, β, π, δ are terms, and μ_m is an entity variable occurring free in C. The terms are then as follows: (i) any entity variable is a term; (ii) if A is a wff and if n is the order of the highest-order variable (bound or free) or term in A, then $\{A\}_{n+1}$ is a term if there is a bound variable of order n in A; otherwise $\{A\}_n$ is a term; (iii) there are no other terms.

The axiom system for the calculus is now largely what one would expect as the outcome of grafting a system of order indices onto the calculus for the

2. See, for instance, "The Paradox of the Liar" (*PL*, p. 73), and "On Types" (*T*, p. 1).

substitutional system of Chapter 4. It is convenient to again follow our practice of dropping nominalizing braces where wffs occur in subject positions. Thus, for instance

$$\alpha_n \;.\!\supset\!.\; \beta_m \supset \alpha_n$$

is more convenient than

$$\alpha_n \supset \{\beta_m \supset \alpha_n\}_z.$$

(Note that $z = m$ if $m > n$, and $z = n$ if $n > m$, and $z = m$ if $m = n$.) The convenience must be used with care, however. Wffs are not assigned order indices; only singular terms are. If we drop the braces the order index is suppressed.

*1907(A_1) $\alpha_n \;.\!\supset\!.\; \alpha_n \vee \beta_m$
*1907(A_2) $\alpha_n \vee \alpha_n \;.\!\supset\!.\; \alpha_n$
*1907(A_3) $\alpha_n \vee \beta_n \;.\!\supset\!.\; \beta_n \vee \alpha_n$
*1907(A_4) $\beta_n \supset \delta_m \;:\!\supset\!:\; \alpha_z \vee \beta_n \;.\!\supset\!.\; \alpha_z \vee \delta_m$
*1907(A_5) $(\mu_n)A\mu_n \supset A\alpha_n|\mu_n.$
where α_n is any term free for μ_n in $A\mu_n$.
*1907(A_6) $(\mu_n)(B \supset A\mu_n) \;.\!\supset\!.\; B \supset (\mu_n)A\mu_n,$
where μ_n does not occur free in B.
Modus ponens (mp)
From A and $\{A\}_n \supset \{B\}_m$, infer B.
Universal generalization (ug)
From A, infer $(\mu_n)A$.
Replacement rule for defined signs
Definiendum may replace and be replaced by definiens wherever it occurs.
*1907df(∃) $(\exists x_n)A = \text{df} \sim (x_n) \sim A$
*1907df(A) $(\mu_{n1}, \ldots, \mu_{mj})A = \text{df } (\mu_n 1), \ldots, (\mu_{mj})a$
*1907df(E) $(\exists\mu_{n1}, \ldots, \mu_{mj})A = \text{df } (\exists\mu_n 1), \ldots, (\exists\mu_{mj})A$
*1907df(⊃) $\alpha_n \supset \beta_m = \text{df} \sim \alpha_n \vee \beta_m,$
*1907df(&) $\alpha_n \;\&\; \beta_m = \text{df} \sim (\sim \alpha_n \vee \sim \beta_m)$
*1907df(=) $\alpha_n \equiv \beta_m = \text{df } \alpha_n \supset \beta_m \;.\&.\; \beta_m \supset \alpha_n.$

Our substitutional axioms, axiom schemata, and definitions will require appropriate modifications for orders. The definitions of "in" and "out" of Chapter 5 are fitted with order indices as follows:

*1907df(out)
α_n out $\beta_m = \text{df } (\mu_n)(\beta_m/\alpha_n;\mu_n!\beta_m)$
*1907(df)in
α_n *in* $\beta_m = \text{df} \sim (\alpha_n \text{ out } \beta_m)$
*1907df(ex_z)
$\alpha_n\ ex_z\beta_m = \text{df} \sim (\exists\mu_z)(\mu_z \textit{ in } \alpha_n \;.\&.\; \mu_z \textit{ in } \beta_m)$
*1907df(ind)
α_n ind $\beta_m = \text{df } \alpha_n \text{ out } \beta_m \;.\&.\; \beta_m \text{ out } \alpha_n$
*1907df(=)

$\alpha_n = \beta_n = \text{df}$
$(p_{n+1}, q_{n+1}, r_{n+1})(a_n)(p_{n+1}/a_n;\alpha_n!q_{n+1}\ \&\ p_{n+1}/a_n;\beta_n!r_{n+1}\ .\supset.\ q_{n+1} \supset r_{n+1})$
$*1907\text{df}(\neq)\quad \alpha_n \neq \beta_n = \text{df} \sim (\alpha_n = \beta_n)$
$*1907(\text{A}_7)\quad \alpha_n\ in\ \{A\alpha_n\}_m$
$*1907(\text{A}_8)\quad \alpha_n\ in\ \{A\}_m\ :\supset:\ \alpha_n\ in\ \beta_{z1}\ .\vee., \ldots, .\vee.\ \alpha_n\ in\ \beta_{zj}$
where $\beta_{z1}, \ldots \beta_{zj}$ are all (j-many) order z_j terms occurring free in A, $n \leq z_j \leq m$.
$*1907(\text{A}_9)\quad (x_n, y_n)(x_n\ in\ y_n\ .\&.\ y_n\ in\ x_n\ :\supset:\ x_n = y_n)$
$*1907(\text{A}_{10})\quad (x_n)(y_m)(v_z)(x_n\ in\ y_m\ .\&.\ y_m\ in\ v_z\ :\supset:\ x_n\ in\ v_z)$
where $n \leq m \leq z$.
$*1907(\text{A}_{11})$
$(p_m, a_n)(q_m)(x_n)(y_n)(p_m/a_n;x_n!q_m\ .\&.\ p_m/a_n;y_n!q_m\ :\supset:$
$a_n\ in\ p_m\ .\supset.\ x_n = y_n)$
$*1907(\text{A}_{12})\quad (p_m, a_n)(x_n)(q_m)(p_m/a_n;x_n!q_m\ .\&.\ a_n\ in\ p_m\ .\&.$
$a_n \neq p_m\ :\supset:\ x_n\ in\ q_m\ .\&.\ x_n \neq q_m).$
$*1907(\text{A}_{13})\text{a}\quad (x_n, y_n)(x_n/x_n;y_n!y_n)$
$*1907(\text{A}_{13})\text{b}\quad (x_m)(y_n)(y_n/x_m:z_m!y_n),$
where $m > n$.
$*1907(\text{A}_{13})\text{c}\quad \sim (p_m/a_n;x_n!q_z),$
where $m \neq z$.

By the definition of "*out*," principle $*1907(\text{A}_{13})$b ensures that for all x_m and y_n such that $m > n$, x_m out y_n.

$*1907(\text{A}_{14})\quad (p_m, a_n)(p_m/a_n;a_n!p_m)$
$*1907(\text{A}_{15})\quad (p_m, a_n)(x_n)(\exists q_m)(p_m/a_n;x_n!q_m\ .\&.$
$(r_m)(p_m/a_n;x_n!r_m\ .\supset.\ r_m = q_m))$

This is our $*1907$(p).

$*1907(\text{A}_{16})\quad (p_m)(\exists q_n)(q_n\ \text{ex}_z\ p_m),$
where $z \leq n$ and $n \leq m$.

This is our $*1907$(q).
$*1907(\text{A}_{17})\quad (\exists\mu_n)(\alpha_n\ \text{out}\ \{A\mu_n|\nu_n\}_m) \supset (\mu_n)(\{A\alpha_n|\nu_n\}_m/\alpha_n;\mu_n!\{A\mu_n|\nu_n\}_m),$
where α_n and μ_n are free for ν_n in A.

This is our $*1907$(r).

$*1907(\text{A}_{18})$
$(\exists\mu_{z1}, \ldots, \mu_{zn})(\alpha_m\ out\ \{A\mu_{z1}|v_{z1}, \ldots, \mu_{zn}|v_{zn}\}_m)\ :\&:$
$\alpha_m \neq \{A\sigma_{z1}|\nu_{z1}, \ldots, \sigma_{zn}|v_{zn}|\}_m\ .\&.\ \alpha_m \neq \beta_{m1}\ .\&., \ldots, .\&.\ \alpha_m \neq \beta_{mk}\ .:\supset:.$
$(\mu_m)(\exists\mu'_{z1}, \ldots, \mu'_{zn})(\{A\sigma_{z1}|\nu_{z1}, \ldots, \sigma_{zn}|\nu_{zn}\}_m/\alpha_m;\mu_m!\{A\mu'_{z1}|\nu_{z1}, \ldots,$
$\mu'_{zn}|\nu_{zn}\}_m$
$:\&:\ \sigma_{z1}/\alpha_m;\mu_m!\mu'_{z1}\ .\&., \ldots, .\&.\ \sigma_{zn}/\alpha_m;\mu_m!\mu'_{zn}),$
where each μ_{zi}, σ_{zi}, $1 \geq i \geq n$, $m \in \mathbb{N} - \{0\}$, is free for its respective v_{zi} in the wff A, and $\beta_{m1}, \ldots, \beta_{mk}$, $k \in \mathbb{N}$, are all the singular terms occurring free in the wff $A\sigma_{z1}|\nu_{z1}, \ldots, \sigma_{zn}|\nu_{zn}$ excepting $\sigma_{z1}, \ldots, \sigma_{zn}$.

This is our $*1907$(s).

$*1907(A_{19})$ $\{(\mu_n)A\mu_n\}_m = \{(\nu_n)A\nu_n|\mu_n\}_m$, where ν_n is free for μ_n in A
$*1907(A_{20})$ $\{(\mu_n)A\mu_n\}_m = \{(\mu_n)B\mu_n\}_m$.⊃. $(\mu_n)(\{A\mu_n\}_m = \{B\mu_n\}_m)$
$*1907(A_{21})$ $\{(\mu_z)A\mu_z\}_{z'} \neq \{\alpha_m \supset \beta_n\}_{z'}$
$*1907(A_{22})$ α_n *in* $\{(\mu_m)A\mu_m\}_z$.&. $\alpha_n \neq \{(\mu_m)A\mu_m\}_z$:⊃:
$(\mu_m)(\alpha_n$ *in* $\{A\mu_m\}_z)$, where μ_m is not α_n and not free in α_n and $n \leq z$ and $m < z$.
$*1907(A_{23})$ $\supset \neq \{\alpha_m \supset \beta_n\}_{z'}$.&. $\supset \neq \{(\mu_z)A_{\mu z}\}_{z'}$

This completes the revisions needed to graft orders on to the substitutional system set out in Chapter 4. In the substitutional theory cum orders, however, we shall need reducibility schemata as well:

$*1907(\text{Reduc})^1$
$(q_m b_n)(\exists p_{n+1}, a_n)(p_{n+1}/a_n \equiv q_m/b_n)$,
$*1907(\text{Reduc})^2$
$(s_m t_{n+1} w_n)(\exists q_{n+2}, p_{n+1}, a_n)(q_{n+2}/p_{n+1}a_n \equiv s_m/t_{n+1}w_n)$,

and more are needed as one ascends types. The nature of these schemata will be discussed in the sequel.

The grafting is rather straightforward. The grammatical rules on forming singular terms by nominalizing wffs are supposed to track the orders of entities. A singular term formed by nominalizing a wff containing a variable of order index m will have an order index at least as high as m. If an entity of order m is a constituent of an entity y, then y will have to have an order at least as high as m.

It is quite natural as well (if one is to assign orders at all) to exclude a sequence such as

$$p_m/a_n;x_{n'}!q_{m'}$$

where n' is not equal to n. The idea is that there is a structural difference between propositions having constituents of different order. For example, if the structure of the proposition

'$(x)(x = x) \supset p$'

differs from the structure of

'$q \supset p$'

(with p and q elementary), then substitution of the proposition '$(x)(x = x)$' for q violates structural integrity. "Les paradoxes" attempted to get at this by putting:

$$(x)(Ax) \supset p = \text{df } (\exists x)(Ax \supset p).$$

This definition clearly establishes the difference in structure in a way that makes the substitution of the proposition '$(x)(x = x)$' for q in '$q \supset p$' impossible. But getting at the idea through such definitions abandons the ontology of general propositions entirely. This plan did not succeed, but its spirit remains in "Mathematical Logic."

The assignment of order indices to substitutional principles is seriously threatened if one maintains that the sequence, $\mu_m/\alpha_n{:}\beta_{n'}!\delta_{m'}$, is well-formed only if $m = m'$ and $n = n'$ and $m \geq n$. In this case, an expression such as

$$p_1/a_2;b_2!q_1$$

is meaningless. On the present proposal it is meaningful and, if true, simply implies that $p_1 = q_1$. We just employ $*1907(\mathrm{A}_{13})\mathrm{b}$. Moreover, had we required $m \geq n$, the definition of "out" would make, for example, "x_2 out y_1," meaningless. Matters are even worse, had we put $m > n$. Here "x_n *out* x_n" would be meaningless too. This would destroy many of the proofs of Chapter 4 concerning propositional identities. The present formulation, however, preserves the theorems of Chapter 4 fitted with appropriate order indices.

For definite descriptions the adjustments due to orders are straightforward. We need not pause here. An order indexed formulation of Chapter 4's proof of $*(\mathrm{CP_{sub}})^1$ goes through as before. We have:

$*1907(\mathrm{CP_{sub}})^1$
$\vdash\ (\exists p_m, a_n)(p_n \textit{ in } a_m \text{ .\&. } (u_n)(p_m/a_n;u_n!\{Au_n|\nu_n\}_m),$

where p_m and a_n are not free in A.

1. $(\exists x_n)(x_n \text{ ex}_n\{A\nu_n\}_m)$ 1907(q), ui.
2. a_n *out* $\{A\nu_n\}_m$ 1, EI, $*1907(\mathrm{T}_{4.5})$b, df(ind).
3. $(\exists \nu_n)(a_n \textit{ out } \{A\nu_n\}_m)$ 2, eg.
4. $(\exists \nu_n)(a_n \text{ out } \{A\nu_n\}_m) \text{ .}\supset\text{. } (\mu_n)(\{Aa_n|\nu_n\}_m/a_n;\mu_n!\{A\mu_n|\nu_n\}_m)$ $*$ 1907(r)
5. $(\mu_n)(\{Aa_n|\nu_n\}_m/a_n;\mu_n!\{A\mu_n|\nu_n\}_m)$ 3, 4, mp.
6. a_n *in* $\{Aa_n|\nu_n\}_m$ $*$ 1907(A_7).
7. $a_n \textit{ in } \{Aa_n|\nu_n\}_m \text{ .\&. } (\mu_n)(\{Aa_n|\nu_n\}_m/a_n;\mu_n!\{A\mu_n|\nu_n\}_m)$ 5, 6, conj.
8. $(\exists p_m, a_n)(a_n \textit{ in } p_m \text{ .\&. } (\mu_n)(p_m;\, a_n;\mu_n!\{A\mu_n|\nu_n\}_m))$ 7, eg, (2ei).

With definite descriptions at hand, $*1907(\mathrm{CP_{sub}})^1$ yields:

$*(1907)(\mathrm{CP})^1$
$\vdash\quad (\exists p_m, a_n)(x_n)(p_m/a_n;x_n \text{ .}\equiv\text{. } \{A\}_m),$

where p_m and a_n are not free in A, and A is any wff containing x_n free. Our definition of "dual substitution" is easily fitted with order indices. We get:

$*1907$df(dual)
$p_m/a_nb_z;x_ny_z!q_m = \text{df}$
$\qquad (\exists h_m e_n t_m)(p_m/a_n;e_n!h_m \text{ .\&. } h_m/b_z;y_z!t_m \text{ .\&. } t_m/e_n;x_n!q_m)).$

Consider the following example:

$q_2/p_1a_o;r_1c_o!s_2 = \text{df}$
$\qquad (\exists h_2e_1t_2)(q_2/p_1;e_1!h_2 \text{ .\&. } h_2/a_o;c_o!t_2 \text{ .\&. } t_2/e_1;r_1!s_2).$

It is essential that a substitutional formula can allow the prototype order index to be more than next higher than the order index of its argument. As in Chapter 5, $*1907$df(dual) yields $*1907(\mathrm{CP_{sub}})^2$.

$$\vdash (\exists p_m,a_n,b_i)(a_n \textit{ in } p_m \text{ .\&. } b_i \textit{ in } p_m \text{ .\&. } (x_n,y_i)(p_m/a_nb_i;x_ny_i!\{Ax_ny_i\}_m),$$
where p_m,a_n,b_i are not free in Ax_ny_i.

This, in turn, results in:

$*1907(\text{CP})^2$
$$\vdash (\exists q_z, p_m, a_n)(r_m, c_n)(q_z/p_ma_n;r_mc_n \text{ .≡. } \{A\}_z),$$

where q_z and p_m and a_n are not free in A, and A is any wff containing r_m and c_n free. The process continues for higher types.

It will be observed that these comprehension principles are weaker than is provided by the reducibility schemata. But with them, we shall see that all is well for recovery of an order stratified version of the proxy (set out in our Chapter 5) of a type-stratified language with bindable predicate variables allowed in subject as well as predicate positions.

Predicativity and Reducibility

9.5 In the system of "Les paradoxes", there is no hiearchy of orders of entities (and of course no hierarchy of types of entities). Accordingly, although the system required a "reducibility" schema—viz., $*1906(\text{Reduc})^{\text{n}}$, the system was not, properly speaking, "ramified." In the substitutional system of 1907, we have a genuine ramification. The introduction of order indices on variables and the accompanying restriction in $*1907(\text{A}_5)$, must be offset by reducibility axiom schemata else the theory of classes and relations-in-extension will not be recovered within the 1907 "ramified" substitutional theory.

To uncover the schema of reducibility for the 1907 substitutional theory, we must first understand the needs of the substitutional proxy for the theory of classes and relations-in-extension. The meaning of "predicativity" is then to be tailored to these needs. Classes are to be constructed so as to be extensional—the orders of the propositions characterizing them should not matter to their identity conditions. For classes of order-0 entites, this means any adequate proxy for classes within the ramified substitutional theory must preserve:

$$(z_o)(Az_o \equiv \beta z_o) \text{ .⊃. }$$
$$\imath(p_n/a_o)[p_n/a_o \approx z_o \ \{Az_o\}_v] = \imath(r_m/a_o)[r_m/a_o \approx z_o \ \{Bz_o\}_i],$$

when index v is not equal to i. Now by definition the above is:

$$(\exists p_na_o)(p_n/a_o \approx z_o \ \{Az_o\}_v \text{ .\&. } (\exists r_mc_o)(r_m/c_o \approx z_o \ \{Bz_o\}_i$$
$$\text{ .\&. } p_n/a_o = r_m/c_o)).$$

The last clause reveals that we must have $m = n$, even though $v \neq i$. Moreover, the natural choice then is to put n as "1". The new 1907 reducibility schema is, therefore, as follows:

$*1907(\text{Reduc})^1$
$$(q_mb_n)(\exists p_{n+1}a_n)(p_{n+1}/a_n \equiv q_m/b_n).$$

When $n = \text{o}$, we have:

$$(q_m b_\text{o})(\exists p_1 a_\text{o})(p_1/a_\text{o} \equiv q_m/b_\text{o}).$$

This is corroborated in Russell's manuscript "On Types." There Russell stated his axiom of reducibility for the first type as follows (*T*, p. 3):

$$(\exists p^1)\colon\ p^n/a = p^1/a \quad \text{Pp.}$$

The same version occurs in "The Paradox of the Liar" (*PL*, p. 73). He uses the letter "*a*" without an order index for an order 0 entity variable, and (as we have seen) his double bar sign is our "≡" for coextensivity.

With $*$1907(Reduc)[1] we can easily arrive at the following theorem schema:

$\vdash\ (\exists p_{n+1} a_n)(x_n)(p_1/a_n;x_n \equiv \{A\}_m)$.

Proof:

1. $(\exists q_m b_n)(x_n)(q_m/b_n;x_n \equiv \{A\}_m)$ $\quad *$ 1907(CP)[1].
2. $(x_n)(q_m/b_n;x_n \equiv \{A\}_m)$ $\quad$ 1, EI.
3. $(\exists p_{n+1} a_n)(p_{n+1}/a_n \equiv q_m/b_n)$ $\quad *$ 1907(Reduc)[1], ui.
4. $(x_n)(p_{n+1}/a_n;x_n \equiv q_m/b_n;x_n)$ $\quad$ 3, EI, df(≡.a)[1].
5. $(x_n)(p_{n+1}/a_n;x_n \equiv \{A\}_m)$ $\quad$ 2, 4, log.
6. $(\exists p_{n+1} a_n)(x_n)(p_{n+1}/a_n;x_n \equiv \{A\}_m)$ $\quad$ 5, eg, (4ei), (2ei).

An instance of this is:

$$(\exists p_1 a_\text{o})(p_1/a_\text{o};x_\text{o} \equiv \{A\}_m).$$

In some respects it is more convenient to employ this than having to pair $*$1907(Reduc)[1] with $*$1907(CP)[1].

What are the substitutional reducibility schema for higher types? Russell gives few clues in the extant manuscripts of 1906 and 1907. The needs of the theory of classes speak loudly enough, however. Is there a class defined by use of

$$\imath(s_3/t_2 w_\text{o})[s_3/t_2 w_\text{o} \approx r_2 c_\text{o}\ \ B(r_2 c_\text{o})]?$$

Suppose so. Then the following would be well-formed:

$$\imath(p_1/a_\text{o})[p_1/a_\text{o} \approx z_\text{o}\ \ Az_\text{o}] \in \imath(s_3/t_2 w_\text{o})[s_3/t_2 w_\text{o} \approx r_2 c_\text{o}\ \ B(r_2 c_\text{o})].$$

But following our earlier definitions, this would be

$$(\exists s_3 t_2 w_\text{o})[s_3/t_2 w_\text{o} \approx r_2 c_\text{o}\ \ B(r_2 c_\text{o})\ \&$$
$$(\exists p_1 a_\text{o})(p_1/a_\text{o} \approx z_\text{o}\ \ Az_\text{o}\ .\&.\ p_1/a_\text{o} \in s_3/t_2 w_\text{o})).$$

And yet the clause,

$$p_1/a_\text{o} \in s_3/t_2 w_\text{o}$$

is defined as,

$$(\exists q_3)(s_3/t_2 w_\text{o};p_1 a_\text{o}!q_3),$$

and this is ill formed. Only entities of like order are intersubstitutable.

In using matrices to proxy the theory of classes, we see that only matrices such as

$$\text{“}p_1/a_o\text{”}, \quad \text{“}q_2/p_1a_o\text{”}, \quad \text{“}j_3/k_2u_1v_o\text{”}, \quad \text{etc.,}$$

are appropriate. Accordingly, this is to be the meaning of "predicative."

$*1907(\text{Reduc})^2$
$(s_mt_{n+1}w_n)(\exists q_{n+2}p_{n+1}a_n)(q_{n+2}/p_{n+1}a_n \;.\!\equiv\!.\; s_m/t_{n+1}w_n).$

From this and $*1907(\text{CP})^2$, we arrive at:

$$\vdash (\exists q_{n+2}p_{n+1}a_n)(r_{n+1}c_n)(q_{n+2}/p_{n+1}a_n;r_{n+1}c_n \;.\!\equiv\!.\; \{A\}_z),$$

where q_{n+2}, p_{n+1}, a_n are not free in A, and r_{n+1}, c_n occur free in A. An instance of this is:

$$\vdash (\exists q_2p_1a_o)(r_1c_o)(q_2/p_1a_o;r_1c_o \;.\!\equiv\!.\; \{A\}_z),$$

The process continues for higher types.

As we see, the needs of the theory of classes and relations-in-extension, coupled with substitution's grammatical demand that the order indices of subsituend and substituendum match, force a particular notion of "predicativity" and reducibility. To illustrate, consider the definition df(=.a) which permits the identity sign to be flanked by matrices:

df(=.a)
$p/a = r/c = \text{df } (stw)(s/tw;pa \equiv s/tw;rc).$

How shall order indices be grafted on to this? In a case such as,

$$p_{n+1}/a_n = r_{n+1}/c_n = \text{df}$$
$$(s_{n+2}t_{n+1}w_n)(s_{n+2}/t_{n+1}w_n;p_{n+1}a_n \equiv s_{n+2}/t_{n+1}w_n;r_{n+1}c_n),$$

all is well, for this is supported by $*1907(\text{Reduc})^2$. For any higher-order attribute ("propositional function") proxied by a matrix, $h_m/d_{n+1}e_n$, we can find a coexemplifying one that is in the range of the quantifiers. But consider this:

$$p_2/a_o = r_2/c_o = \text{df } (s_3t_2w_o)(s_3/t_2w_o;p_2a_o \;.\!\equiv\!.\; s_3/t_2w_0;r_2a_o).$$

This won't work as a definition. Indiscernibility here is restricted to order three, and higher orders are left out because there is *no* reducibility principle of the form:

$$(s_mt_2w_o)(\exists q_3p_2a_o)(q_3/p_2a_o \;.\!\equiv\!.\; s_m/t_2w_o).$$

Fortunately, substitution can rectify this problem. As we saw in Chapter 5, there were two ways to proxy the identity of attributes. The second way is to put:

$$p_2/a_o = r_2/c_o = \text{df } p_2 = r_2 \;.\&.\; a_o = c_o.$$

So the 1907 theory can scotch the original df(=.a) adopting,

1907df(=.a)^1
$p_m/a_n = r_m/a_n = \text{df } p_m = r_m \text{ .\&. } a_n = c_n,$

and similar definitions for higher types. As we shall see, this problem, imposed by the grammatical demand that the order indices of substituend and substituendum match, will arise again in the context of *Principia Mathematica*.

9.6 The changes to the theory of classes of our Chapter 6 imposed by orders are now straightforward. From ∗1907(Reduc)^n, we have the following comprehension theorem schemata for classes:

∗1907(CP$_\in$)^1 $\vdash (\exists p_{m+1}a_m)(p_{m+1}/a_m \approx z_m\ Az_m)$
∗1907(CP$_\in$)^2 $\vdash (\exists s_{m+2}t_{m+1}w_m)(s_{m+2}/t_{m+1}w_m \approx r_{m+1}c_m\ B(r_{m+1}c_m))$

and so forth. Note that A and B can be any wffs. Note as well that we might have written $\{Az_m\}_v$, and $\{B(r_{m+1}c_m)\}_v$ in these schemata. The system is quite unlike that of "Les paradoxes" in allowing nominalization of general wffs. It is convenient, however, to drop braces as has been Russell's practice. We can go on to affix order indices to the definitions of Chapter 6 without trouble. For instance, we put:

1907df(C≈)^1
$p_{m+1}/a_m \approx r_{m+1}/c_m = \text{df } (z_m)(z_m \in p_{m+1}/a_m \text{ .}\equiv\text{. } z_m \in r_{m+1}/c_m)$
1907df(C≈.a)^1
$p_{m+1}/a_m \approx z_m\ Az_m = \text{df } (z_m)(z_m \in p_{m+1}/a_m \text{ .}\equiv\text{. } Az_m)$
1907df(C≈.b)^1
$Az_m \approx z_m\ Bz_m = \text{df } (z_m)(Az_m \text{ .}\equiv\text{. } Bz_m)$
1907df(C≈)^2
$q_{m+2}/p_{m+1}a_m \approx s_{m+2}/t_{m+1}w_m = \text{df}$
$(r_{m+1}c_m)([r_{m+1}/c_m][r_{m+1}/c_m \in q_{m+2}/p_{m+1}a_m] \text{ .}\equiv\text{.}$
$[r_{m+1}/c_m][r_{m+1}/c_m \in s_{m+2}/t_{m+1}w_m])$
1907df(C≈.a)^2
$q_{m+2}/p_{m+1}a_m \approx r_{m+1}c_m\ A(r_{m+1}c_m) = \text{df}$
$(r_{m+1}c_m)([r_{m+1}/c_m][r_{m+1}/c_m \in q_{m+2}/p_{m+1}a_m] \text{ .}\equiv\text{.}$
$[r_{m+1}/c_m]A(r_{m+1}c_m))$
1907df(C≈.b)^2
$A(r_{m+1}c_m) \approx r_{m+1}c_m\ B(r_{m+1}c_m) = \text{df}$
$(r_{m+1}c_m)([r_{m+1}/c_m]\ A(r_{m+1}c_m) \text{ .}\equiv\text{. } [r_{m+1}/c_m]\ B(r_{m+1}c_m))$

and so on. As before, we have:

1907df(A)^1
$[r_{m+1}/c_m]\ A(r_{m+1}c_m) = \text{df } (\exists q_{m+1}b_m)(q_{m+1}/b_m \approx r_{m+1}/c_m \text{ .\&. } A(q_{m+1}b_m))$
1907df(A)^2
$[q_{m+2}/p_{m+1}a_m]\ A(q_{m+2}p_{m+1}a_m) = \text{df}$
$(\exists h_{m+2}d_{m+1}e_m)(h_{m+2}/d_{m+1}e_m \approx q_{m+2}/p_{m+1}a_m \text{ .\&. } A(h_{m+2}d_{m+1}e_m)).$

The existence theorem schemata for classes follow straightforwardly from the definitions and the reducibility schemata.

Our contextual definitions for eliminating substitutional class symbols are next modified for orders.

1907df(C)[1]
$x_m \in \imath(p_{m+1}/a_m)[p_{m+1}/a_m \approx z_m\ Az_m] = \mathrm{df}$
$(\exists p_{m+1}a_m)(p_{m+1}/a_m \approx z_m\ Az_m\ .\&.\ x_m \in p_{m+1}/a_m)$
1907df(C.a)[1]
$x_m \in p_{m+1}/a_m = \mathrm{df}\ (\exists q_{m+1})(p_{m+1}/a_m;x_m!q_{m+1}\ \&\ q_{m+1}).$
1907df(C =)[1]
$\imath(p_{m+1}/a_m)[p_{m+1}/a_m \approx z_m\ Az_m] = \imath(p_{m+1}/a_m)[p_{m+1}/a_m \approx z_m\ Bz_m] = \mathrm{df}$
$(\exists p_{m+1}a_m)(p_{m+1}/a_m \approx z_m\ Az_m\ .\&.$
$(\exists r_{m+1}c_m)(r_{m+1}/c_m \approx z_m\ Bz_m\ .\&.\ p_{m+1}/a_m = r_{m+1}/c_m)$
1907df(C)[2]
$\imath(p_{m+1}/a_m)[p_{m+1}/a_m \approx z_m\ Az_m] \in$
$\imath(s_{m+2}/t_{m+1}w_m)[s_{m+2}/t_{m+1}w_m \approx r_{m+1}c_m\ B(r_{m+1}c_m)] = \mathrm{df}$
$(\exists s_{m+2}t_{m+1}w_m)(s_{m+2}/t_{m+1}w_m \approx r_{m+1}c_m\ B(r_{m+1}c_m)\ .\&.$
$(\exists p_{m+1}a_m)(p_{m+1}a_m \approx z_m\ Az_m\ .\&.\ p_{m+1}/a_m \in s_{m+2}/t_{m+1}w_m))$
1907df(C.a)[2]
$p_{m+1}/a_m \in s_{m+2}/t_{m+1}w_m = \mathrm{df}\ (\exists t_{m+2})(s_{m+2}/t_{m+1}w_m;p_{m+1}a_m!t_{m+2}\ \&\ t_{m+2})$
1907df(C =)[2]
$\imath(s_{m+2}/t_{m+1}w_m)[s_{m+2}/t_{m+1}w_m \approx r_{m+1}c_m\ A(r_{m+1}c_m)] =$
$\imath(s_{m+2}/t_{m+1}w_m)[s_{m+2}/t_{m+1}w_m \approx r_{m+1}c_m\ B(r_{m+1}c_m)] = \mathrm{df}$
$(\exists s_{m+2}t_{m+1}w_m)(s_{m+2}/t_{m+1}w_m \approx r_{m+1}c_m\ A(r_{m+1}c_m)\ .\&.$
$(\exists h_{m+2}d_{m+1}e_m)(h_{m+2}/d_{m+1}e_m \approx r_{m+1}c_m\ B(r_{m+1}c_m)$
$.\&.\ s_{m+2}/t_{m+1}w_m = h_{m+2}/d_{m+1}e_m)).$

Like adjustments apply to the other definitions of Chapter 6. We shall not pause for them since the changes should now be clear.

To illustrate the affectations orders produce, consider again the principle of abstraction for classes of classes (of individuals). We want to capture the theorem

$$\vdash\ \hat{z}Az \in \hat{\alpha}B\alpha\ .\equiv.\ B(\hat{z}Az),$$

where A and B are any wffs. We have:

$*$1907(Abs)[2]
$\vdash\ \imath(p_1/a_o)[p_1/a_o \approx z_o\ A(z_o)] \in \imath(s_2/t_1w_o)[s_2/t_1w_o \approx r_1c_o\ B(r_1c_o)]$
$.\equiv.\ (\exists p_1a_o)(p_1/a_0 \approx z_o\ Az_o\ .\&.\ B(p_1a_o))$
Proof:
1. $\rightarrow\ \imath(p_1/a_o)[p_1/a_o \approx z_o\ Az_o] \in \imath(s_2/t_1w_o)[s_2/t_1w_o \approx r_1c_o\ B(r_1c_o)]$
2. $(\exists s_2t_1w_o(s_2/t_1w_o \approx r_1c_o\ B(r_1c_o)\ .\&.$ $(\exists p_1a_o)(p_1/a_o \approx z_o\ Az_o\ .\&.\ p_1/a_o \in s_2/t_1w_o)),$ 1, df(C)[2].
3. $s_2/t_1w_o \approx r_1c_o\ B(r_1c_o)\ .\&.$ $(\exists p_1a_o)(p_1/a_o \approx z_o\ Az_o\ .\&.\ p_1/a_o \in s_2/t_1w_o)$ 2, EI.
4. $s_2/t_1w_o \approx r_1c_o\ B(r_1c_o)$ 3, simp.
5. $(\exists p_1a_o)(p_1/a_o \approx z_o\ Az_o\ .\&.\ p_1/a_o \in s_2/t_1w_o)$ 3, simp.
6. $q_1/b_o \approx z_o\ Az_o\ .\&.\ q_1/b_o \in s_2/t_1w_o$ 5, EI.

7. $(\exists p_1 a_o)(p_1/a_o \approx q_1/b_o \ .\&.\ p_1/a_o \in s_2/t_1 w_o))$ 5, 6, df(C≈)^1, log.
8. $(\exists p_1 a_o)(p_1/a_o \approx q_1/b_o \ .\&.\ B(p_1 a_o))$ 4, 7, df(C≈.a)^2, log.
9. __ $(\exists p_1 a_o)(p_1/a_o \approx z_o\ Az_o \ .\&.\ B(p_1, a_o))$ 6, 8, df(C≈)^1, log, (6ei), (3ei).
10. $1 \supset 9$ cp.
11. → $(\exists p_1 a_o)(p_1/a_o \approx z_o\ Az_o \ .\&.\ B(p_1 a_o))$
12. $(\exists s_2 t_1 w_o)(s_2/t_1 w_o \approx r_1 c_o\ B(r_1 c_o))$ $*(\mathrm{CP}_{\in})^2$
13. $s_2/t_1 w_o \approx r_1 c_o\ B(r_1 c_o)$ 12, EI.
14. $q_1/b_o \approx z_o\ Az_o \ .\&.\ B(q_1 b_o)$ 11, EI.
15. $(\exists p_1 a_o)(p_1/a_o \approx q_1/b_o \ .\&.\ B(p_1 a_o))$ 11, 14, df(C≈)^1, log.
16. $(\exists p_1 a_o)(p_1/a_o \approx q_1/b_o \ .\&.\ p_1/a_o \in s_2/t_1 w_o)$ 13, 15, df(C≈.a)^2.
17. $(\exists p_1 a_o)(p_1/z_0 \approx z_o A z_o \ .\&.\ p_1/a_o \in s_2/t_1 w_o)$ 14, 16, df(C≈)^1.
18. 13 & 17 conj.
19. $(\exists s_2 t_1 w_o)18$ eg, (14ei), (13ei).
20. __ $\iota(p_1/a_o)[p_1/a_o \approx z_o\ Az_o] \in \iota(s_2/t_1 w_o)[s_2/tw_o \approx r_1 c_o\ B(r_1 c_o)]$ 9, df(C)^2.
21. $11 \supset 20$ cp.
22. 10 & 21 conj.
22. $*(\mathrm{Abs})^2$ 22, df(≡).

As we see, the addition of orders does not despoil the theory of classes so long as the reducibility schemata are in place.

The substitutional treatment of relations-in-extension set forth in Chapter 7 can next be affixed with order indices. The issue of predicativity arises here as well. But the order-indexed comprehension theorem schemata for relations-in-extension pose no problems. They will follow from $*1907(\mathrm{Reduc})^n$. We need only parallel the definitions for relations-in-extension of Chapter 7. Since the pattern is clear, we need not take up the changes here. With reducibility in place, mathematical induction is recovered in the 1907 ramified substitutional theory, and substititutional proofs of the Peano postulates go through.

Paradoxes of Propositions Avoided

9.7 Orders of propositions avoid the paradoxes of propositions. Consider first the *liar*. With order indices added, the *liar* is not formulable. We now have:

$$\mathrm{Bel}^{\mathrm{at}}\{(\exists p_n)(\mathrm{Bel}^{\mathrm{at}} p_n \ \&\ \sim p_n)\}_{n+1}.$$

The paradox cannot arise because the quantifier in the proposition

$$\{(\exists p_n)(\mathrm{Bel}^{\mathrm{at}} p_n \ \&\ \sim p_n)\}_{n+1}$$

does not range over all propositions. Its range is restricted to propositions of order *n*. In short,

"$\{(\exists p_n)(\mathrm{Bel}^{\mathrm{at}} p_n \ \&\ \sim p_n)\}_{n+1}$"

is not a term which is an admissible substituend for "p_n" in a universal instantiation allowed by $*1907(A_5)$.

Consider next the paradoxes relevant to substitution. The Appendix B (PoM) paradox is not expressible. The following is not an instance of $*1907(CP)^1$,

$$(\exists p_m, a_n)(x_n)(p_m/a_n;x_n \ .\!\equiv\!. \\ \{(\exists r_n, c_n)(x_n = \{(z_n)(r_n/c_n;z_n \ .\!\supset\!. \ z_n)\}_{n+1} \ .\&. \sim (r_n/c_n;x_n))\}_m).$$

From the definition of the identity sign, the expression

$$\text{“}x_n = \{(z_n)(r_n/c_n{:}z_n \ .\!\supset\!. \ z_n)\}_{n+1}\text{”}$$

is not well-formed.

Consider next the troublesome p_o/a_o paradox. The schema $*1907(CP)^1$ does yield:

$$(\exists p_m a_n)(x_n)(p_m/a_n;x_n \ .\!\equiv\!. \\ \{(\exists r_n c_n)(x_n = \{r_n/c_n;b_n!q_n\}_n \ \& \sim (r_n/c_n;x_n))\}_m).$$

But now m has to be $n+1$. So the term "$\{p_m/a_n;b_n!q_m\}_m$" cannot be put for "x_n" in a *universal instantiation*. The same result occurs in the following simple variant of the p_o/a_o. The schema $*1907(CP)^1$ yeilds:

$$(\exists p_m a_n)(x_n)(p_m/a_n;x_n \ .\!\equiv\!. \ \{(\exists r_n c_n)(x_n = \{r_n \supset c_n\}_n \ \& \sim (r_n/c_n;x_n))\}_m).$$

Again m has to be $n+1$. Accordingly, the grammar prevents the term "$\{p_m \supset a_n\}_m$" from being put for "x_n" in a *universal instantiation*. The paradoxes of propositions are dodged by the hierarchy of orders.

The hierarchy of orders is quite unlike the situation of "Les paradoxes." Both systems eliminate the paradoxes. But the reducibility schema $*1906(Reduc)^1$ of "Les paradoxes" revives the p_o/a_o. In contrast, the system of "Mathematical Logic" supports reducibilities. Reducibility $*1907(Reduc)^1$ does not yield

$$(\exists p_n a_n)(x_n)(p_n/a_n;x_n \ .\!\equiv\!. \ (\exists r_n c_n)(x_n = \{r_n \supset c_n\}_n \ .\&. \sim (r_n/c_n;x_n))).$$

Rather we only get:

$$(\exists p_{n+1} a_n)(x_n)(p_{n+1}/a_n;x_n \ .\!\equiv\!. \ (\exists r_n c_n)(x_n = \{r_n \supset c_n\}_n \ .\&. \sim (r_n/c_n;x_n))).$$

No contradiction can result from this.

Moreover, "Mathematical Logic" differs from "Les paradoxes" when it comes to matters of cardinality. In the earlier work, there was no proof that the cardinality of any class of *entities* was nondenumerable. In "Mathematical Logic," it is possible to show that for any n the class of all $n+1$ order entities is larger in cardinality than the class of n-order entities. The p_o/a_o and variants, after all, are diagonal constructions akin to that used in proving Cantor's power-class theorem that the cardinality of any class (finite or infinite) is smaller than that of its power-class. From, $*1907(Reduc)^2$ we have:

$$(\exists p_{n+1} a_n)(x_n)(p_{n+1}/a_n;x_n \equiv (\exists r_{n+z} c_n)(A(x_n, r_{n+z}, c_n) \ \& \sim (r_{n+z}/c_n;x_n))),$$

assuming that there is some formula A that is functional. That is, assume

$$(x_n r_{n+z} c_n q_{n+z} b_n)(A(x_n, r_{n+z}, c_n) \ \& \ A(x_n, q_{n+z}, b_n) \ .\supset. \ r_{n+z}/c_n \equiv q_{n+z}/b_n),$$

and

$$(r_{n+z} c_n)(\exists x_n) A(x_n, r_{n+z}, c_n),$$

where z is not 0. If there is indeed a formula A that proxies a function from all the classes of n-order entities into n-order entities we get a contradiction. Accordingly, there must be no such function and there are more $n+1$ order entities than n-order entities.

On our reconstruction of the 1907 substitutional theory we kept elementary propositions as 0-order entities and were able to show an infinity theorem. Assuming that the cardinality of 0-order entities is $\aleph_0$, then the cardinality of the class of 1-order entities is greater than $\aleph_0$. Is its cardinality $2^{\aleph_0}$? This is an interesting question that we must leave unresolved.

9.8 We now see how order indexed variables are to be grafted on to the calculus of substitution of Chapter 4. "Mathematical Logic" intends just such a graft, but Russell's unfortunate attitude concerning use and mention tends to obscure what is going on. Speaking of "Mathematical Logic" Quine asks: "Variables, in the easiest sense, are letters; and what contain them are notational expressions. Is Russell then assigning types to his objects or to his notations?" Believing that "Mathematical Logic" is committed to a hierarchy of types and orders of "attributes" (i.e., Russell's "propositional functions" in an ontological sense), Quine quite rightly dismisses ramification as ill-conceived. Quine sees ramification as the product of Russell's sloppy use of expressions such as "propositional function," "proposition," and "variable." Attributes in intension are independent of thought and linguistic representation. They do not contain variables or quantifiers; their expressions do. Quine is quite correct that *if* ramified type theory is committed to an ontology of "propositional functions," ramification is ill-motivated.

Quine's premise was wrong, however. "Mathematical Logic" does not assume that every open wff stands for an attribute ("propositional function"). The language of propositional functions is taken up as a matter of convenience. An ontology of propositions was to be assumed. Russell's cavalier attitude concerning use and mention is innocuous so long as readers understand him. Russell concurs with Quine that the variable is simply a single letter—its "determination" at this or that entity being primitive and unanalyzable (to us "meta-linguistic"). Nonetheless, there are general propositions—entities which (according to "Mathematical Logic") come in orders depending on the nature of the generality they contain. Formulae, on the other hand, contain quantifier *symbols*. The symbols used as variables are to have type (order) indices (except in virtue of order ambiguity). Russell was not confused at all.

What is the effect of the orders of propositions on the substitutional proxy for a type regimented theory of attributes? The effect was that substitution now proxies a ramified type theory of attributes. In this respect, substitution gives us new insight into why Russell found himself forced to provide a solution of paradoxes which, by modern lights, seem not germane to the matter of establishing logicism. The paradoxes of propositions are a mixed bunch. Some, such as the *liar*, involve the introduction of terms which are semantic, intentional, or "epistemological" (in Ramsey's sense)—terms such as "believes," "asserts," and "denotes." Substitution could be "Ramisified" by setting these aside. But others, like the p_0/a_0, remain. When applied to "Mathematical Logic," Quine's quip that the ramified theory "calls for amputation" is misguided. Due to the p_0/a_0 paradox, ramification and the substitutional theory are connected at the chest, and it is not clear that amputation is possible without killing the patient.

Indeed, it is precisely because of this that the patient died in Russell's hands! The substitutional system with orders avoids the paradoxes of propositions, but it left Russell with serious problems. Order indices seem flatly incompatible with the *raison d'être* of the original substitutional theory—viz., that any calculus for logic must adopt only one style of variables (entity variables). In the 1907 substitutional theory, "entity" is no longer univocal in meaning. This seemed philosophically unacceptable to Russell. How can logic support the claim that there are "entities" which are 0-order, entities that are 1st-order, and so on? This very statement violates order stratification. In the next chapter we examine the system of *Principia Mathematica*. It is commonly believed that Russell gave up on the unrestricted variable in *Principia*, and the work there is viewed as paradigmatic of a system with restricted variables. We shall find that quite the contrary is the case. In *Principia*, Russell felt he had found the final solution—a solution of the paradoxes which enables the demonstration of logicism and the *preservation* of its doctrine of the unrestricted variable.

CHAPTER TEN

The Logic of *Principia Mathematica*

10.0 There are many different renditions of the formal system of *Principia Mathematica*. To name just a few, there are Church (1978), Hatcher (1968), Copi (1971), and Chihara (1973). The source of the contention between them lies in the degree of significance each attributes to certain features of the philosophical explanations set out in the Introduction of the work. All regard the explanations as inconsistent and concur that pruning is necessary for coherence. This presentation offers a new rendition of the formal system—a rendition which at last shows that the real *Principia* comports with its Introduction.

The Formal System of *Principia* (*cum* $*10$)

10.1 Let us use "A", "B", "C" as meta-linguistic letters for wffs. The primitive signs of the formal language are "$\sim$" and "$\vee$", brackets and parentheses. (We no longer have braces for nominalizing a wff; in *Principia* there are no propositions and so no singular terms for them.) The signs "$\sim$" and "$\vee$" are now to be *statement* connectives. The individual and predicate variables come with order\type superscripts. The variables are x, y, z with any positive numeral subscript and any order\type superscript. (For convenience we shall often omit the subscript and use further lower-case letters of the English alphabet.) To get at the notion of an order\type symbol let us render a definition.

> An order\type symbol is any expression which satisfies the following recursive definition:

(i) "o" is an order\type symbol.
(ii) If $t_1, \ldots, t_m$ are order\type symbols, then the expression $(t_1, \ldots, t_m)$ is also an order\type symbol.
(iii) These are the only order\type symbols.

Examples of our variables are then x^{o}, $x^{(\text{o})}$, $x^{((\text{o}),\text{o})}$, and so on. The intent is that the variable x^{o} is an individual variable (i.e., a variable of type o). The variable $x^{(\text{o})}$ is a predicate variable whose type is such that its admissible arguments will be individual variables; $x^{((\text{o}),\text{o})}$ is a dyadic predicate variable whose first argument is a predicate variable of type (o) and whose second argument is an individual variable. The aspect of *order* of the order\type symbol is to be understood by means of our next definition:

> The *order* of an order\type symbol is given by the following recursive definition:
>
> (i) the order\type symbol "o" has order O.
> (ii) An order\type symbol $(t_1, \ldots, t_m)$ has order $n+1$, if the highest order of the order\type symbols $t_1, \ldots, t_m$ is n.

One can give a parenthesis-counting algorithm for determining the order of an order\type symbol. Count parentheses from left to right, adding +1 for each left parenthesis and −1 for each right. The order of the order\type symbol is the highest positive integer obtained in the counting process.

Now our order\type symbols are formally convenient but it is useful to adopt an alternative notation with is handy when it comes to the demonstration of certain important points about *Principia*'s order\types. We can put superscripts on the right for the order\type of the arguments to a variable and superscripts on the left of the variable for the order of the whole order\type symbol. The individual variables then look like: ${}^{\text{o}}x^{\text{o}}$, ${}^{\text{o}}y^{\text{o}}$, ${}^{\text{o}}z^{\text{o}}$, and so forth. The predicate variables of the form $x^{(\text{o})}$ can be written as ${}^{1}\phi^{(\text{o}\backslash\text{o})}$; the predicate variables of the form $x^{((\text{o}))}$ can be written as ${}^{2}\phi^{(1\backslash(\text{o}\backslash\text{o}))}$; variables such as $x^{((\text{o}),\text{o})}$ can be written as ${}^{2}\phi^{(1\backslash(\text{o}\backslash\text{o}),\text{o}\backslash\text{o})}$, and so on. In this way we can more easily see the significance of what will come to be a rule governing wffs of the system—viz., that arguments to a predicate variable must match in order\type.

The terms of the language are as follows:

(i) all variables are terms;
(ii) there are no other terms.

The "atomic wffs" of the language are of the form,

$$x^{(t_1,\ldots,t_j)}(y_1^{t_1}, \ldots, y_j^{t_j}).$$

The wffs of the language are determined inductively as the smallest set K containing all atomic wffs of the language and such that $\sim A$, $(A \vee B)$, and $(\forall x^t)C$ are in K if A, B, and C are in K and C contains the variable x^t free. We use $\ulcorner A(y^t|x^t)\urcorner$ for the wff exactly like $A(x^t)$ except for having free occurrences of the term y^t for every free occurrence of x^t in A.

Set within this background grammar the deductive system of *Principia* with $*10$ comes to the following:

$*1.2$ Pp $A \vee A$ $.\supset.$ A
$*1.3$ Pp B $.\supset.$ $A \vee B$
$*1.4$ Pp $A \vee B$ $.\supset.$ $B \vee A$
$*1.5$ Pp $A \vee (B \vee C)$ $:\supset:$ $B \vee (A \vee C)$[1]
$*1.6$ Pp $B \supset C$ $:\supset:$ $A \vee B$ $.\supset.$ $A \vee C$
$*1.1$ *Modus ponens*
From A and $A \supset B$, infer B
$*1.01$ $A \supset B = \text{df} \sim A \vee B$
$*3.01$ $A \,\&\, B = \text{df} \sim (\sim A \vee \sim B)$
$*4.01$ $A \equiv B = \text{df}\ A \supset B\ .\&.\ B \supset A$.

We have here used our meta-linguistic "A", "B", and "C" since on the method of Principia's $*10$ these apply to any wffs, elementary or otherwise. For the identity sign, we have the following definition:

$$*13.01 \quad x^t = y^t = \text{df}\ (\forall z^{(t)})(z^{(t)}(x^t) \equiv z^{(t)}(y^t)).$$

This definition is tied to the reducibility schema discussed below.

The quantification theory developed in *Principia*'s $*10$ is next given (with minor modifications). We have the schemata:

$*10.01$ $(\exists x^t)A = \text{df} \sim (\forall x^t) \sim A$
$*10.1$ Pp $(\forall x^t)\, A(x^t)$ $.\supset.$ $A[y^t|x^t]$,
where the variable y^t is free for x^t in A.
$*10.11$ From A infer $(\forall x^t)A$,
where the variable x^t occurs free in A.
$*10.12$ Pp $(\forall x^t)(B \vee A(x^t))$ $.\supset.$ $B \vee (\forall x^t)\, A(x^t)$,
where x^t does not occur free in B.

To complete the system, we need a general formulation of *Principia*'s schemata of reducibility.[2]

$*1910$(Reduc)
$$(\exists x^{(t_1,\ldots,t_j)})(\forall y_1^{t_1}),\ldots,(\forall y_j^{t_j})(x^{(t_1,\ldots,t_j)}(y_1^{t_1},\ldots,y_j^{t_j}) \equiv A),$$

where $x^{(t_1,\ldots,t_j)}$ does not occur free in A.

1. Of course $*1.5$ was redundant. See Bernays (1926).

2. *Principia* gave only two reducibility axioms, one for monadic functions and one for dyadic functions. The intent, however, was that further axioms for any finite number of variables would be introduced where needed.

The Perils of Typical Ambiguity

10.2 Compared with the baroque systems of ramified types often attributed to *Principia* the present austere system above must appear shocking. It is time then that we offer arguments for its historical accuracy.

Principia's logical connectives are statement connectives rather than predicate constants as in Russell's earlier theories (and in particular as in his substitutional theory). To be sure, one will find the following in *Principia* (*PM*, p. 93):

> If p and q are any propositions, the proposition "p or q" i.e., "either p is true or q is true," where the alternatives are not mutually exclusive, will be represented by "$p \vee q$".

At first blush this seems to make "$\vee$" into a dyadic relation. But as such it should be able to be flanked by any singular terms—not just propositional terms. Worse, *Principia* has explicitly abandoned an ontology of propositions (*PM*, p. 44). When Whitehead and Russell use the term "proposition" they must mean "well-formed formula." A charitable interpretation, then, would be that the above passage intends to render the truth conditions for a wff "$p \vee q$", where "p" and "q" are meta-linguistic letters for wffs. This is corroborated in a letter to Jourdain of 2 January 1911:

> I no longer think it significant to deny $x \supset p$, where x is not a proposition. I think that, strictly, one ought not to use a single letter for a proposition, but always some such symbol as ϕx (Quoted from Grattan–Guinness (1977b, p. 136)

Thus, "$p \vee q$" is not properly read "either p is true or q is true." This is its truth-condition. The case is similar when Whitehead and Russell read "$\sim p$" as "p is false." The logical constants are statement connectives; they are flanked by wffs to form wffs.[3] Whitehead and Russell are using "p", "q", etc., as schematic letters for wffs.[4]

Of course, it may be doubted whether the formal system employed axioms with special "sentential variables" together with a rule of uniform replacement or axiom schemata. Sentential variables are also compatible with regarding the logical particles as statement connectives. Some passages suggest schemata. For instance, we find (*PM*, p. 99):

> The proofs of the earlier of the propositions of this number consist simply in noticing that they are instances of the general rules given in $*1$. In such cases these rules are not premises, since they assert any instance of themselves.

3. This point is emphasized by Quine (1967), who thinks Russell simply conflated predicates with statement connectives. This is rather uncharitable. When he had propositions, Russell legitimately took the logical particles as predicates, reading the horseshoe as "implies." When he abandons propositions "$\supset$" is a statement connective even though, in some cases the locution "implies" is retained.

4. When the deductive system is characterized via $*9$ "p", "q" etc., are for quantifier-free wffs.

Indeed, it is even claimed that "noticing that they are instances" cannot be formulated as a rule of replacement (*ibid.*). Church (1956), however, reads this quite oppositely—as adopting axioms and an implicit rule of replacement. Moreover, Church cites Russell's 1919 *Introduction to Mathematical Philosophy* as evidence. Russell seems to admit that *Principia* should have had a replacement rule (*IMP*, p. 151).

With respect to the sentential part of the calculus, it is not clear which is the correct view. Russell was just too sloppy about the meta-linguistic issues. He may well have thought of axiom schemata as also needing some rule governing what wffs are to be proper instances. In any case, the quantificational part of ramified type theory militates against the approach of axioms and a rule of replacement. Schematic order\type indices and the distinction between predicate variables and schematic letters for wffs are essential to the grammar and to the reducibility principles.

Principia is infamously poor in distinguishing its statements about its formal language from statements made within that language. Of course there are those who argue that the very idea of a meta-language within which we set out the formal system is out of sorts with Russell's conception of logic. We addressed this matter in Chapter 1 and found the arguments wanting. The distinction is absolutely indispensable for an adequate understanding of the formal system of *Principia*.

Consider, for instance, the status of statements setting out the system itself. All too often it is assumed that *allowing* no meta-language, Russell must violate type theory to state type theory. As Hylton recently put it (Hylton 1990, p. 317):

> A single propositional function ($\hat{\phi}$ is *predicative*) is applicable (truly or falsely) to entities of different type; this contravenes type restrictions. Another example concerns the notion of type itself. Proposition $*9.131$ is a definition of 'being the same type as'. . . . The relation is one which itself violates type restrictions.

This is mistaken. The notions "is predicative," and "is same type as" are meta-linguistic; they concern the formal grammar of *Principia* and the conditions for well-formedness of symbols.

There are, to be sure, a number of "principles" in *Principia* that speak to the question of sameness of type. But care is needed in understanding them. Consider the following (*PM*, p. 140):

$*10.121$ If "ϕx" is significant, then if a is of the same type as x, then "ϕa" is significant, and vice versa.

$*10.122$ If, for some a, there is a proposition ϕa, then there is a function $\phi\hat{x}$ and vice versa.

Observe how these are used. If ϕa is a proposition, Whitehead and Russell tell us, then (by $*10.122$) $\phi\hat{x}$ is a function; and since "$\phi\hat{x}$ always" is a primitive, there is a proposition $(x)\phi x$. (See the proof of $*9.62$.) The intent here is just to say that if there is a wff $A\nu^t$, we shall have a wff $(\mu^t)A\mu^t$. The issue is about order\type indices and not about propositional functions in an ontological sense at all. The principles arise because Whitehead and Russell adopt a technique they call "typical ambiguity"—the omission of order\type indices on their expressions. They employ this technique with a vengeance. Properly done, the formal grammar should have been rendered with schematic order\-type indices and then typical ambiguity adopted as a convention of omitting them. But instead the formal grammar is set out without any indices. This necessitates "principles" which govern the identification of variables of "the same type" (i.e., variables which, were they to have had order\type indices, would have the same indices). Consider another example. Whitehead and Russell write (*PM*, p. 136):

> If $\phi\hat{x}$ and $\psi\hat{x}$ are functions of the same type, then (by ∗10.121) there is an a for which "ϕa" and "ψa" are significant. So "$\phi a \vee \psi a$" is significant. So (by ∗10.122) there is a function $\phi\hat{x} \vee \psi\hat{x}$. Hence "$(x)(\phi x \vee \psi x)$" is significant.

(See also the proof of ∗9.61.) The problem is that since the μ of $A\mu$ is typically ambiguous, it does not follow that $(\mu)(A\mu \vee B\mu)$ is a wff simply because $A\mu$ and $B\mu$ are wffs. The μ of $A\mu$ may have a different order\type index than the μ of $B\mu$. So Whitehead and Russell require a means of assuring that we can generate the wffs we want in appropriate cases.

It is quite clear, then, that the "principle" ∗9.131 for "sameness of type" is just a statement of grammatical well-formedness that governs the wffs of the system. It and principles like it above are necessitated only by a quite deplorable misuse of typical ambiguity. They are entirely avoided if the formal grammar is set out (as we have) with schematically ambiguous order\type indices and typical ambiguity is recast as the technique of omitting indices under conventions of restoration.

Curiously, Whitehead was actually on to this point. In May of 1910, with the printing of Volume 1 of *Principia* well underway, he wrote to Russell:[5]

> I don't feel sure that ∗3.03 is right as it stands—or at least as it is explained. It appears to me as if two ideas are muddled up together—namely, a true logical premise and a test which supplements the incompleteness of our symbolism.
>
> A true logical premiss must [be] such as would still be required, if our symbolism were complete and adequate. Now in such a case the type is always in evidence. E.g. let every letter representing an individual have i as subscript, then we have $\vdash .\phi x_i$ and $\vdash .\psi x_i$, we do not need any axiom to assure that x_i and y_i are of the same type, and that any possible value of x_i is a possible value of y_i and vice versa

5. See letters #210.057451 and #710.057442 in the Russell Archives.

The analog of $*3.03$ is

> $*10.13$ If $\phi\hat{x}$ and $\psi\hat{x}$ take arguments of the same type, and we have "$\vdash .\phi x$" and "$\vdash .\psi x$", we shall have "$\vdash .\phi x \cdot \psi x$."

Whitehead is quite correct. That the type (order\type) of the variable "*x*" in the first is the same as in the second would be given if the language had not suppressed the indices. There would be no need for this theorem or principles such as $*9.131$, $*9.14$, or the analogs $*10.131$ etc.

10.3 The situation of "sameness of type" is only one of many misunderstandings precipitated by Whitehead and Russell's use of typical ambiguity. The common misconception that for Russell logic must "contain its own meta-language" only compounds the misunderstandings. Consider some examples. *Principia* tells us that "any symbol whose meaning is not determinate is called a *variable*, and the various determinations of which its meaning is susceptible are called the *values* of the variable" (*PM*, p. 4). But then the work goes on to say "variables will be denoted by single letters" (*ibid*., p. 5). Dismissing any meta-language, many explain the second passage by assuming that *Principia*'s "propositional functions" are Plàtonic entities with variables (in an admittedly obscure ontological sense) as constituents. On this view, single letters denote these constituents of propositional functions.[6] This is far away from the truth. The proper reading imposes no archaic ontology of variables as constituents of propositional functions. Certain letters *are* variables; a variable is just a symbol.

What then of the talk about denoting variables? Well the simple point is that we do talk about (denote) symbols, and this happens in the meta-language when setting forth any formal deductive system. For example, (*PM*, p. 140):

> $*10.11$ If ϕy is true whatever possible argument y may be, then $(x).\phi x$ is true.

This is just the inference rule of *universal generalization*; and here the italicized "*y*" is used to denote variables.

Sloppy as their expression is, Whitehead and Russell did embrace a meta-language/object-language distinction. Once we see this, the fog surrounding *Principia* begins to clear. Consider the expression "individual." An individual, we are told, is "something which exists on its own account; it is then obviously not a proposition, since propositions, as explained in Chapter II of the Introduction [*PM*, p. 43] are incomplete symbols" (*ibid*., p. 162). And they continue (*ibid*., p. 51):

> We shall use such letters as *a*, *b*, *c*, *x*, *y*, *z*, *w*, to denote objects which are neither propositions nor functions. Such objects we call *individuals*. Such objects will be constituents of propositions or functions, and will be *gen-*

6. Most recently, Hylton (1990) takes this view.

uine constituents, in the sense that they do not disappear on analysis, as (for example) classes do, or phrases of the form "the so and so."

Well if there are no propositions in the ontological sense, then individuals cannot be their constituents. Moreover, if "x is an individual" is a primitive of the object language, then one may as well have "x exists" which is explicitly excluded from the symbolism (*ibid.*, p. 175). Russell is just sloppy. The word "individual" has two uses. First, it means "individual variable" (i.e., one among the lower case Latin letters "a", "b", "x", "y", etc.) But, of course, on the intended interpretation, individual variables are assigned entities. This is its second use. An individual is any entity, and no entity is a proposition or a class.

So individual variables are certain designated letters of the formal language. Similarly, Whitehead and Russell designate the letters that are to be the predicate variables. They write (*ibid.*, p. 165):

> We shall use the small Latin letters (other than p, q, r, s) for variables of the lowest type concerned in any context. For functions, we shall use the letters ϕ, X, ψ, f, g, F.

This tells what letters the formal system is to adopt as its function terms (i.e., its predicate variables). Unfortunately, the case of letters such as "ϕ" is complicated by a particularly injurious misuse of typical ambiguity in *Principia.* The omission of order\type indices is taken to allow "apparent" (bound) variables to be fixed in order\type while "real" (free) variables *occurring in theses* are to be unrestricted ("ambiguous") as to order (*ibid.*, p. 128). (This is the heart of Whitehead and Russell's distinction between "all" and "any.")[7]

In *Principia*, the symbol "$\phi!x$" as opposed to "ϕx" is used to indicate predicativity. Thus the reducibility principles are stated as follows (*PM*, 167):

$*12.1 \quad (\exists f):\phi x \,.\equiv_x.\, f!x \quad \mathrm{Pp}$
$*12.11 \quad (\exists f):\phi(x, y) \,.\equiv_{x,y}.\, f!(x, y) \quad \mathrm{Pp}$

These use letters such as "ϕ" without the shriek, as special "variables" which because free (and in a principle or thesis) are somehow not to be restricted in order. Whitehead and Russell write (*ibid*, p. 165):

> It is unnecessary to introduce a special notation for non-predicative functions of a given order and taking arguments of a given order. . . . We require, however, a means of symbolizing a function whose order is not assigned. We shall use "ϕx" or "$f(X!\hat{z})$" or etc. to express a function (ϕ or f) whose order, relative to its argument is not given. Such a function cannot be made into an apparent variable unless we suppose its order previously fixed. As the only purpose of the notation is to avoid the necessity of fixing the order, such a function will not be used as an apparent variable; the only functions which will be so used will be predicative functions.

7. It also arises in expressing the reducibility axioms.

This won't do. Typical ambiguity is only intelligible as a convention of *suppressing* order\type symbols; the formal language and deductive system must be stateable without them. One cannot have indices on bound variables but drop them from free variables in theses, else the ramified type regimented rules of *universal generalization* and principles like *universal instantiation* become impossible to state. All free variables must come with indices. Whitehead and Russell's typically ambiguous variables must be understood to be schematic letters for formulae.

Accordingly, a letter such as "ϕ" must not play a dual role—as predicate variable and as schematic letter. In our reformulation, all variables, bound or free, are alike fixed in order\type. It is schematic letters for wffs, letters such as "*A*", "*B*", "*C*", that realize Whitehead and Russell's use of the typically ambiguous "any."

10.4 Viewing matters this way yields a great conceptual clarification of the text. At once we get the following important result:

(∗) ALL PREDICATE VARIABLES ARE PREDICATIVE.

All bound and free predicate variables must come with order\type indices and the indices are such as to be predicative.

With schematic letters distinguished from object-language variables, we came to thesis (∗). But we are also in a position to see something even more striking. In *Principia* we find (*PM*, p. 132):

> Primitive idea: *Individual.* We say *x* is "individual" when it is neither a proposition nor a function.

The point here is simply that individual variables are primitive symbols of the language. The notion of an individual variable is a primitive idea. Paralleling this, we find (*ibid.*, p. 164):

> "Matrix" or "predicative function" is a primitive idea.

We saw that care must be taken with *Principia*'s use of the word "individual"; extra special care must be taken when it comes to Whitehead and Russell's use of words such as "function," "propositional function," and "matrix." The simple fact is that, as used above, a "matrix" is simply a predicate variable. Whitehead and Russell are saying that predicate variables are primitive signs of the formal language. To be sure, *Principia* does go on to write as follows (*ibid.*):

> A function is said to be *predicative* when it is a matrix. . . . The variables *occurring in* the present work, from this point onward, will all be either individuals or matrices of some order in the above hierarchy.

This makes it appear as if there are function terms other than matrices and that some among them are not predicative. But this is wrong. The appearance results from *Principia*'s failure to distinguish schematic letters from genuine variables. Once the distinction is made, we see what Whitehead and Russell are struggling to say. The only terms of the language are to be individual variables and predicate variables; and the predicate variables have order\type indices in accordance with the definition of predicativity. Taken together with (∗) we now have:

(∗∗) ALL AND ONLY PREDICATE VARIABLES ARE PREDICATIVE.

This result is of inestimable importance.

Thesis (∗∗) at once puts to rest perhaps the most often repeated objection to *Principia*—the objection that it does not render an unambiguous meaning of "predicativity." Church (1976) maintains that *Principia* contains two different accounts of "predicativity"—one in the Introduction that parallels that of the 1907 "Mathematical Logic as Based on the Theory of Types," and quite another at ∗12. In the Introduction we find Russell allowing that if a function has several arguments the highest order of which is n then the function is predicative if it is of order $n + 1$ (*PM*, p. 53):

> We will define a function of one variable as *predicative* when it is of the next order above that of its argument, i.e., of the lowest order compatible with its having that argument. If a function has several variables, and the highest order of function occurring among the arguments is the nth, we call the function predicative if it is of the $(n + 1)$th order, i.e., again of the lowest order compatible with its having the arguments it has.

As we saw, ∗12 has the following (*ibid.*, p. 167):

> A predicative function is one which contains no apparent variables, i.e., a matrix.

Church claims that ∗12 introduces a much more austere notion of predicativity—viz., that predicative function (terms), being matrices, contain *no apparent (bound) variables.*

Now Whitehead and Russell sensed that readers might be confused by the claim that predicative functions are matrices and that they would thereby be led to Church's interpretation. They tried to help with the following (*ibid.*, p. 52):

> Owing to the introduction of the variable first-order function $\phi!\hat{x}$, we now have a new set of matrices. Thus "$\phi!x$" is a function which contains no apparent variables, . . . (It should be observed that when ϕ is assigned, we may obtain a function whose values do involve individuals as apparent variables, for example if $\phi!x$ is $(y).\psi(x,y)$. But so long as ϕ is variable, $\phi!x$ contains no apparent variables.)

But this does little by way of clarification. The important point is that the only predicate terms are predicate variables, and all predicate variables are predicative.

Of course, the Introduction and ∗12 would not occur on the meaning of "predicativity" if there are predicate terms other than predicate variables—for instance, if circumflexion is a predicate term forming operator in *Principia*. So our (∗∗) naturally leads to another thesis:

> (∗ ∗ ∗) CIRCUMFLEXION WAS NOT A TERM FORMING OPERATOR IN *PRINCIPIA*

If one can form the term

$$\text{“}(\forall^{o}x^{o})^{1}\phi^{(o\backslash o,o\backslash o)}({}^{o}x^{o},{}^{o}\hat{z}^{o})\text{”}$$

then there are predicative terms that are not predicate variables. Accepting circumflexion, Church could not escape from his conclusion. But circumflexion was never intended as a predicate term forming operator. Whitehead and Russell themselves say as much, writing (*ibid.*, p. 19):

> We have found it convenient and possible—except in the explanatory portions—to keep the explicit use of symbols of the type "$\phi\hat{z}$," either as constants [e.g., $\hat{x} = a$] or as real variables, almost entirely out of this work.

Aside from its use to mark subject positions of predicate variables,[8] circumflexion is used only as a heuristic device in introducing and explaining the system.

Of course, even those who accept circumflexion admit that it is poorly set forth in *Principia*. For one thing, as Quine (1963, p. 249) points out, it would require *concretion* principles. Such principles are nowhere to be found in the work.[9] Another problem is that the device of circumflexing variables would have to be improved to avoid ambiguities.[10] These problems are often attributed to Russell's sloppiness, however, and circumflexion has been adopted by every interpreter of *Principia* to date. It will come as something of a shock, then, to be told that circumflexion was not a predicate term forming operation. But we can show that it is actually incompatible with *Principia*'s formal grammar and destroys its treatment of class symbols.

To demonstrate the deleterious effects of circumflexion, we need only reflect on three features of *Principia*'s grammar:

(i) Arguments to a predicate variable must agree in order\type.
(ii) All predicate variables are predicative.
(iii) (Predicative) predicate variables must take predicative arguments.

Suppose circumflexion was a term forming operation of the system. It follows from (ii) that since circumflexion always involves capping a variable, there can be no predicate terms which take a nonpredicative term as an argument. Circumflexion would then allow nonpredicative terms and from (ii) there are no variables appropriate to them. So UG and UI are not possible with respect

8. This role is quite minor and is easily omitted as we have in 10.1.
9. Goldfarb (1989, p. 39, fn 20) finds something at ∗9.131, but is mistaken as we have seen.
10. See Hatcher (1982), p. 109.

to such terms. Moreover, *Principia*'s definition of the identity sign cannot apply to such purported terms. *Principia* has:

$$*13.01 \quad x = y = \text{df } (\phi)(\phi!x \equiv \phi!y).$$

By clause (i) no predicate term can take arguments differing in order\type. If we can formulate terms by circumflexion then *Principia* would include terms in its language which are not admissible substituends for quantified variables and which cannot flank the identity sign. This is an intolerable situation.

This matter has been suppressed[11] because interpreters, seeking to "improve' *Principia*, have commonly ignored grammatical requirement (i) that arguments to a function term must have the same fixed order\type. Accordingly, (ii) and (iii) are ignored as well, and non-predicative variables are allowed. But (i) cannot be dropped if the system of the historical *Principia* is to be respected. Indeed, we shall see in subsequent sections that (i) plays a central role in Russell's explanation and justification of the order\type indices of the system.

Now given (i) Russell has an important motivation for (ii) and (iii). With (i) in place, clauses (ii) and (iii) are required when Russell's treatment of class and relation-in-extension symbols is regimented by ramification. Suppose (i) is true and (ii) and (iii) false. Suppose there are predicate variables such as "${}^{2}\phi^{(o\backslash o)}$", and "${}^{3}\Sigma^{(2\backslash(o\backslash o))}$", where the second is regarded as predicative. Next consider the following instance of Russell's contextual definitions for classes:

$$\begin{aligned}&{}^{o}\hat{z}^{o} A \in {}^{3}\hat{\alpha}^{(2\backslash(o\backslash o)} B = \text{df} \\ &(\exists^{3}\Sigma^{(2\backslash(o\backslash o))})((\forall^{2}\alpha^{(o\backslash o)})({}^{3}\Sigma^{2\backslash(o\backslash o))}({}^{2}\alpha^{(o\backslash o)}) \equiv B({}^{2}\alpha^{(o\backslash o)})) \\ &\qquad \& \ (\exists^{1}\phi^{(o\backslash o)})((\forall^{o}x^{o})({}^{1}\phi^{(o\backslash o)}({}^{o}x^{o}) \equiv A^{o}x^{o}) \ .\&. \ {}^{3}\Sigma^{(2\backslash(o\backslash o))}({}^{1}\phi^{(o\backslash o)}))).\end{aligned}$$

As we can see, given constraint (i) the last clause is not well-formed. *Principia* has no nonpredicate class variables such as "${}^{2}\alpha^{(o\backslash o)}$", and no predicate variables such as "${}^{2}\phi^{(o\backslash o)}$". The text is explicit. The reason is that their presence would wreak havoc for the contextual definition of classes—given constraint (i). The needs of the theory of classes and relations-in-extension are clear. There must be no nonpredicative predicate variables so that reducibility can afford a notion of predicativity according to which a monadic predicate variable is "predicative" only when the arguments it can take are themselves predicative.[12]

11. Hatcher (1982) is aware of it.

12. This dissolves Böer's problem with *Principia*'s treatment of class symbols (Böer, 1973). The predicativity requirement, he maintains, destroys the treatment of class symbols for it is not in general true that an intensional context A of a nonpredicative propositional function ${}^{2}\phi^{(o/o)}$ is such that:

$$\begin{aligned}&A^{2}\phi^{(o/o)} \ .\supset. \\ &\qquad (\exists^{1}\psi^{(o/o)})((\forall^{o}x^{o})({}^{1}\psi^{(o/o)}({}^{o}x^{o}) \equiv {}^{2}\phi^{(o/o)}({}^{o}x^{o})) \ .\&. \ A^{1}\psi^{(o/o)}).\end{aligned}$$

Böer is quite correct. We need only consider the case where A is the context "$\ldots = {}^{2}\phi^{(o/o)}$". But what Böer doesn't see is that there are no nonpredicative predicate variables (and accordingly no nonpredicative propositional functions in the ontological sense) in *Principia* in the first place.

Observe that it would not be possible to respond to the above argument by claiming that classes of individuals should be differentiated in terms of the formulae that individuals must satisfy to belong to them. That is, one cannot say that ${}^{3}\hat{\alpha}^{(2\backslash(o\backslash o)}B$ is a class of all classes of individuals *characterizable in terms of second-order properties of individuals*. Being extensional, classes are not differentiated in terms of the sorts of formulae of which they are extensions. That is,

$$(\forall^{o}x^{o})(A^{o}x^{o} \equiv B^{o}x^{o}) \;.\supset.\; {}^{o}\hat{z}^{o}A = {}^{o}\,\hat{z}^{o}B,$$

holds even when A contains no bound variables of higher order\type than o\o, and B is some wff that contains, say, a bound variable of order\type 2\(1\(o\o)). If arguments to a predicate variable must match in order\type, there must be no nonpredicative variables or predicate terms. Circumflexion is not a term forming operator in *Principia.*

Wait, wait! It will be objected that without circumflexion as a predicate term forming operator we have left *Principia* without comprehension principles. Common as it is, the view that *Principia* had no comprehension principles and instead achieved comprehension though circumflexion is quite mistaken. *Principia* has comprehension principles—its comprehension principles are precisely its reducibility schemata!

Orders within Types or Types within Orders?

10.5 The present interpretation dissolves Church's objection that there are two definitions of "predicativity" in *Principia*. It also settles the longstanding dispute over whether orders are within types or types are within orders. The dispute has generated many different interpretations of *Principia*'s ramified hierarchy.

Copi's (1971) rendition of *Principia*'s ramified-type theory claims that types came first and were split, in an effort to avoid new "semantic" paradoxes, into orders. In contrast, Chihara (1973), claims that *Principia* was ramified from the onset. Orders come first and are split into "types." Now the historical development of ramified types shows that types came first and later were split into orders. This follows because we must include the substitutional theory as itself proxying a theory of types.[13] So Copi is partly right. Nonetheless, both the Copi and Chihara accounts are unhistorical.

13. Linsky holds that "type theory was effectively ramified from its earliest formulations around 1906–1907" (Linsky 1990, p. 129). Linsky has "Mathematical Logic" in mind and dismisses the substitutional theory as not properly a type theory since there are no types of entities. But as we have seen, "Mathematical Logic" has no types of entities either! The work explicitly advocates substitution, though Russell was unsatisfied with orders. Now it can be said of *Principia* that it was formulated from the beginning with order\types so that orders were not grafted on to types. But this is unilluminating since it divorces the work from its historical development.

Copi does not have order\type indices. He has separate order indices and type indices. Arguments to a predicate variable need only match in their type index, so long as the order index of the argument does not exceed the order index of the predicate variable or term to which it is argument. The individual variables look like "x^{o}"; and in the monadic cases the predicate variables look like $\ulcorner {}^{n}\phi^{t} \urcorner$, where "$n$" is an order index and "t" a type index. A predicate variable is *predicative* when its order index is "1" (Copi 1971, p. 95). Moreover, Copi allows, "${}^{1}\phi^{((o))}$," an order 1 predicate variable whose arguments are type (o) predicate variables of order 1. We can see the motivation by noting that if one allows circumflex as a predicate term forming operator, one wants to say that a term such as

$$(\forall x^{o})\,{}^{1}\hat{\phi}^{(o)}(x^{o}),$$

should have order index "1". Surely, circumflexing a variable to construct a term from a wff should not raise the order. Only the presence of a quantifier binding a predicate variable raises order. Suppose the order index were "2." Then by a principle of *concretion* for circumflex terms, the second-order wff would be equivalent to a first-order wff. This just seems wrong. So Copi concludes that circumflexion does not raise the order index. He admits that in *Principia* one can find the claim that a first-order function is one whose arguments are individuals (*PM*, p. 167). Moreover, he recognizes that *Principia* says that the order of a function must be "compatible" with its type—so that the lowest order a predicate variable of type t can have is the order of its type symbol. But he glosses this, writing that "this suggestion seems designed to foster the idea that there is a closer connection between types and orders than really obtains" (Copi 1971, p. 86, fn1).

Chihara has order\types and the matching requirement. He takes seriously *Principia*'s statement that order must be compatible with type. Unfortunately, he labors under the error of thinking there are circumflex predicate terms in *Principia*. Hoping to find a closer fit to the text than Copi, Chihara decides to regard the circumflex term,

$$\text{"}(\forall^{o}x^{o})\,{}^{1}\hat{\phi}^{(o\backslash o)}({}^{o}x^{o})\text{"}$$

as having order "2." He writes that "the order of a propositional function is the least integer greater than the order of all its bound variables (quantified or circumflex)" (Chihara 1973, p. 20). Chihara maintains that the hierarchy begins from orders, and split into "types." (Let us use the expression "types∗" to distinguish Chihara's use of the term.) The type∗ of a monadic function, Chihara explains, depends on more than simply whether it is a function of individuals, a function of functions of individuals, etc. The order of the arguments it can take is also relevant. All first-order functions are functions of individuals. They are T1.0, where T0 is the type of individuals. But second-order functions are split into two types∗: those of type∗ T2.0 which can take individuals, and those of type∗ T2.1.0 that can take as arguments only first-order functions (of individuals). Accordingly, third-order functions are split into four types∗: those of type∗ T3.0 that take individuals, those of type∗

T3.1.0 that take first-order functions of individuals, those of type T3.2.0 that take second-order functions of individuals, and those of type* T.3.2.1.0 that take second-order functions of first-order functions of individuals (Chihara 1973, p. 20).

In the monadic case, Chihara tells us, a propositional function is *predicative* if its type is either T1.0 or Tk.k $-$ 1.σ (for some natural number k, with σ a sequence of the required sort) (*ibid.*, p. 22). That is, the predicative monadic propositional functions are those whose indices descend numerically from left to right, e.g., T1.0 T2.1.0 T3.2.1.0 T4.3.2.1.0, etc. (*ibid.*, p. 45).

Chihara's types* are based upon his recognition that *Principia* has order\-type indices and not Copi's separate order and type indices. That is, arguments to a predicate variable must match in order and type. Unfortunately, he fails to appreciate the consequences of having order\types for the notion of "predicativity" and for *Principia*'s treatment of class and relation-in-extension symbols. As we saw, if arguments to a predicate variable must match in order\type, then the system mustn't allow circumflexion as a term-forming operator; and moreover, the system had better not have nonpredicative variables. Chihara recognizes that the predicative predicate variables will have order indices in accordance with T1.0, T2.1.0, T3.2.1.0, etc. But he seems unaware that allowing the other variables and circumflex terms brings ruin.

Hatcher's presentation of the ramified types of *Principia* agrees with Chihara in many respects. The notion of an order\type that we have used in our characterization of *Principia*'s ramified types (and that generates Chihara's types*) is Hatcher's notion of the "order of a type symbol." But Hatcher attributes to *Principia* what he calls "order-types":

$$(t_1, \ldots, t_m)/k/(y_1, \ldots, y_m),$$

where the t_i are type symbols and k and y_i are numerals for natural numbers subject to the restriction that the number indicated by y_i is for each i, greater than or equal to the order of the corresponding type symbol t_i, and the number indicated by k is greater than the greatest value of the number indicated by y_i (Hatcher 1968, p. 124). If t_i is o, then y_i is 0. Let k be the highest order of all variables in a circumflex predicate term. Then the circumflex term is of order-type $(t_1, \ldots, t_m)\,/r/(y_1, \ldots, y_m)$, where r is k if the term has no bound variables of order k, and is $k + 1$ otherwise. Accordingly, like Chihara, Hatcher takes the circumflex term,

$$(\forall x^{o})^{1}\hat{\phi}^{(o)}(x^{o}),$$

to have order type ((o))/2/1. (Recall that Chihara regarded it as having order 2.) On Hatcher's view, a predicate variable or term of order-type $(t_1, \ldots, t_m)$ $/k/(y_1, \ldots, y_m)$ is *predicative* if and only if k is $y + 1$, where y is the maxim value of the orders $y_1, \ldots, y_m$ (*ibid.*, p. 125).

Hatcher has no requirement that arguments to a predicate variable must match in order-type. If x is a term of order-type

$$(t_1, \ldots, t_m)/k/(y_1, \ldots, y_m)$$

and $z_1, \ldots, z_m$ are terms of type t_i, respectively, and of order less than or equal to y_i, respectively, then $x(z_1, \ldots, z_m)$ is a wff (*ibid.*, p. 124). In this respect, Hatcher's system is a *cumulative* hierarchy; universal instantiation is permitted in accordance with the restrictions on the well-formed formulae.

Church (1976) offers a recursive definition of *Principia*'s ramified-types (r-types) as well. Church's recursive definition is framed so as to include propositions in the ontology. For convenience, Church represents propositions of order N as 0-ary propositional functions of level n; that is, as propositional functions of r-type $(\ )/n$, where '( )' represents the null-sequence. Church is well aware that the Introduction of the work advocates a "multiple-relation theory of judgment" according to which an ontology of propositions is to be rejected. But noting that Russell's purpose was to devise an intensional logic, he dismisses the multiple-relation theory as probably a late addition to the Introduction which plays no significant role. The ontology of propositions, he says, is "clearly demanded by the background and purpose of Russell's logic" (Church 1976, p. 291).

In subsequent sections, we shall see that, while Russell's intent was indeed to formulate an intensional logic, Church is quite mistaken about the multiple-relation theory. For our present purposes, however, we need only revise Church's recursive characterization of r-types (rendered below) so as to drop commitment to propositions. Simply exclude all r-types of the form $()/n$, where $n \in \mathbb{N} - \{0\}$. That is, we should require in Church's clause (2) below that $m \in \mathbb{N} - \{0\}$ (Cocchiarella, 1980, p. 56).

Church renders a recursive definition of r-types as follows (*ibid.*, 290):

1. There is an r-type to i which all and only individual variables belong.
2. If $m \in \mathbb{N}$, $n \in \mathbb{N} - \{0\}$, and $\beta_1, \ldots, \beta_m$ are given r-types, then there is an r-type $(\beta_1, \ldots, \beta_m)/n$ to which belong all and only m-ary predicate variables of *level* n and with arguments of r-types $\beta_1, \ldots, \beta_m$, respectively.
3. The *order* of an individual variable is 0. The *order* of a variable of r-type $(\beta_1, \ldots, \beta_m)/n$ is $N + n$, where N is the greatest of the orders corresponding to the r-types $\beta_1, \ldots, \beta_m$ (and $N = 0$ if $m = 0$).

Illustrations will make the recursion clear. The r-type $(i)/1$ will have order 1, because by stipulation the order of an individual variable i is 0 and $0 + 1 = 1$. Similarly, $(i)/2$ will be order 2. The order of $((i)/1)/1$ will be 2, since the order of $(i)/1$ is 1 and $1 + 1 = 2$. A predicate term or variable will be *predicative* if and only if its level n equals 1.

Church recognizes[14] that historically *Principia*'s ramified type theory was not cumulative (Church 1976, p. 289). But like Hatcher, Church is interested in a cumulative hierarchy. Accordingly, Church allows that $f(x_1, \ldots, x_m)$ is a well-formed formula if and only if f is a predicate variable (or predicate term) of

14. Nonetheless, Church does not address the problems.

r-type $(\beta_1, \ldots, \beta_m)/n$ where $m > 0$, and x_i is a variable (or term) of r-type β_i or directly lower than β_i for $1 \leq i \leq m$.

An r-type $(\beta_1, \ldots, \beta_m)/n$ is *directly lower* than r-type $(\alpha_1, \ldots, \alpha_m)/k$, if $\alpha_i = \beta_i$ for $1 \leq i \leq m$, and $k > n$. The levels are cumulative in that the range of a variable of given r-type shall include the range of every variable of directly lower r-type.

For purposes of comparison, it will be useful to present the differing views in the following chart:

Copi	Hatcher	Chihara	Church
${}^1\phi^{(o)}$	(o)/1/o	T1.0	$(i)/1$
${}^2\phi^{(o)}$	(o)/2	T2.0	$(i)/2$
${}^3\phi^{(o)}$	(o)/3	T3.0	$(i)/3$
${}^4\phi^{(o)}$	(o)/4	T4.0	$(i)/4$
${}^1\phi^{((o))}$	none	none	none
${}^2\phi^{((o))}$	((o))/2/1	T2.1.0	$((i)/1)/1$
${}^3\phi^{((o))}$	((o))/3/1	T3.1.0	$((i)/1)/2$
	((o))/3/2	T3.2.0	$((i)/2)/1$
${}^4\phi^{((o))}$	((o))/4/1	T4.1.0	$((i)/1)/3$
	((o))/4/2	T4.2.0	$((i)/2)/2$
	((o))/4/3	T4.3.0	$((i)/3)/1$
${}^1\phi^{(((o)))}$	none	none	none
${}^2\phi^{(((o)))}$	none	none	none
${}^3\phi^{(((o)))}$	(((o)))/3/2	T3.2.1.0	$(((i)/1)/1)/1$
${}^4\phi^{(((o)))}$	(((o)))/4/3	T4.3.1.0	$(((i)/1)/2)/1$
	none	T4.3.2.0	$(((i)/2)/1)/1$

As we can see, Hatcher is sympathetic to Chihara. But Hatcher counts variables with order-type ((o))/3/2, corresponding to Chihara's T3.2.0, as being predicative. Chihara does not; only ((o))/2/1, i.e., T2.1.0, is predicative in this type. Moreover, Hatcher's order-types are not discriminating enough. They cannot distinguish Chihara's T4.3.2.0 from his T4.3.1.0. Both have Hatcher's order-type (((o)))/4/3. We see, however, that like Chihara, Copi's predicate variables ${}^1\phi^{((o))}$, ${}^1\phi^{(((o)))}$, ${}^2\phi^{(((o)))}$, etc., have no counterparts in Hatcher's system. They violate the restrictions Hatcher places on his indices. The order of the type symbol ((o)) is 2, and the order of the type symbol (((o))) is 3.

Though Church's recursion does not include circumflex predicate terms, it would seem that Church (like Hatcher) concurs with Chihara that the circumflex term,

$$(\forall x^o)^1\hat{\phi}^{(o)}(x^o),$$

has order 2, its r-type being $((i)/1)/1$. There is no level 0, so it would not be possible for it to have r-type $((i)/1)/0$ and order 1. Moreover, Copi's variables ${}^1\phi^{((o))}$, ${}^1\phi^{(((o)))}$, ${}^2\phi^{(((o)))}$, and the like, have no analogs in Church's system. The lowest level being 1, Church's system has an analog of the notion of the "order of a type symbol." So Church's system is in the spirit of Chihara's account.

Indeed, Church's system is more finely discriminating than Hatcher's. We saw that Hatcher cannot distinguish Chihara's T4.3.2.0 from his T4.3.1.0. Church's system can. The former has r-type $(((i)/2)/1)/1$; the latter has r-type $((i)/1)/2/1$. Church counts *both* as predicative; Chihara, on the contrary, counts neither as predicative. Only T3.2.1.0 is predicative in this type, and T4.3.2.1.0 is predicative in the next higher type.

Church captures all the fine discriminations set forth by Chihara's types∗; Hatcher captures only some. All the same, in adopting cumulative grammars, both Church and Hatcher have abandoned the very basis of the distinctions—that is, the criterion that arguments to a predicate variable must match in order and type. Perhaps, then, we can see why Hatcher's system ignores some of the distinctions in Chihara's types∗. In a cumulative hierarchy, it seems to serve no purpose to distinguish (e.g.,) T4.3.2.0 from T4.3.1.0. Chihara, on the other hand, has the matching requirement, and so must distinguish them.

Copi, Chihara, Hatcher, and even Church, have lost sight of the formal system of the historical *Principia.* It is not surprising then to find them remarking that they cannot understand features of the philosophical justifications offered for the system. Chihara goes so far as to say that "many of the distinctive features of his [Russell's] system receive no real justification at all. Why cannot a single variable range over the propositional functions of, say T4.0, T3.0, T2.0, and T1.0?" (Chihara 1973, p. 43). The first step in understanding the philosophical justification, however, lies in accepting that *Principia*'s formal system is designed to embrace the consequences of its requirement that arguments to a predicate variable must match in *both* order and type. *Principia*'s philosophical justifications do track its formal theory, once one has hold of that theory.

The Docrine of the Unlimited Variable

10.6 Having set out and defended a new characterization of the formal system of *Principia*, we are in a position to examine the philosophical justification of the system set forth by Whitehead and Russell. We shall argue that the justification was intended to preserve the doctrine of the unrestricted variable—a doctrine that dates all the way back to Russell's 1903 *Principles of Mathematics*.

Principia adopts a language with order\type indexed individual and predicate variables. For this reason, it is easy to altogether miss the fact that the system was intended to preserve the doctrine of the unrestricted variable in *Principles.* To understand how this is possible requires that we first observe that Russell's plan for the preservation of unrestricted variables in the system of *Principia* departs from the plan adopted in its predecessor theory—the substitutional theory. We can see this by comparing the two plans side-by-side.

In "Les paradoxes," Russell motivated the substitutional language with the following argument (*InS*, p. 205):

> The difficulty of applying the vicious-circle principle arises from the argument by which, as it seems, we can prove that our variables must be capable of *all* values. . . . There is one way in which we might seek to evade this conclusion. We may say that 'ϕx is always true' means 'ϕx is true whenever it is significant' or 'ϕx is never false.' We might then say that a given function ϕx will always have a certain range of *significance* which will either be *individuals*, or *classes*, or *classes of classes*, or *dual relations of individuals*, or, etc.

> The difficulty of this view [i.e., the appeal to ranges of significance] lies in the proposition (say) 'ϕx is only significant when x is a class.' This proposition must not be restricted, as to its range, to the case when x is a class; for we want it to imply 'ϕx is not significant when x is not a class.' We thus find that we are brought back after all to variables with an unrestricted range. If this is to be avoided, the range of significance must be somehow given with the variable; this can only be done by employing variables having some internal *structure* for such as are to be of some definite logical type other than individuals. . . . But then we have to assume that a single letter, such as x, can only stand for an individual; and that can only be the case if individuals are really all entities, and classes, etc., are merely *façons de parler*.

The founding idea of the language of substitution was to have a formal grammar with only one style of entity/individual variables. Then, through syntactic constructions, type stratified languages which use different sorts of variables—predicate variables, class variables, relation$_e$ variables, and so on—were proxied in such a way that different ranges of significance for these variables are built into the formal grammar of substitution.

In contrast, consider the following passage from "Mathematical Logic" (*ML*, p. 73):

> The difficulty which besets attempts to restrict the variable is that restrictions naturally express themselves as hypotheses that the variable is of such or such a kind, and that, when so expressed, the resulting hypothetical is free from the intended restriction. . . . Thus a variable can never be restricted within a certain range if the propositional function in which the variable occurs remains significant when the variable is outside that range. But if the function ceases to be significant when the variable goes outside a certain range, then the variable is *ipso facto* confined to that range, without the need of any explicit statement to that effect.

This suggests a new plan. Russell is now willing to allow many sorts of predicate variables as primitive signs of his new formal grammar, rather than only one "entity" variable as in substitution. Nonetheless, all such predicate variables are to be *semantically* interpreted in such a way that they are not genuine!

In "Mathematical Logic" Russell explains that in reading '$(x)\phi x$' as 'ϕx always' we witness an "*internal* limitation upon x, given by the nature of the function; and this is a limitation which does not require explicit statement, since it is impossible for a function to be true more generally than for all its values" (*ML*, p. 72). It is now the intended semantic interpretation governing the introduction of predicate variables that is to preserve the doctrine of the unrestricted variable. In *Principia Mathematica* the doctrine of the unrestricted variable has become the doctrine of the "internally limited" variable.

The doctrine of the "internally limited" variable seems easy to ignore, for unlike the "unrestricted" variable it allows many sorts of variables in the formal calculus. The straightforward semantic interpretation of such variables is a hierarchy of sorts of entities for each sort of variable. Of course, if this is the case, Russell is deprived of any reason for abandoning substitution—save the convenience of using indexed predicate variables instead of substitutional formulae. This is not sufficient reason, and scholars have therefore been perplexed by Russell's abandonment of substitution.

Offering help to the perplexed, Grattan-Guinness (1974, 1977b) maintains that the *liar* paradox, in conjunction with Russell's growing dislike of false propositions, led to the abandonment of propositions. This, he says, undermined the ontology of the substitutional system. Grattan-Guinness quotes from a letter to Jourdain of 1 June 1907:

> Consideration of the paradox of the *liar* and its analogs has led me to be chary of treating propositions as entities. I therefore no longer regard as valid the proof of $\aleph_0$ entities. . . . I now think that existence-theorems beyond the finite require a definite assumption that the number of entities is not finite.

Russell makes a similar comment in his 1907 "On the Nature of Truth" (*OT*, p. 48). The era of substitution was over, for Russell could no longer quantify over propositions. As Russell himself explained in the 1907 manuscript "On Fundamentals" (*OF*, p. 6):

> A value of an apparent variable must be something, and thus the no-classes theory won't work. It worked while we had propositions because then they became apparent variables where a variable matrix was wanted.

In Grattan-Guinness's view, *Principia* abandons propositions for a new theory of truth—and a calculus of restricted predicate variables ranging over a ramified hierarchy of propositional functions.

Hylton (1980) has a variant of this interpretation. Instead of the *liar*, he places the weight upon a formulation in substitutional terms of the Appendix B paradox of propositions in *Principles*. Hylton argues that "Les paradoxes" had to adopt a hierarchy of orders of propositions to solve this paradox. Once adopted, the *raison d'être* of the substitutional theory was lost. Variables had to be restricted, so Russell decided in "Mathematical Logic" that he may as well adopt an ontology of propositional functions and restricted predicate variables for them.

Hylton was wrong about "Les paradoxes." No hierarchy of propositions is adopted in the work. The 1907 "Mathematical Logic" did adopt such a hierarchy. But Russell is explicit that the notation of predicate variables is but a "convenience" eliminable in favor of the language of substitution (*ML*, p. 77). Substitution was not abandoned until after this work was written.

The paradoxes of propositions did, as Grattan-Guinness observed, result in Russell's abandonment of propositions and ended substitution. The new correspondence theory of truth was generated from the ashes. But Grattan-Guinness misses the full significance of the new theory of truth. It is the centerpiece of Russell's new philosophical explanation of *Principia*'s introduction of a language with order\type indexed predicate variables.

Russell practically says this outright. In his 1910 article "The Theory of Logical Types," a précis of the ramified type theory of *Principia*, he explains that since "Mathematical Logic" appeared in a mathematical journal, he could not devote more space to philosophical issues. He admits that as a consequence certain "obscurities" resulted from his endeavour to be brief. "These obscurities," he goes on, "I shall try to remove in the following pages, of which the purpose is explanatory rather than controversial" (*TT*, p. 215). "Mathematical Logic" is a substitutional theory with an ontology of propositions. It left its readers with a philosophical obscurity indeed—a language with order stratified variables. To avoid this, Russell had been working out a new "no-propositions" theory of truth. The work appeared in the 1907 paper "On The Nature of Truth," but (as Russell notes in the paper) he was not then confident enough to endorse it.

In *Principia* he was. It is an error to think that in abandoning substitution Russell abandoned the doctrine of the unrestricted variable of *Principles*, adopting predicate variables restricted to entities (attributes or "propositional functions") of a fixed order\type. The fact is that Russell abandoned the substitutional theory's *syntactic* preservation of the doctrine in favor of a *semantic* preservation. In *Principia*, predicate variables are introduced so as to be "internally limited" by their semantic conditions of significance. The conditions of significance are rendered (where the order part of the order\type indices are concerned) by none other than the new theory of truth and an informal nominalistic semantics governing the introduction of predicate variables.

Poincaré's Vicious Circle Principle

10.7 The system we advanced in 10.1 (*Principia* cum $*10$) can be interpreted with a realist semantics, so that predicate variables range over attributes ("propositional functions" in the ontological sense). This, however, leaves Russell with no reason to have abandoned substitution for the no-propositions theory of *Principia*. The Introduction to *Principia* would have looked like Russell's 1907 "Mathematical Logic." It would reference the substitutional theory, not the no-propositions theory; and predicate variables (and subsequently class symbols) would be regarded as notational conveniences eliminable via translation into the substitutional language.

Interpretative tradition, of course, misses this. It takes Russell to have assigned Poincaré's Vicious Circle Principle (VCP) the role of justifying a realist interpretation of *Principia*'s ramified type theory—the fact that it never had this justificatory role in Russell's earlier works notwithstanding. The VCP cannot bear the weight.

In the Schilpp (1944) volume on Russell, Gödel famously pointed out that there are three distinct variants of the VCP in *Principia* and that, in every variant, it should be dismissed as a diagnosis of the paradoxes germane to logicism. Following Ramsey, he argues that it is simply a mistake to think that all paradoxes, logical/mathematical (dealing with the logical form of predication) and semantic/epistemological (dealing with relations of truth, reference, designation, etc.), stem from a common source. Quine agrees, proclaiming that Whitehead and Russell conflate the propositional function—a mind and language independent attribute—with the open wff involved in its comprehension. Attributes are not "simple" or "complex," linguistic expression are. If the comprehension principles mark a genuine ontological commitment to mind and language independent attributes, then the wffs expressing their exemplification conditions do not reflect anything about the ontological nature of the attribute itself. Ramification simply "calls for amputation." Gödel and Quine are quite right. Conceived platonistically, attributes are independent of their linguistic or mental representation. The VCP is empty.

There is a way to make the VCP relevant, however. If we take the fact that *Principia*'s Introduction references the new recursive (and, therefore, "no-propositions") definition of "truth" and "falsehood" seriously, then we see that the order part of order\type indices on predicate variables tracks the meaning of "truth" as applied to wffs. If the semantics for *Principia*'s predicate variables is nominalistic, then all variants of the VCP apply. Consider, for instance, the following formulation of the VCP (*PM*, p. 37):

> Given any set of objects such that, if we suppose the set to have a total, it will contain members which presuppose this total, then such a set cannot have a total.

In applying this formulation of the VCP, Whitehead and Russell have to claim that a propositional function is an "ambiguity awaiting determination" and as such "presupposes that its values are determinate" (*PM*, p. 39):

> A function is not a well-defined function unless all of its values are already well-defined. It follows from this that no function can have among its values anything that presupposes the function, for if it had, we could not regard the objects ambiguously denoted by the function as definite until the function was definite, while conversely, as we have just seen, the function cannot be definite until its values are definite.

The notion of "presupposition" is quite out of place if predicate variables are interpreted as ranging over a Platonic realm of attributes. But if the calculus of *Principia* is motivated by a nominalistic interpretation, the introduction of a given predicate variable requires that the wffs its semantic interpretation assigns it be determinate in advance. The introduction of the predicate variable, as it were, "presupposes" the wffs that are its values. When interpreted nominalistically, the order indices on predicate variables can be seen to be appropriate to the truth conditions of the wffs the variable dummies. The process of generating new languages may continue as high as one likes. But it cannot be extended to a "limit." Impredicativity is ruled out.

The idea of a nominalistic reading of Whitehead and Russell's notion of a "propositional function" is not new.[15] Many have found passages of *Principia* which would seem explicit. Consider the following (*PM*, p. 38):

> By a "propositional function" we mean something which contains a variable x, and become a *proposition* as soon as a value is assigned to x. That is to say, it differs from a proposition solely by the fact that it is ambiguous: It contains a variable of which the value is unassigned.

Since there are no propositions in an ontological sense, it seems clear that a propositional function is just an open wff. The view is repeated by Russell in the 1911/2 article "On The Notion of Cause" (*ONC*, p. 133, fn1). Indeed, similar passages occur in the 1919 *Introduction to Mathematical Philosophy* (pp. 155 and 157) and in the 1959 *My Philosophical Development* (pp. 62 and 92). In these later works, Russell explicitly advocates a nominalistic interpretation.

All too often, however, interpreters construct the idea of a nominalistic semantics in straw and summarily reject it. It would not, for instance, be appropriate to construe *Principia*'s individual variables nominalistically. *Principia*'s infinity "axiom" for individuals must retain its objectual reading. Moreover, the nominalistic semantics for the predicate variables does not mean Russell adopted nominalism. There are universals; and indeed, universals have both an individual and a predicable nature. But universals are values of the *individual* variables, not the predicate variables of *Principia*. The predicate variables alone are given a nominalistic interpretation in the intended semantics.

The nature of a nominalistic semantics for predicate variables must not be misunderstood. Whitehead and Russell offer a "direct inspection" argument, according to which a propositional function is essentially an "ambiguity awaiting determination." Since the ambiguity must be eliminated, they say, e.g., that "$\phi\hat{x}$ is a man" is meaningless. Chihara construes this as follows (Chihara 1973, p. 26):

> If we interpret the authors of PM to be thinking here of propositional functions as being open sentences, then the sentence "$\phi\hat{x}$ is a man" is neither meaningless nor indefinite. . . . Thus if "$\phi\hat{x}$" stands for (denotes)

15. For instance, Sainsbury (1979, p. 287) saw it clearly and so did Kripke (1976) in his discussion of a substitutional semantics for quantification.

> "x is a man," then "$\phi\hat{x}$ is a man" says that the open sentence "x is a man" is a man, and hence what the sentence says is false, not meaningless or indeterminate.

This misunderstands the situation.[16] Chihara takes an objectual reading of predicate variables—with open wffs in their range. But an appropriate nominalistic semantics takes predicate variables to be dummy schematic letters for wffs in a modern "substitutional" interpretation of quantification such as that afforded by Kripke (1976) or by Cocchiarella (1986).[17] In a nominalistic semantics, wffs must "fit within" one another. Such a semantics cannot support "ϕ is a man." In the above example, it would render "is a man is a man" and this is indeed meaningless as Whitehead and Russell said.

If predicate variables are to be interpreted as dummy schematic letters for wffs of a fixed language, the recursive truth-definition of the semantics (the definition of satisfaction) becomes essential to the intelligibility of the predicate variables. Accordingly, the variables of quantification are limited by the meaning of "truth" as applied to the wffs in which they occur. The limitations are not "restrictions" because they are a part of the significance conditions. This philosophically justifies the (suppressed) order indices of the formal language. The order index of a term must take into consideration the order index of the term it can take as argument. Were it not for this, the definition of "truth" would not be foundational and would not depend upon the sort of quantifier symbols involved.

Now Sainsbury has seen this. But he argues that attractive as a nominalistic semantics is, it cannot help. Russell would have to argue that "any coherent function quantifier has to meet the threat of vicious circularity and so has to be set in a ramified hierarchy" (Sainsbury 1979, p. 287). Well, the VCP is certainly irrelevant to any theory which posits an ontology of attributes according to unabashedly contingent and non-logical principles. The VCP is equally irrelevant if (as in Zermelo–Frankel set theory) an ontology of classes is assumed outright in accordance with ad hoc principles for avoiding paradoxes. But it is relative to the context of logicism that the question as to the relevance of the VCP has to be assessed. The paradoxes confronting logicism have to be *solved*, not dodged by some ad hoc device. Indeed, in Russell's view, they must be solved within a calculus for logic whose *genuine* variables are wholly unrestricted. The VCP is empty if one abandons the doctrine of the unrestricted variable and logicism. But this cannot be a criticism of Russell who had to work with both.

16. There is another similar misunderstanding when Chihara remarks that "$\phi\hat{x}$ is a propositional function" is meaningful in *Principia* as a "true statement about an ambiguity" (Chihara 1973, p. 24). Recall that "is a propositional function" is not a legitimate expression of the formal language of *Principia*. When Whitehead and Russell say "$\phi\hat{x}$" is a propositional function they usually mean to say that "ϕ" is a predicate variable.

17. See Cocchiarella (1986, p. 42). Cocchiarella does not think a nominalistic semantics is appropriate for *Principia*, however.

A nominalistic semantics validates the VCP. Observe, however, that on the present interpretation the VCP is not itself the solution of anything. In *Principia*, Russell viewed the VCP as he had in earlier articles—as a guiding principle toward finding a solution within logic that *preserves* the doctrine of the unrestricted variable of *Principles*. It is the "direct inspection" argument and the informal nominalistic semantics that is doing the work, not the VCP.

The Philosophical Justification of the Type Part of an Order/Type Index

10.8 Whitehead and Russell separate the matter of justifying the grammar of the formal calculus into two parts: the justification of the type part of an order\type index; and the justification of the order part of an order\type index. The type part is rendered in the section of *Principia*'s Introduction entitled "Why a Given Function Requires Arguments of a Certain Type" and shall be the concern of this section. Orders will be examined in the next.

The philosophical justification of the type part of the order\type indices on predicate variables lies in the so-called "direct inspection" argument of *Principia*. We are to glean the need for the type part of the order\type index by inspection of the nature of "propositional functions" themselves.

According to the argument, "propositional functions" are said to be "ambiguities awaiting determination." Now we have seen that Whitehead and Russell often use the phrase "propositional function" equivocally. Sometimes the intent is speak of an open wff of the object language, other times a predicate variable of that language. What is intended by saying a propositional function is an "ambiguity"?

The "direct inspection" argument offers a philosophical justification of type indices on lines similar to Frege's hierarchy of "levels" of functions—an ontological hierarchy introduced on a linguistic paradigm. For Frege, a function is unsaturated. Second-level functions *mutually* saturate with first-level functions. First-level functions do not *fall under* second-level functions; they *fall within* them in such a way that the first-level function always preserves (as it were) a predicate position. For this reason, Frege's formal grammar did not permit function expressions occurring in subject positions. The direct inspection argument is akin to Frege's view that *falling within* must be distinguished from *falling under*. But it diverges from Frege in allowing predicative variables in "subject positions" insofar as the variables are fixed with type indices that represent (in Frege's words) *falling within*.

The passages in *Principia* concerning the "ambiguity of a function" can be interpreted so as to support this reading. Whitehead and Russell write (*PM*, p. 47):

> A direct consideration of the kinds of functions which have functions as arguments and the kinds of functions which have arguments other than functions will show, if we are not mistaken, that not only is it impossible

> for a function $\phi\hat{z}$ to have itself or anything derived from it as argument, but that if $\psi\hat{z}$ is another function such that there are arguments a with which both "ϕa" and "ψa" are significant, then $\psi\hat{z}$ and anything derived from it cannot significantly be arguments to $\phi\hat{z}$. This arises from the fact that a function is essentially an ambiguity, and that, if it is to occur in a definite proposition, it must occur in such a way that the ambiguity has disappeared, and a wholly unambiguous statement has resulted.

The idea is that a "function" (i.e., a predicate variable) can occur in a subject position (argument position) of another predicate variable only if this position represents a predicate position in the semantics. Whitehead and Russell go on to generalize (*PM*, p. 48):

> When a function can occur significantly as argument, something which is not a function cannot occur significantly as argument. But conversely, when something which is not a function can occur significantly as an argument, a function cannot occur significantly.

If the semantics is nominalistic (with predicate variables interpreted as dummy schematic letters for formulae) this idea is validated.

To see this, let us take the illustration that Whitehead and Russell used to make the direct inspection argument. They write (*PM*, p. 48):

> Take, e.g., "x is a man," and consider "$\phi\hat{z}$ is a man." Here there is nothing definite which is said to be a man. A function, in fact, is not a definite object which could be or not be a man; it is a mere ambiguity awaiting determination, and in order that it may occur significantly it must receive the necessary determination.

Taking a nominalistic semantics for predicate variables, consider "$\phi(\psi)$". If the semantics assigns "ϕ" to the formula " . . . is a man", and assigns "ψ" to " . . . is mortal" the semantic interpretation of "$\phi(\psi)$" would be the ungrammatical ". . . is mortal is a man." However, if "ϕ" is to be assigned "Every thing is such that . . . it . . .", then the result is "Everything is such that it is mortal". Putting "$\phi^{((o))}$" and "$\psi^{(o)}$" to track the semantic requirement that the formula assigned to "$\psi^{(o)}$" must occupy a predicate position in the formula assigned to "$\phi^{((o))}$, type indices on the predicate variables of the formal grammar are thereby philosophically justified.[18]

18. This interpretation has the merit of being testable. For disconfirmation, we have only to find an occasion where Russell allows a predicate variable to occur in a context which cannot represent a *falling within*. Interestingly, the Grelling paradox requires the two-place predicate "denotes" and it would not be possible to have "denotes$^{(o,(o))}$" for the type index "(o)" would not represent a *falling within* in this case. The Grelling paradox appeared in 1908. Curiously, Russell did not mention it. Perhaps we see why.

The Philosophical Justification of the Order Part of an Order/Type Index

10.9 As is well known, *Principia* offers two renditions of quantification theory. In our first section we set out the system they call ∗10. This system is regarded by Whitehead and Russell as something of a formal convenience. The system of quantification theory which reflects the philosophical foundations of the system of order\types is *Principia*'s ∗9.

The system of ∗9 can be stated as follows. The individual variables and predicate variables are just those of ∗10. The set of *elementary* wffs of the language is next determined inductively as the smallest set K containing all the atomic wffs and such that $\sim A$, $(A \vee B)$ are in K if A and B are in K. The set of *Q-formulae* of the language is then determined inductively as the smallest set K' containing $(\forall\mu^t)A$ and $(\exists\mu^t)A$ where A is an elementary wff containing μ^t free and such that $(\forall\mu^t)C$ and $(\exists\mu^t)C$ are in K' whenever C is in K' and contains μ^t free. The wffs of the language are then the elementary and Q-formulae.

Using the letters "p", "q", "r", etc., as schematic letters for *elementary* (quantifier-free) wffs, the deductive system for elementary wffs is then:

∗1.2 Pp $p \vee p \;.\!\supset\!.\; p$
∗1.3 Pp $q \;.\!\supset\!.\; p \vee q$
∗1.4 Pp $p \vee q \;.\!\supset\!.\; q \vee p$
∗1.5 Pp $p \vee (q \vee r) \;:\!\supset\!:\; q \vee (p \vee r)$
∗1.6 Pp $q \supset r \;:\!\supset\!:\; p \vee q \;.\!\supset\!.\; p \vee r$
∗1.1 *Modus ponens*$_1$
From p and $p \supset q$, infer q
∗1.01 $p \supset q = \mathrm{df} \sim p \vee q$
∗3.01 $p \,\&\, q = \mathrm{df} \sim (\sim p \vee \sim q)$
∗4.01 $p \equiv q = \mathrm{df}\ p \supset q \;.\&.\; q \supset p.$

This is then extended to quantified wffs in virtue of the following definitions (which have been modernized below):

∗9.01 $\sim (\forall\mu^t)A = \mathrm{df}\ (\exists\mu^t) \sim A$
∗9.02 $\sim (\exists\mu^t)A = \mathrm{df}\ (\forall\mu^t) \sim A$
∗9.03 $(\forall\mu^t)A(\mu^t) \vee p = \mathrm{df}\ (\forall\mu^t)(A(\mu^t) \vee p)$
∗9.04 $p \vee (\forall\mu^t)A(\mu^t) = \mathrm{df}\ (\forall\mu^t)(p \vee A(\mu^t))$
∗9.05 $(\exists\mu^t)A(\mu^t) \vee p = \mathrm{df}\ (\exists\mu^t)(A(\mu^t) \vee p)$
∗9.06 $\mathrm{p} \vee (\exists\mu^t)\mathrm{A}(\mu^t) = \mathrm{df}\ (\exists\mu^t)(\mathrm{p} \vee \mathrm{A}(\mu^t))$

These definitions require that p is a quantifier free wff in which μ does not occur free. New definitions are needed for other cases. As we saw in our discussion of Russell's article "Les paradoxes," we shall have to stray from *Principia* here, adding ∗9.XX and ∗9.YY.

∗9.07 $(\forall\mu^t)A(\mu^t) \vee (\exists\nu^n)B(\nu^n) = \mathrm{df}$
$(\forall\mu^t)(\exists\nu^n)(A(\mu^t) \vee B(\nu^n))$

$$*9.08 \quad (\exists\mu^t)A(\mu^t) \vee (\forall\nu^n)B(\nu^n) = \text{df} \quad (\forall\nu^n)(\exists\mu^t)(A(\mu^t) \vee B(\nu^n))$$

$$*9.XX \quad (\exists\mu^t)A(\mu^t) \vee (\exists\nu^n)B(\nu^n) = \text{df} \quad (\exists\mu^t)(\exists\nu^n)(A(\mu^t) \vee B(\nu^n))$$

$$*9.YY \quad (\forall\mu^t)A(\mu^t) \vee (\forall\nu^n)B(\nu^n) = \text{df} \quad (\forall\mu^t)(\forall\nu^n)(A(\mu^t) \vee B(\nu^n)).$$

For the theory of deduction for quantified wffs, we find:

$*9.1$ Pp $A[\alpha^t|\mu^t]$.⊃. $(\exists\mu^t)\ A(\mu^t)$
$*9.11$ Pp $A[\alpha^t|\mu^t] \vee A[\beta^t|\mu^t]$.⊃. $(\exists\mu^t)\ A(\mu^t)$.

From $*9.1$ Russell proves *universal instantiation*:

$*9.2$ $(\forall\mu^t)\ A(\mu^t)$.⊃. $A[\alpha^t|\mu^t]$,

where α^t is free for μ^t in A.

The system continues with inference rules for wffs that are not elementary:

$*9.12$ Pp *Modus ponens*$_2$
From A and $A \supset B$, infer B.
$*9.13$ Pp *Universal generalization*$_2$
From $A(\mu^t)$, infer $(\forall\mu^t)A$.

Absent but implicit is the following rule:

Variable rewrite
From $X[A]$, infer $X[B|A]$,
where B results from replacing all bound occurrences of the individual variable μ in A by occurrences of the individual variable ν, and ν does not occur in A.

Also absent and needed[19] is the rule allowing switching the order of quantifiers:

Switch
From $X[(\forall\mu^t)(\exists\nu^t)A]$, infer $X[(\exists\nu^t)(\forall\mu^t)A]$,
where all free occurrences of μ^t in A are on one side of a logical connective and all free occurrences of the variable ν^t in A are on the other.

Finally, reducibility handles comprehension:

$*1910$(Reduc)
$$(\exists x^{(t_1,\ldots,t_j)})(\forall y_1^{t_1}), \ldots, (\forall y_j^{t_j})(x^{(t_1,\ldots,t_j)}(y_1^{t_1}, \ldots, y_j^{t_j}) \equiv A),$$

where $x^{(t_1,\ldots,t_j)}$ does not occur free in A.

10.10 The system of $*9$ is designed to be sensitive to the philosophical distinctions relevant to Whitehead and Russell's explanation of the order part of the order\type indices on predicate variables. Central to that explanation is the abandonment of Russell's former ontology of propositions. With a no-

19. See Chapter 8.

propositions theory, "truth" and "falsehood" no longer need be regarded as primitive properties. Rather, there are now to be different senses of "truth" and "falsehood" as applied to statements differing in structure. The different senses, in turn, explain and philosophically justify the order part of order\type indices on a predicate variable.

The different senses of "truth" and "falsehood" are generated by a recursive analysis which defines "truth"and "falsehood" in terms of correspondence of belief with fact. Whitehead and Russell write (*PM*, p. 42):

> That the words 'true' and 'false' have many different meanings according to the kind of proposition to which they are applied, is not difficult to see. Let us take any function $\phi\hat{x}$, and let ϕa be one of its values. Let us call the sort of truth applicable to ϕa '*first* truth'. (This is not to assume that this would be first truth in another context; it is merely to indicate that it is the first sort of truth in our context.) Consider now the proposition $(x)\phi x$. If this has truth of the sort appropriate to it, that will mean that every value ϕx has 'first-truth'. . . . Thus if we call the sort of truth that is appropriate to $(x)\phi x$ '*second* truth,' we may define '$\{(x).\phi x\}$ has second truth' as meaning 'every value for $\phi\hat{x}$ has first truth,' i.e., '(x). (ϕx has first truth)'. . . . Similar remarks apply to falsehood.

This is unquestionably a recursive definition. But to fully understand how it is used to explain the order part of an order\type index, we must go a bit further than this rough sketch.

For the time being, let us introduce very finely grained indices. Instead of variables such as

$$^{1}\phi^{(\mathrm{o}\backslash\mathrm{o})}, \quad ^{2}\phi^{(1\backslash(\mathrm{o}\backslash\mathrm{o}))},$$

we put,

$$^{1.s;e.a}\phi^{(\mathrm{o}\backslash\mathrm{o})}, \quad ^{2.t;1.s;e.a}\phi^{(1.n;e.m\backslash(\mathrm{o}\backslash\mathrm{o}))},$$

and so on. Here $m \leq a$ and each $n \leq s$. Atomic wffs now look like this: $^{1.0;e.1}\phi^{(\mathrm{o}\backslash\mathrm{o},\ldots,\mathrm{o}\backslash\mathrm{o})}\,({}^{\mathrm{o}}x_1^{\mathrm{o}},\ldots,{}^{\mathrm{o}}x_n^{\mathrm{o}})$. The recursive characterization of "truth" "and falsehood" for quantified wffs would then be defined with respect to our finely indexed predicate variables as follows:

(0) where A is an atomic wff $^{1.\mathrm{o};e.1}\phi^{(\mathrm{o}\backslash\mathrm{o},\ldots,\mathrm{o}\backslash\mathrm{o})}\,({}^{\mathrm{o}}x_1^{\mathrm{o}},\ldots,{}^{\mathrm{o}}x_n^{\mathrm{o}})$ and R is an n-place universal and $c,\ldots,c_n$ are entities, and the semantic interpretation assigns R to $^{1.\mathrm{o};e.1}\phi^{(\mathrm{o}\backslash\mathrm{o},\ldots,\mathrm{o}\backslash\mathrm{o})}$, and $c_1.\ldots,c_n$ to ${}^{\mathrm{o}}x_1^{\mathrm{o}},\ldots,{}^{\mathrm{o}}x_n^{\mathrm{o}}$ respectively, then $[A]$ is true$_{\mathrm{o}}$ iff there is a complex (fact) consisting of c_1, and , . . . , and, c_n (in proper order) and R, with R occurring as concept.

(i) For any atomic wff A,

 $\sim [A]$ is true$_{\mathrm{e.1}}$ iff $[A]$ is not true$_{\mathrm{o}}$

(ii) For any elementary wffs A and B,

 $[A \vee B]$ is true $_{\mathrm{e.a}}$ iff

 Either $[A]$ is true$_{\mathrm{e.b}}$ or $[B]$ is true$_{\mathrm{e.c}}$, for some b, c such that $b + c = a$.

(iii)a For any wff A containing ${}^{\mathrm{o}}x^{\mathrm{o}}$ free,

$[(\forall^{o}x^{o})A]$ is $\text{true}_{z.w;\ldots;k.m;\ldots;1.s;e.a}$ iff
$(\forall^{o}x^{o})(\,[A]$ is $\text{true}_{z.w;\ldots;k.m;\ldots;1.(s-1);e.a})$

(iii)b For any wff A containing ${}^{o}x^{o}$ free,
$[(\exists^{o}x^{o})A]$ is $\text{true}_{z.w;\ldots;k.m;\ldots;1.s;e.a}$ iff
$[\exists^{o}x^{o})(\,[A]$ is $\text{true}_{z.w;\ldots;k.m;\ldots;1.(s-1);e.a})$.

We also need:

(iv)a for any wff A,
$[A]$ is $\text{true}_{z.o;\ldots;k+1.o;k.m;\ldots;1.s;e.a}$ iff
$[A]$ is $\text{true}_{k.m;\ldots;1.s;e.a}$.

(iv)b For any wff A,
$[A]$ is $\text{true}_{1.o;e.a}$ iff $[A]$ is $\text{true}_{e.a}$.

Each of the indices indicates the *number* and the order of quantifiers that occur in the wff A in question. For predicate variables, we have:

(v)a For any wff A containing ${}^{k.m;\ldots;1.s;e.a}\phi^{t}$ free,
$[(\forall^{k.m;\ldots;1.s;e.a}\phi^{t})A]$ is true_{q} iff
$(\forall^{k.m;\ldots;1.s;e.a}\phi^{t})([A]$ is $\text{true}_{\text{p}})$.

(v)b For any wff A containing ${}^{k.m;\ldots;1.s;e.a}\phi^{t}$ free,
$[\exists^{k.m;\ldots;1.s;e.a}\phi^{t})A]$ is true_{q} iff
$(\exists^{k.m;\ldots;1.s;e.a}\phi^{t})([A]$ is $\text{true}_{p})$.

(vi) For any wff A,
$[A]$ is false_{n} iff $[A]$ is not true_{n}.

In (v)a and (v)b the index q is to a sequence of numerals appropriate to,

$$z.w;\ldots;k+1.n;k.m^{\circ};\ldots;1.s^{\circ};e.a^{\circ}.$$

The index p is to be a sequence of numerals appropriate to,

$$z.w;\ldots;k+1.n-1;k.m^{\circ}+(mxd);\ldots;1.s^{\circ}+(sxd);e.a^{\circ}+(axd),$$

where d is the number of occurrences of the bound predicate variable in predicate positions in A when A is in primitive notation.

It should be noted that in the above recursion the wff A is in prenex normal form. This is needed else the recursion would not apply to wffs with quantified wffs occurring subordinate to a propositional connective. For example, where p is a quantifier-free statement not containing μ free. The meaning of "truth" for a statement such as $\ulcorner(\forall\mu)A\mu \vee p\urcorner$ is to be the same as for $\ulcorner(\forall\mu)(A\mu \vee p)\urcorner$, because the former shall be defined in terms of the latter.

As we can see, the order indices of the predicate variables reflect the recursive definition of "truth" and "falsehood." Whitehead and Russell's idea is that this philosophically explains them. The order part of an order\type index is designed to reflect the nature of the recursion.

Of course, such finely grained indices are not adopted in *Principia*. Nor could they have been, else reducibility would have to fix the suborders as well as the orders. And where would they be fixed? Certainly not at 1. Fortunately, *Principia*'s claim that the philosophical justification of orders lies in the recursive truth definition is not committed to the finely grained order indices above. The order indices on predicate variables reflect only the order of truth in the recursion, first, second, third, etc., and not also the suborders within first, second, third, etc., which reflect the *number* of quantifiers of the given order. Whitehead and Russell do maintain that the number is relevant to the meaning of "truth" and "falsehood" (*PM*, p. 162):

> First-order propositions are not all of the same type, since as was explained in ∗9, two propositions which do not contain the same number of apparent variables cannot be of the same type.

All the same, Whitehead and Russell downplay this feature on the grounds that "no reflexive fallacies result" (*ibid.*). More exactly, the suborders are not reflected in the order indices of the variables because the suborders do not prevent the recursion from reaching its base case. An increase in the suborder will not violate the recursive definition, but increase the order and the recursion cannot work. Accordingly, although the senses of "truth" and "falsehood" change relative to the number of quantifiers, and although the predicate variables of the formal language have order indices to reflect the recursive truth conditions of wffs, the finely grained order indices would not be warranted by the constraints of the recursion.

Returning then to the usual order and order\type indices on the variables as we find in *Principia*, we can see the relationship between the indices and the truth-conditions as follows. For the moment, put aside reducibility comprehension principles in favor of a fully predicative type theory. (Recall that in *Principia* reducibility principles are the only comprehension principles.) The present comprehension principles (which will soon be obviated by the more general reducibility principles) are now:

$\mathrm{Pp}_1\ (\exists^1\phi^{(o\backslash o)})(\forall^o z^o)({}^1\phi^{(o\backslash o)}({}^o z^o)\ .\!\equiv\!.\ A)$,
where (i) ${}^1\phi^{(o\backslash o)}$ is not free in A; and (ii) A has first-order truth or first-order falsehood.
$\mathrm{Pp}_2\ (\exists^2\phi^{(1\backslash(o\backslash o))})(\forall^1\psi^{(o\backslash o)})({}^2\phi^{(1\backslash(o\backslash o))}({}^1\psi^{(o\backslash o)})\ .\!\equiv\!.\ A)$,
where (i) ${}^2\phi^{(1\backslash(o\backslash o))}$ is not free in A; and (ii) A has second-order truth or second-order falsehood.[20]

20. It might be objected that Russell has:

$$(\exists^2\phi^{(1\backslash(o\backslash o))})(\forall^1\psi^{(o\backslash o)})({}^2\phi^{(1\backslash(o\backslash o))}({}^1\psi^{(o\backslash o)})\ .\!\equiv\!.\ (\forall^o y^o)\,{}^1\psi^{(o\backslash o)}({}^o y^o))$$

as an instance of comprehension where A is $(\forall^o y^o)\,{}^1\psi^{(o\backslash o)}({}^o y^o)$ and yet here A has truth$_1$, not truth$_2$. Shouldn't there be a variable ${}^1\phi^{(1\backslash(o\backslash o))}$? The answer is that no such variables are needed. It is always possible get an equivalent. Just put for A,

$$(\forall^1\phi^{(o\backslash o)})(\forall^o z^o)({}^1\phi^{(o\backslash o)}({}^o z^o) \supset {}^1\phi^{(o\backslash o)}({}^o z^o))\ .\&.\ (\forall^o y^o)\,{}^1\psi^{(o\backslash o)}({}^o y^o)$$

and this wff has truth$_2$.

And so on. Whitehead and Russell are offering a philosophical explanation and justification of the order part of the order\type indices on predicate variables by appeal to an informal semantics. The comprehension principles employ wffs A with fixed truth-conditions. The order part of order\type indices reflect the recursive truth-conditions of the wffs A used in the comprehension principles.

As we see, the hierarchy of senses of "truth" and "falsehood" are generated in such a way that they step downward, one level at a time, until the base notions of "truth" and "falsehood" for atomic wffs are reached. Since the truth-definition is offered as a semantic justification for the introduction of predicate variables with order\type indices, we get a philosophical explanation of the "matching requirement"—that is, the formal requirement that arguments to a predicate variable must match in their order indices as well as their type incides.[21] The definition defines "truth$_n$" as applied to wffs containing at least one quantifier binding variables of order $n-1$ and no quantifiers binding variables of higher order; and "truth$_n$" means that in removing this quantifier (or all such if there are many) the resulting wff has "truth$_{n-1}$." If we could instantiate the variables of order $n-1$ to, say, $n-2$, the definition of "truth$_n$" would be lost. Higher senses of "truth" (and "falsehood") are recursively defined in such a way that they collapse step-by-step to the base of atomic "truth" (and "falsehood"). The matching requirement, therefore, is a product of the semantic justification—the recursive definition of "truth" set out in the Introduction of *Principia*.

Many interpretations of *Principia* have ignored or been perplexed by the matching requirement. Why not have a cumulative hierarchy, allowing variables (and circumflex terms) of the same type but differing in order be arguments to the same predicate variable? A nominalistic semantics for predicate variables with a cumulative recursive truth-definition is possible. In a cumulative hierarchy, universal instantiation would be:

$$(\forall \mu^t)A\mu^t \ .\supset.\ A(\nu^n|\mu^t),$$

where ν^n is a term and $n \leq t$. All that seems needed to ensure that the recursion reaches its base case (for quantifier free wffs) is the demand that the order index of a predicate variable be greater than the highest-order arguments it can take. Why then not a cumulative hierarchy? The answer is that Whitehead and Russell are offering a semantics which they take to be generated from a *philosophically correct* definition of "truth" and "falsehood" in terms of correspondence. Not any semantics that turns the trick will do. This takes us to the controversial and much maligned "multiple-relation theory of judgment," for this theory was the centerpiece of Whitehead and Russell's new correspondence definition of "truth."

21. In the 1907 ramified substitutional theory, "$p/a;b!q$" was not a wff unless the order index of a matched that of b. The justification was unabashedly ad hoc; it avoids the "p_o/a_o" paradox with minimal damage to the system.

The Multiple-Relation Theory of Judgment

10.11 In *Principia*'s recursive definition of "truth," "truth" (and "falsehood") predicates flank statements and not names of statements. This may at first seem a use–mention confusion.[22] But on the correspondence theory Russell envisions, when a genuinely atomic statement is flanked by "is true" it *is* to be viewed as a sort of name—viz., a disguised definite description.

That a statement flanked by ". . . is true" is a disguised definite description is suggested by Russell in several passages. In "On the Nature of Truth," we find Russell saying that his new theory of "truth" is "an extension of the principle applied in my article, 'On Denoting' [*Mind*, October, 1905], where it is pointed out that such propositions as 'The present King of France is bald' contain no constituent corresponding to the phrase 'the present King of France' " (*OT*, p. 48). Compare the following from *Principia* (*PM*, p. 48):

> A statement in which a proposition appears as subject will only be significant if it can be reduced to a statement about the terms which appear in the proposition. A proposition, like such phrases as "the so-and-so," where grammatically it appears as subject, must be broken up into its constituents if we are to find the true subject or subjects.

A reference is then made to *Principia*'s Chapter III, where the theory of definite descriptions is introduced. Moreover, Whitehead and Russell go on as follows (*PM*, p. 44):

> A "proposition," in the sense in which a proposition is supposed to be *the* object of a judgment, is a false abstraction, because a judgment has several objects not one. . . . The phrase which expresses a proposition is what we call an "incomplete" symbol; it does not have meaning in itself, but requires some supplementation in order to acquire a complete meaning. This fact is somewhat concealed by the circumstance that judgment in itself supplies a sufficient supplement and that judgment in itself makes no *verbal* addition to the proposition.

So it seems clear that when an atomic statement flanks ". . . is true" it is to be taken as a disguised definite description. But what definite description is it to be? Help comes when we examine Russell's new view that with propositions abandoned, judgment (belief) is to be regarded as a multiple relation and not a dyadic relation between a mind and a proposition.

On the new correspondence theory, "truth" (in the simplest case) is defined as the correspondence between a belief complex and another complex composed of the objects of the belief complex. For example, when a person judges (believes) that a is R to b, there is a structured entity composed of the mind m of the person in a belief relation to the entity a and the relation R and the entity b. Accordingly,

22. See Quine (1963).

[*a* is *R* to *b*] has *elementary* (i.e., "first" or "atomic") truth

is to be analyzed in terms of a belief complex corresponding to a complex composed solely of *a* and *R* and *b*. Russell uses hyphens to nominalize a formula to make a name of a fact; for instance, he uses "*a*-*R*-*b*" or "*a*-in-the-relation-*R*-to-*b*" to name the corresponding complex. (The expression "*a*-*R*-*b*" is a term, not a statement.) For example, when Othello (whose mind is *m*) believes that Desdomona loves Cassio, there is a complex mental state,

m's believing with respect to Desdemona, loves, and Cassio.

This complex (fact) is true insofar as there is a corresponding complex (fact),

Desdemona's loving Cassio.

It is false if there is no such complex.

What then is to be the definite description that the statement "*a* is *R* to *b*" disguises when flanked by ". . . is true"? Russell's idea might be that the disguised description is:[23]

"the complex (fact) corresponding to *m*'s understanding with respect to *a* and *R* and *b*, for some mind *m*."

Russell takes 'supposing' and 'understanding' as multiple relations akin to 'believing' and these seem more appropriate for the general recursive definition.[24] Whitehead and Russell explain that

[$(x)(\text{Man}(x) \supset \text{Mortal}(x))$] has *second* truth,

is to be analyzed with,

(x)(if 'x is a man' has elementary truth then 'x is mortal' has elementary truth),

where *elementary truth* was previously defined (*PM*, 46). If we plug in the definition of "elementary truth" in terms of the multiple-relation theory, we get:

[$(x)(\text{Man}(x) \supset \text{Mortal}(x))$] has *second* truth iff for every x, if there is a fact corresponding to *m*'s understanding with respect to x and Humanity, for some mind *m*, then there is a fact corresponding to *m*'s understanding with respect to x and Mortality, for some mind *m*.

We might assume with Russell that to define "truth" as correspondence between a mental state and fact one must assume at least one mind, but surely not one in a believing relation and not one for every entity. 'Understanding,' on the other hand, turns the trick so long as it is not viewed as an occurrent state. Put generally, then, Russell and Whitehead have:

23. I took this approach in Landini (1991).

24. See *KAKD*, p. 159, and also *TK*.

> $[R(c_1, \ldots, c_n)]$ has atomic truth iff there is a fact corresponding to *m*'s understanding with respect to c_1, and, . . . , and, c_n and *R*, for some mind *m*.

Of course, it is assumed by the theory that there is a fact corresponding to this understanding complex just when there is a fact consisting of c_1, and, . . ., and c_n (in proper order),[25] and *R*, with *R* occurring as concept. Recursion then handles the connectives and quantifier cases, showing how to build up truth-conditions for quantified wffs from the atomic cases.

If this is correct, we can reply to one of the most intractable criticisms of Russell's theory. Anscombe put the problem as follows: "What happens when I judge that *A* is *not* to the right of *B*? Do I stand in the judging relation to *A*, *B*, *to the right of*, and *not*? Similar questions arise for the other logical constants, 'if,' 'and,' 'or.' "(Anscombe 1959, p. 46). Anscombe's problem properly arises when one attempts to apply the multiple-relation theory to render truth conditions for ascriptions of propositional attitudes. What are the constituents of a belief complex when one believes that all men are mortal? *Principia* never says. In *Principia*, however, the multiple-relation theory had only the more narrow task of facilitating the recursive definition of "truth" as applied to statements not involving propositional attitude verbs. On the present construal, the problem of the mental content before one's mind when the truth conditions for a sentence involving logical particles and quantifiers is rendered need never arise.

10.12 Attention to "truth" as correspondence between a mental state and fact, however, has led many to hold that the multiple-relation theory was surely intended to introduce *one* belief complex for every statement that is believed.

It is possible to preserve *Principia*'s recursive "truth" definition and yet construe the multiple-relation theory so that there is always one belief complex. One could take the statement "*a* is *R* to *b*" to disguise the following definite description when it is flanked by ". . . is true":

> "the belief complex consisting of *m* related to *R* and c^1, and, . . . , and c^n."[26]

On this construal, the disguised description purports to refer to a belief complex rather than the purported complex (fact) that would correspond to it. Whose belief complex? The person making the discursive judgment. Thus if we unguise the description in,

> "[Desdemona loves Cassio] has atomic truth,"

when asserted by Othello, we get,

> "the belief complex consisting of Othello related to Desdemona and loves and Cassio."

25. This encounters the problem of order that Russell would struggle to overcome shortly after the appearance of *Principia*'s first volume. For details, see Landini (1991).

26. This description would also face the infamous problem of order. See Landini (1991).

Accordingly,

> [Desdemona loves Cassio] is ^{m}true iff the belief complex consisting of *m* related to Desdemona, and loves, and Cassio, is true.

This then holds iff there is a complex consisting of Desdemona and loves and Cassio (in proper order).

The present alternative preserves the recursive "truth" definition as follows. We have:

> [All men are mortal] is ^{m}true iff the general belief complex consisting of *m* related to Humanity and Mortality is true.

This, in turn, holds iff for every *x*, if there is a fact consisting of *x* and Humanity, then there is a fact consisting of *x* and Mortality. This construal avoids the "too few minds" problem that Russell's recursive truth definition faces unless 'understanding' is employed as the multiple relation and is taken to be a nonoccurrent mental state.

This alternative has a certain appeal. But it faces the vexing question as to what are to be the constituents severally before the mind in a molecular or general belief. The judgment "All men are mortal," Whitehead and Russell tell us, "collects together a number of elementary judgments," though it is not composed of them, and is of a "radically new kind" (*PM*, p. 45). Just what sort of judgment is a general judgment?

Whitehead and Russell never answer this question in *Principia*. But it is very important not to answer it for them by undermining the recursive nature of *Principia*'s truth (correspondence) definition. That definition is, as we argued, the centerpiece of their philosophical justification of the order part of the order\type indices on predicate variables.

Consider, for example, Cocchiarella's innovative answer to Ansombe's question. Cocchiarella begins from the interpretation that *Principia* assumes that every open wff stands for a Platonic attribute (i.e., a "propositional function" in the ontological sense). Cocchiarella takes a belief as a multiple relation between a mind and the constituents, including propositional functions, of the would-be proposition being judged (Cocchiarella 1980, p. 102). Accordingly, he offers the following as the mental belief complex for the judgment "All men are mortal,"

> '$J\{m, (x)(\hat{\phi}x \supset \hat{\psi}x), \hat{x} \text{ is a man}, \hat{x} \text{ is mortal}\}$.'

In Cocchiarella's view, judgments (mental states) are arranged into a hierarchy of orders. The order of the judgment's truth or falsehood is the maximum of the orders of the propositional functions occurring as objects of the judgment. "Instead of taking propositions of different orders as being single entities that are themselves objectively true or false," Cocchiarella writes, "Russell now assumes that judgments as particular occurrences, or statements as potential judgments, are vehicles of a hierarchically ordered system of truth and falsehood" (Cocchiarella 1980, p. 104).

Cocchiarella then determines the notion of "correspondence" appropriate to the judgment in accordance with the order of the judgment. The judgment (above) contains a second-order propositional function, says Cocchiarella, so the judgment has *second-order* truth or *second-order* falsehood. The judgment is true when it corresponds to several complexes—viz., the-mortality-of-x_1, the-mortality-of-x_2, etc, where all the x's are men.

Cocchiarella has things top-down, establishing orders of judgments (complexes) on the basis of an independently established hierarchy of orders of attributes (propositional functions). Faced with the problems of the contents of belief complexes, Cocchiarella allows propositional functions as objects of belief complexes. Indeed, he says that "without including propositional functions among the single entities contained in a judgment or belief complex, there would simply be no multiple-relation theory of belief at all" (Cocchiarella 1989, p. 189). This cannot be correct if, as we have argued, Whitehead and Russell have things "bottom-up." Unlike Cocchiarella's interpretation, we find no hierarchy of orders of belief complexes ("judgments") at all; and no independently conceived ontological hierarchy of orders of propositional functions used to generate them. All complexes are on a par with individuals. The multiple-relation theory is meant to comport with *Principia*'s recursive definition of "truth" (and "falsehood"). It is recursion that generates the meanings of "truth" (i.e., "correspondence") and these are the very explanation of "order." In an effort to resolve the problem of the constituents of general judgments, one must not construe the multiple-relation theory in a way that dislodges recursion as the centerpiece of Whitehead and Russell's philosophical explanation of orders.[27] Cocchiarella's interpretation does just this.

What Is Logic?

10.13 Once we see that it was a recursive truth definition that generates orders, we get an entirely new perspective on *Principia*. Far from being a system of genuinely restricted variables ranging over a ramified hierarchy of Platonic entities (propositional functions), the system aims at capturing (at least in spirit) the philosophical idea that had exercised Russell ever since his *Principles of Mathematics*—viz., the doctrine that the unrestricted variable is essential to any calculus for logic. Whitehead and Russell are endeavoring to philosophically explain order\types in a way that *preserves* unrestricted variables. The "direct inspection" argument and the recursive truth definition (of

27. Perhaps the simplest way to do this is to hold that the multiple-relation theory is not part of the recursive truth (correspondence) definition. On this view, the multiple-relation theory is only intended as a sketch which shows how it is possible to disarm a central argument for an ontology of propositions. It is intended to show, that is, how one might deny that belief (judgment) must be a relation between a mind and a proposition.

which the multiple-relation theory is a component) are their attempts to achieve this goal. If the order\type indices on predicate variables are philosophically explained through their nominalistic significance conditions, then *Principia* has only one style of *genuine* variables—individual/entity variables. As Russell would put matters in 1911, "the disjunction 'universal–particular' includes all objects" (*KAKD*, p. 155). All entities, universal or particular, are values of the individual/entity variables of the *Principia*. Particulars include complexes (i.e., facts) such as *a*-in-the-relation-*R*-to-*b*, and belief complexes such as *m*-in-the-relation-believes-to-*a*-and-humanity; they also include non-complex existent individuals. Universals include sensible qualities such as 'redness,' 'roundness,' and 'hardness,' as well as relations of space and time. But there is no assumption that open wffs always stand for single entities—i.e., propositional functions in an ontological sense. Russell's calculus for logic treats all entities, universal and particular, on a par. There are no types or orders of entities.

Now in *Principles*, logic is a synthetic a priori science whose subject matter is structure, and structures are identified as propositions. What according to *Principia* is logic the science of? On the present construal, it cannot be the science of propositional functions (in an ontological sense).

Russell's ideas here began to surface in the 1912 *Problems of Philosophy* and his 1913 (unfinished book) *The Theory of Knowledge*. His ideas were sketchy. In a little manuscript entitled "What is Logic?," Russell maintains that logic is the synthetic a priori science of "logical forms." Not all particulars are concrete, it turns out. Logical forms are abstract particulars, and these particulars are the subjects of the science of logic. Russell was not certain as to the nature of such particulars, but in *Theory of Knowledge* he decides that they are special "abstract facts" that have no constituents. In the *Problems of Philosophy*, Russell tells us that "all a priori knowledge deals exclusively with the relations of universals" (*Pr*, p. 103). In particular, knowledge of arithmetic and logic lies in that "we have the power of sometimes perceiving such relations between universals" (*ibid.*, p. 105). This is repeated in *Theory of Knowledge*. Russell maintains that "perceptions of facts (as opposed to corresponding judgments) must be included as data . . . and perceived facts are not always particular; general logical facts, for example, are often such as can be perceived" (*ibid.*, p. 47).

Observe that perception, here, is regarded as a dyadic relation. Logical forms are not proper objects for a correspondence relation. Belief states cannot be made true by corresponding to logical forms. But the relation that holds between universals that one perceives in recognizing a "truth" of logic, is grounded in an intuition of a logical form. Logical forms are objects of a special form of "pure logical intuition," a form of immediate acquaintance acquired through a process of abstraction and generalization to a *summum genus* (*TK*, p. 114). In Russell's view, such intuition is involved in our understanding of logical words like "particular," "universal," "relation," "dual complex," and "predicate" (*ibid.*, p. 97).

10.14 But does *Principia* truly preserve the fundamental conception of logic as a science that is found in *Principles*? In *Principles*, "logical truth" is logical necessity, and both are founded upon the universality of the science. In the calculus of propositions that Russell originally envisioned, any *fully* closed formula that is true is a logical truth. Truths of logic apply to every entity—their genuine variables being wholly unrestricted. But in *Principia*, there are (presumably) true, fully closed formulae, which (for all that) cannot be regarded as truths of logic. For instance, there is

$$(\phi)((x)\phi x \supset (\exists x)\phi x).$$

The basic calculus of *Principia* is committed to at least one individual. Though on the present interpretation universals are counted among individuals, there is no logical assumption that would ground an infinity of them. This feature, Russell would later write, "I now view as a defect in logical purity" (*IMP*, p. 203, fn). *Principia*'s axiom of infinity is another example. Though Whitehead and Russell applied it only as an antecedent where needed, it undermines the demonstration that the concept "natural number" has been conceptually analyzed in purely logical terms.

Moreover, there are instances of reducibility that are fully general and (presumably) true, and yet are not, in the context of ramified types, proper candidates for truths of pure logic. In the 1910 *Principia*, Russell tried to gloss reducibility with a pragmatic argument. Since classes are extensional entities, they are not differentiated in terms of the orders of the wffs of which they are extensions. Accordingly, Russell maintains that if within a context of predicate variables stratified by order\types, a theory of classes had been assumed outright, with "∈" a primitive relation, variables for classes would always be of the lowest possible order compatible with their type. As Russell put it: with classes assumed as entities outright, reducibility would be true (*PM*, pp. 58, 166). Comprehension of a class of individuals would be rendered by:

$$(\exists {}^1c^{(o\backslash o)})(\forall {}^oz^o)(\in^{(o\backslash o,1\backslash(o\backslash o)} ({}^oz^o, {}^1c^{(o\backslash o)}) .\equiv. B),$$
where ${}^1c^{(o\backslash o)}$ is not free in B.

Here there are no restrictions on the bound variables that may occur in B. The comprehension principles of such a predicative order\type theory would include:

$$(\exists {}^1\phi^{(o\backslash o)})(\forall {}^oz^o)({}^1\phi^{(o\backslash o)}({}^oz^o) \equiv A),$$
where ${}^1\phi^{(o\backslash o)}$ is not free in A, and no variables in A have higher order\type than 1\(o\o).

Thus, for any existential instantiation of the comprehension schema for classes, we would get the following instance of comprehension:

$$(\exists {}^1\phi^{(o\backslash o)})(\forall {}^oz^o)({}^1\phi^{(o\backslash o)}({}^oz^o) .\equiv. \in^{(o\backslash o,1\backslash(o\backslash o))} ({}^oz^o, {}^1c^{(o\backslash o)}).$$

But then we shall have as well:

$$(\exists {}^1\phi^{(o\backslash o)})(\forall {}^oz^o)({}^1\phi^{(o\backslash o)}({}^oz^o) .\equiv. B),$$

and here there is no restriction on the bound variables occurring in *B*. That is, we have a reducibility theorem schema.

Russell seems to reason as follows. If the theory of extension is part of logic, and the calculus for logic is properly *Principia*'s ramified type theory, then the comprehension principles for the calculus for logic should reflect this fact. Reducibility principles, being the comprehension principles for the calculus, are thus regarded as truths of logic. By 1919 he thinks better. The argument begs the question in favor of logicism. The comprehension principles that are philosophically warranted are those of predicative type theory. Reducibility schemata would then be grafted on to predicative type theory to generate (together with the contextual definition for class and relation$_e$ terms) the theory of extension. Confronted with the paradoxes, logicism has to show that the theory of extension is a part of logic, not assume it. There are object-language instances of reducibility that are fully general and true, and yet not truths of logic.

Curiously, Russell still clung to the *Principles* view on logical necessity in his 1919 Logical Atomism lectures. Russell maintains that a *first-order wff* is logically necessary if, replacing all individual and predicate constants with individual and predicate variables, its universal closure is true. The idea behind such a replacement is to capture the idea that a logical truth (a logical necessity) is "true in virtue of logical form." One reaches the logical form via such a replacement—or as Russell put it, by "abstraction and generalization to a *summum genus*." This process reaches a limit with wffs that are second-order and fully general; it requires the notion in *Principles* of "logical truth" as fully general truth. Yet this cannot be maintained in *Principia*; and Russell is therefore at a loss as to the general nature of logical truth (necessity).

What Logic is Not

10.15 In the 1919 *Introduction to Mathematical Philosophy*, Russell acknowledged the problem of logical truth, and he suggests the propositions of logic and (non-applied) mathematics all have the characteristic of being "tautologies" (*IMP*, p. 204). The importance of "tautology," he explained, is due to Wittgenstein (*ibid.*, p. 204 fn). The influence of Wittgenstein is apparent. It is no less apparent that Wittgenstein had a very different notion of "tautologyhood" in mind!

For Russell, logic is a synthetic *a priori* science; it has a subject matter (logical forms) which are objects of pure logical intuition. Wittgenstein disagreed heartily. Logic is not a science at all; it has no subject matter or "content," and is not governed by any intuition of "objects." In his *Tractatus*, he writes:

> 6.1231 The mark of a logical proposition is *not* general validity. To be general means no more than to be accidentally valid for all things. An ungeneralized proposition can be tautological just as well as a generalized one.

In a similar vein, Wittgenstein wrote:

> 5.525 It is incorrect to render the proposition "$(\exists x).fx$" in the words "*fx* is *possible*," as Russell does. The certainty, possibility, or impossibility of a situation is not expressed by a proposition, but by an expressions being a tautology, a proposition with sense, or a contradiction.

If logic were a science of certain "objects," whether they be logical forms, propositions, or some other, its truths would (in Wittgenstein's view) be "accidental" truths about these objects. This challenges Russell's idea that logical necessity lies in full generality and truth. Accordingly, Wittgenstein maintains, logic is not a science of objects; it is not about anything at all. He writes:

> 6.111 All theories that make a proposition of logic appear to have content are false.
>
> 6.124 The propositions of logic describe the scaffolding of the world, or rather they represent it. They have no 'subject-matter.'

Principia's very conception that there is a calculus for logic, where one begins from some epistemically primitive truths and deduces others, is misguided in Wittgenstein's view:

> 5.452 In Russell and Whitehead's *Principia Mathematica* there occur definitions and primitive propositions expressed in words. Why this sudden appearance of words? It would require a justification, but none is given, or could be given, since the procedure is in fact illicit.

Russell's calculus for logic must be set out in a meta-language where rules for the manipulation of signs are given in words. But, Wittgenstein argues, this raises the question as to the legitimacy of the rules. Since the meta-language uses logic, it presupposes the very ideas it endeavors to set forth. Because logic is the precondition for thought (and structure generally) it cannot be made into a proper object of study.

As we can see, Russell's idea that there is a logical intuition of certain objects (logical forms) which grounds our understanding of logical notions such as "particular," "universal," "dyadic complex," "relation," "identity," and "fact," was odious to Wittgenstein. These are pseudo-concepts. The same holds for semantic notions such as "believes," "knows," "is true," "is possible," "is necessary," "designates," and "means." In Wittgenstein's view, Russell should have seen that these are pseudo-predicates to be banished with "exists." There can be no "theory" of truth; the truth-conditions of a statement are given with the very syntax of the statement. The very idea of epistemology—a "theory of knowledge"—was grotesque, and the multiple-relation theory of judgment particularly so.[28] Russell's 1912 *Problems of*

28. See Landini (1991) for a detailed discussion of Wittgenstein's objections to Russell's multiple-relation theory. For a dissenting opinion, see Griffin (1980).

Philosophy errs by pretending there is a subject matter to discuss, all such "theories" make logic into a science, as if there are logical objects (logical forms).

Wittgenstein's replacement was his notion of "tautologyhood." This is tied to his *Doctrine of Showing* from which it follows that a logically perfect language must be extensional so that such pseudo-concepts are not expressible. Logical (and semantic) notions must be given with the symbolism. The "truths" of logic and mathematics would show that they are part of the scaffolding for thought and the world through the syntax itself. Logic, in Wittgenstein's view, is by its very definition decidable; and this is connoted by his use of the word "tautology." Here are a few passages from the *Tractatus*:

> 6.126 One can calculate [determine] whether a proposition belongs to logic by calculating the logical properties of the *symbol*.
>
> 6.1265 It is always possible to construe logic in such a way that every proposition is its own proof.
>
> 6.127 All propositions of logic are of equal status; it is not the case that some of them are essentially primitive propositions and others are essentially derived propositions. Every tautology itself shows that it is a tautology.

Decidability was to supplant Russell's conception of logic as a science of structure. Logic is not a science of "objects" (structures) of any sort, be they reified as propositions, logical forms, or otherwise.

Mathematics is in the same category. Wittgenstein was not a "logicist" in the sense of advocating a "reduction" of (non-applied mathematics) to logic via the theory of extension (classes and relations-in-extension). Nonetheless, mathematics, like logic, is not a science, Wittgenstein writes:

> 6.031 The theory of classes is completely superfluous in mathematics. This is connected with the fact that the generality required in mathematics is not accidental generality.

This does not challenge the idea of logicism; it challenges Whitehead, Russell's, and Frege's methods. If one takes (non-applied) mathematics to be about classes it makes it appear as if it has a subject matter—classes; and its "truths" seem as if they are accidental truths of these objects. There is no subject-matter to mathematics in Wittgenstein's view. Each equation stands by itself, revealing its necessity in the symbolism alone.

> 6.2 Mathematics is a logical method. The propositions of mathematics are equations, and therefore pseudo-propositions.
>
> 6.22 The logic of the world, which is shown in tautologies by the propositions of logic, is shown in equations by the equations of mathematics.

This is surely in the spirit of logicism, for both logic and mathematics are conceptualized in the same way. The logical/mathematical structure of the world (and of thought) would be shown in the syntax of a perfect language.

We now see that Wittgenstein used the word "tautology" (and "equation") in a sense quite different from what Russell had realized. For Wittgenstein, they connote decidability. When, in 1919 Russell agreed that Wittgenstein's notion of a "tautology" might show the way toward a new criterion for logical and mathematical truth, he missed this connotation entirely. Furthermore, it is quite clear that, since he continued to view both logic and mathematics as a science, he rejected both it and the doctrine of showing upon which it is based.[29]

Wittgenstein had been enthralled by Russell's endeavors to build logical distinctions in to logical syntax—endeavors motivated by Russell's tireless struggle to preserve unrestricted variables in his calculus for logic. At first, Wittgenstein appeared to Russell as if a protégé—someone who would carry the torch. Distilled by Wittgenstein, however, Russell's quest to build logical distinctions into logical grammar was transformed into the *Doctrine of Showing*. The doctrine undermines Russell's conception of logic as a science. In its sweeping dismissal of logical and semantic notions as pseudo-concepts, it was destructive of Russell's positive contributions. In short, the doctrine swept away the baby with the bath water!

History proved Russell the master. Church showed that no predicate calculus is decidable. The doctrine of showing cannot be substantiated, and its call for extensionality does not hold the key to the philosophy of logic and mathematics.[30] Indeed, as Church (1974) aptly noted, extensionality is antithetical to Russell's very conception of logic. A Russellian calculus for logic, he admonishes, should be an intensional calculus which embraces propositions, for this is what is clearly demanded by the background and purpose of Russell's logic (Church 1976, p. 291).

Church is quite correct. But Church turns to a realist construal of *Principia Mathematica*—embellished with an ontology of propositions as values of a type-theory of attributes ("propositional functions"). Then, attending to the paradoxes of propositions (such as the *Principles* "Appendix B," and the *liar*) he is led to a dark conclusion (Church 1984, p. 521):

> If, following early Russell, we hold that the object of an assertion or belief is a proposition and then impose on propositions the strong conditions of identity which this requires, while at the same time undertaking to formulate a logic that will suffice for classical mathematics, we therefore find no alternative except for ramified type theory with axioms of reducibility.

29. In his 1921 introduction to the *Tractatus*, Russell explicitly rejects the doctrine of showing, noting that it is avoidable by means of a hierarchy of languages.

30. Russell worked out some of the consequences of accepting extensionality in his Appendix B and Introduction to the second edition of *Principia Mathematica*. Mathematical induction can be recovered without reducibility in the new system. Nonetheless, the success is limited. See Landini (1996).

Church draws this conclusion on the basis of the "Appendix B" paradox (instead of the *liar*). But Church missed the fact that *Principia* is not the proper background for Russell's logic of propositions. The proper background is the substitutional theory—a theory which genuinely solves the paradoxes of classes and predication. The *liar* can be dismissed altogether, and the Appendix B found a solution in *Les paradoxes*. Substitution has only begun to be explored. Though logicism is commonly thought to be dead, the pronouncement is premature. A genuine solution of the p_o/a_o paradox (and variants) may yet be found; and if it is, the conception of logic in *Principles* will be saved by substitution just as Russell had originally hoped.

Appendix A: Proof of the Peano Postulates

A.0 So that the reader can get a firm understanding of the technique of the substitutional theory, we shall prove the Peano Postulates. Our plan is to first prove postulates 1, 2, 3 and 5. We then discuss the matter of proving that the universal class of individuals is not a member of any natural number. Finally, we return to Peano Postulate$_4$. In the theorems that follow, we shall continue our practice of first writing the familiar class-theoretic formulation and then the substitutional formulation where the type becomes a salient feature of the grammar. We shall also continue our practice of introducing convenient definitions to help the reader in understanding the theorems and also to aid in their expression.

Finite Cardinals

A.1 Finite cardinal numbers are constructed in the usual Frege/Russell manner—though now regimented as to type.

df(0)

$$0(rc) = \text{df } (y) \sim (y \in \imath(l/m)[l/m \approx r/c]).$$

We next introduce definitions for 'successor' and 'natural number' in the substitutional theory. Recall that the symbol for the successor of a class n is defined as follows:

$$S(n) = \text{df } \hat{\alpha}(\ (\exists y)(y \in \alpha \ \&\ \alpha - \iota y \in n)).$$

Here we have:

$$\alpha - \iota y = \text{df } \hat{z}(z \in \alpha \ \&\ z \neq y).$$

In the language of substitution we shall put:

df(S)

$$\text{S}(hde)(rc) \;=\text{df}\; (\exists y)(y \in \imath(q/b)[q/b \approx r/c] \;\&\; \imath(p/a)[p/a \;.{\approx_z}.\; (r/c - \iota y)(z)] \in \imath(s/tw)[s/tw \approx h/de]).$$

The existence theorem for successor may now be stated:

Successor$_1$
$\vdash \;(hde)(\exists stw)(s/tw \approx_{rc}\; S(hde)(rc)).$

The theorem follows directly from $(\text{CP}_\in)^2$. The Frege/Russell definition of the property of being a natural number was (in class-theoretic notation) as follows:

$$N = \text{df}\; \hat{\delta}(\alpha)(0 \in \alpha \;\&\; (\beta)(\beta \in \alpha \supset S(\beta) \in \alpha) \;.\!\supset\!.\; \delta \in \alpha).$$

That is, the set of natural numbers is the smallest set which contains 0 and the successor of every element it contains. To put this in substitutional terms, we first give the definition schema:

df(N)
$\text{N}(qpa) \;= \text{df}$
$(lmno)(\;\imath(s/tw)[s/tw \approx_{rc} 0(rc)] \in \;\imath(j/kuv)[j/kuv \approx l/mno]$
$\quad\&\; (hde)(\;\imath(s/tw)[s/tw \approx h/de] \in \;\imath(j/kuv)[j/kuv \approx l/mno] \;.\!\supset\!.$
$\quad\quad\imath(s/tw)[s/tw \approx_{rc} S(hde)(rc)] \in \;\imath(j/kuv)[j/kuv \approx l/mno])$
$\quad\quad.\!\supset\!.\, \imath(s/tw)[s/tw \approx q/pa] \in \;\imath(j/kuv)[j/kuv \approx l/mno])$

Thus, the existence theorem for the class of natural numbers is:

$\vdash \;(\exists jkuv)(j/kuv \approx_{qpa}\; N(qpa)).$

This follows from $*(\text{CP}_\in)^3$.

The Peano Postulates (Excepting Peano$_4$)

A.2 We are now in a position to prove the Peano postulates in the substitutional theory. We begin with the following:

$(\text{T}_{\text{A.1}})\quad \vdash\; \Lambda^{(o)} \in 0^{((o))}$
$\vdash\; \imath(p/a)[p/a \approx_z \Lambda z] \in \;\imath(s/tw)[s/tw \approx_{rc} 0(rc)].$

Proof:
1. $(\exists pa)(p/a \approx_z \Lambda z)\quad *(\text{CP}_\in)^1.$
2. $p/a \approx_z \Lambda z\quad$ 1, EI.
3. $y \in \imath(q/b)[q/b \approx p/a] \;.\!\equiv.\; y \in p/a\quad *(\text{Abs})^1.$
4. $y \in p/a \;.\!\equiv.\; \Lambda y\quad$ 2, $\text{df}(\approx.\text{a})^1$.
5. $y = y\quad (\text{T}_{4.8})\text{a}$, ui.
6. $\sim (y \neq y)\quad$ 5, dneg, df($\neq$).
7. $\sim \Lambda y\quad$ 6, df(Λ).
8. $\sim (y \in p/a)\quad$ 4, 7, df($\equiv$), mt.
9. $\sim (y \in \imath(q/b)[q/b \approx p/a])\quad$ 8, $*(\text{Abs})^1$.
10. (y)9 UG.
11. $0(pa)\quad$ 10, df(0).
12. 2 & 11 conj.
13. $(\exists pa)[p/a \approx_z \Lambda z \;.\&, 0(pa)]\quad$ 12, eg, (2ei).
14. $\imath(p/a)[p/a \approx_z \Lambda z] \in \imath(s/tw)[s/tw \approx_{rc} 0(rc)]\quad$ 13, $*(\text{Abs})^2$.

$(T_{A.2})$
$\vdash (\forall\alpha^{(o)})(\alpha^{(o)} \in 0^{((o))} .\supset. \alpha^{(o)} = \Lambda^{(o)})$
$\vdash (lm)(\imath(p/a)[p/a \approx l/m] \in \imath(s/tw)[s/tw \approx_{rc} 0(rc)] .\supset.$
$\imath(p/a)[p/a \approx l/m] = \imath(p/a)[p/a \approx_z \Lambda z])$.

Proof:

1. $\rightarrow \imath(p/a)[p/a \approx l/m] \in \imath(s/tw)[s/tw \approx_{rc} 0(rc)]$
2. $(\exists pa)(p/a \approx l/m .\&. 0(pa))$ 1, $*(\text{Abs})^1$.
3. $d/e \approx l/m .\&. 0(de)$ 2, EI.
4. 0(de) 3, simp.
5. $(y) \sim (y \in \imath(p/a)[p/a \approx d/e])$ 4, df(0).
6. $\imath(p/a)[p/a \approx d/e] = \imath(p/a)[p/a \approx_z \Lambda z]$ 5, $(T_{6.24})^1$.
7. $d/e \approx_z \Lambda z$ 6, $*(V)^1$, df($\approx$).
8. $l/m \approx_z \Lambda z$ 3, 7, simp, df$(\approx.\text{a})^1$.
9. __ $\imath(p/a)[p/a \approx l/m] = \imath(p/a)[p/a \approx_z \Lambda z]$ 8, $*(V)^1$.
10. $1 \supset 9$ cp.
11. $(lm)10$ ug.

The class $0^{((o))}$ is the class with only the empty class in it.

Of course, these theorems will repeat at higher types. That is, we have a new empty class, Zero, One, etc., for classes of classes of individuals.

df$(\Lambda)^2$
$\Lambda(rc) = \text{df} \sim (\imath(p/a)[p/a \approx r/c] = \imath(p/a)[p/a \approx r/c])$
df$(0)^2$
$0(hde) = \text{df } (lm) \sim (\imath(p/a)[p/a \approx l/m] \in \imath(s/tw)[s/tw \approx_{rc} hde])$

We then have the theorem:

$(T_{A.2})^2 \quad \vdash \imath(s/tw)[s/tw \approx_{rc} \Lambda(rc)] \in \imath(j/kuv)[j/kuv \approx_{qpa} 0(qpa)]$

The repetition of cardinal numbers at each type level in the so-called simple (non-ramified) theory of types is commonly attacked as counterintuitive. But we can now see that it is the perfectly natural consequence of Russell's logical construction of propositional functions (and classes) through the theory of propositions. To be sure, a proof for a lower type does not, of itself, give us a proof of a higher type. This is so even though we can see from the pattern that proofs for higher types will go through.

For the present purposes we shall deduce the Peano postulates within the lowest possible type and simply observe, as should be clear, that extension to higher types is readily forthcoming.

Peano_1 says that Zero is a member of the class of natural numbers.

$(T_{A.3}) \quad \vdash 0^{((o))} \in N^{(((o))}$
$\vdash \imath(s/tw)[s/tw \approx_{rc} 0(rc)] \in \imath(j/kuv)[j/kuv \approx_{qpa} N(qpa)]$

Proof:

1. $(lmno)(\imath(s/tw)[s/tw \approx_{rc} 0(rc)] \in \imath(j/kuv)[j/kuv \approx l/mno] .\&.$
$(hde)(\imath(s/tw)[s/tw \approx h/de] \in \imath(j/kuv)[j/kuv \approx l/mno] .\supset.$
$\imath(s/tw)[s/tw \approx_{rc} S(hde)(rc)] \in \imath(j/kuv)[j/kuv \approx l/mno])$
$:\supset: \imath(s/tw)[s/tw \approx_{rc} 0(rc)] \in \imath(j/kuv)[j/kuv \approx l/mno])$
log.
2. $(\exists s^\circ t^\circ w^\circ)(s^\circ/t^\circ w^\circ \approx_{qpa} 0(rc) .\&.$
$(lmno)(\imath(s/tw)[s/tw \approx_{rc} 0(rc)] \in \imath(j/kuv)[j/kuv \approx l/mno] .\&.$

$(hde)(\imath(s/tw)[s/tw \approx h/de] \in \imath(j/kuv)[j/kuv \approx l/mno]$:⊃:
$\imath(s/tw)[s/tw \approx_{rc} S(hde)(rc)] \in \imath(j/kuv)[j/kuv \approx l/mno])$
:⊃: $\imath(s/tw)[s/tw \approx s^\circ/t^\circ w^\circ] \in \imath(j/kuv)[j/kuv \approx l/mno]))$
1, $*(T_{6.19})^2$.

3. $(\exists s^\circ t^\circ w^\circ)(s^\circ/t^\circ w^\circ \approx_{qpa} 0(rc)$.&. $N(s^\circ t^\circ w^\circ))$ 2, df(N).
4. $\imath(s/tw)[s/tw \approx_{rc} 0(rc)] \in \imath(j/kuv)[j/kuv \approx_{qpa} N(qpa)]$
3, $*(\text{Abs})^3$.

Now on to Peano$_2$. We wish to first prove the following lemma.

$(T_{A.4})$
$\vdash (\forall\alpha^{((o))})(0^{((o))} \neq S(\alpha^{((o))}))$
$\vdash (hde) \sim (\imath(s/tw)[s/tw \approx_{rc} 0(rc)] = \imath(s/tw)[s/tw \approx_{rc} S(hde)(rc)])$

Proof:

1. $\rightarrow \sim (hde) \sim (\imath(s/tw)[s/tw \approx_{rc} 0(rc)] = \imath(s/tw)[s/tw \approx_{rc} S(hde)(rc)])$
2. $(\exists hde)(\imath(s/tw)[s/tw \approx_{rc} 0(rc)] = \imath(s/tw)[s/tw \approx_{rc} S(hde)(rc)])$
1, df(∃).
3. $\imath(s/tw)[s/tw \approx_{rc} 0(rc)] = \imath(s/tw)[s/tw \approx_{rc} S(hde)(rc)]$ 2, EI.
4. $\imath(p/a)[p/a \approx_z \Lambda z] \in \imath(s/tw)[s/tw \approx_{rc} 0(rc)]$ $(T_{A.1})$.
5. $\imath(p/a)[p/a \approx_z \Lambda z] \in \imath(s/tw)[s/tw \approx_{rc} S(hde)(rc)]$
3, 4, $*(\text{Ext})^2$.
6. $(\exists pa)(p/a \approx_z \Lambda z$.&. $S(hde)(pa))$ 5, $*(\text{Abs})^2$.
7. $t/w \approx_z \Lambda z$.&. $S(hde)(tw)$ 6, EI.
8. $S(hde)(tw)$ 7, simp.
9. $(\exists y)(y \in \imath(p/a)[p/a \approx t/w]$ &
$\imath(p/a)[p/a .\approx_z. (t/w - \iota y)(z)] \in \imath(s/tw)[s/tw \approx h/de])$ 8, df(S).
10. $y \in \imath(p/a)[p/a \approx t/w]$ &
$\imath(p/a)[p/a .\approx_z. (t/w - \iota y)(z)] \in \imath(s/tw)[s/tw \approx h/de]$ 9, EI.
11. $y \in \imath(p/a)[p/a \approx t/w]$ 10, simp.
12. $y \in t/w$ 11, $*(\text{Abs})^1$.
13. Λy 7, 12, simp, $\text{df}(\approx.a)^1$.
14. $y \neq y$ 27, $\text{df}(\Lambda)^1$.
15. $\underline{\quad} F$ 14, $(T_{4.8})$a, (7ei), (3ei).
16. $1 \supset F$ 1–15 cp.
17. $(hde) \sim (\imath(s/tw)[s/tw \approx_{rc} 0(rc)] = \imath(s/tw)[s/tw \approx_{rc} S(hde)(rc)])$
16, reductio.

From this lemma we quickly arrive at (Peano$_2$):

$(T_{A.5})$
$\vdash (\forall\alpha^{((o))})(\alpha^{((o))} \in N^{(((o)))}$.⊃. $0^{((o))} \neq S(\alpha^{((o))}))$.
$\vdash (hde)(\imath(s/tw)[s/tw \approx h/de] \in \imath(j/kuv)[j/kuv \approx_{qpa} N(qpa)]$.⊃.
$\sim (\imath(s/tw)[s/tw \approx_{rc} 0(rc)] = \imath(s/tw)[s/tw \approx_{rc} S(hde)(rc)]))$.

The proof is left as an exercise.

Peano$_3$ asserts that the successor of every natural number is a natural number.

$(T_{A.6})$
$\vdash (\forall\alpha^{((o))})(\alpha^{((o))} \in N^{(((o)))}$.⊃. $S(\alpha^{((o))}) \in N^{(((o)))})$
$\vdash (hde)(\imath(stw)[s/tw \approx h/de] \in \imath(j/kuv)[j/kuv \approx_{qpa} N(qpa)]$:⊃:
$\imath(s/tw)[s/tw \approx_{rc} S(hde)(rc)] \in \imath(j/kuv)[j/kuv \approx_{qpa} N(qpa)])$.

Proof:

1. $\rightarrow$ $\imath(s/tw)[s/tw \approx h/de] \in \imath(j/kuv)[j/kuv \approx_{qpa} N(qpa)]$
2. $(\exists fxy)[f/xy \approx h/de\ .\&.\ N(fxy)]$ 1, $*(\text{Abs})^2$.
3. $N(hde)$ 2, $*(\text{T}_{6.19})^2$.
4. $(lmno)(\imath(s/tw)[s/tw \approx_{rc} 0(rc)] \in \imath(j/kuv)[j/kuv \approx l/mno]\ .\&.$ $(hde)(\imath(s/tw)[s/tw \approx h/de] \in \imath(j/kuv)[j/kuv \approx l/mno]\ .\supset.$ $\imath(s/tw)[s/tw \approx_{rc} S(hde)(rc)] \in \imath(j/kuv)[j/kuv \approx l/mno])$ $:\supset:\ \imath(s/tw)[s/tw \approx h/de] \in \imath(j/kuv)[j/kuv \approx l/mno])$ 3, df(N).
5. $\rightarrow$ $\imath(s/tw)[s/tw \approx_{rc} 0(rc)] \in \imath(j/kuv)[j/kuv \approx l/mno]\ .\&.$ $(hde)(\imath(s/tw)[s/tw \approx h/de] \in \imath(j/kuv)[j/kuv \approx l/mno]\ .\supset.$ $\imath(s/tw)[s/tw \approx_{rc} S(hde)(rc)] \in \imath(j/kuv)[j/kuv \approx l/mno])$
6. $\imath(s/tw)[s/tw \approx_{rc} 0(rc)] \in \imath(j/kuv)[j/kuv \approx l/mno]\ .\&.$ $(hde)(\imath(s/tw)[s/tw \approx h/de] \in \imath(j/kuv)[j/kuv \approx l/mno]\ .\supset.$ $\imath(s/tw)[s/tw \approx_{rc} S(hde)(rc)] \in \imath(j/kuv)[j/kuv \approx l/mno])$ $:\supset:\ \imath(s/tw)[s/tw \approx h/de] \in \imath(j/kuv)[j/kuv \approx l/mno])$ 4, ui.
7. $\imath(s/tw)[s/tw \approx h/de] \in \imath(j/kuv)[j/kuv \approx l/mno])$ 5, 6, mp.
8. $(hde)(\imath(s/tw)[s/tw \approx h/de] \in \imath(j/kuv)[j/kuv \approx l/mno]\ .\supset.$ $\imath(s/tw)[s/tw \approx_{rc} S(hde)(rc)] \in \imath(j/kuv)[j/kuv \approx l/mno])$ 5, simp.
9. $\imath(s/tw)[s/tw \approx h/de] \in \imath(j/kuv)[j/kuv \approx l/mno]\ .\supset.$ $\imath(s/tw)[s/tw \approx_{rc} S(hde)(rc)] \in \imath(j/kuv)[j/kuv \approx l/mno]$ 8, ui.
10. $\_\_$ $\imath(s/tw)[s/tw \approx_{rc} S(hde)(rc))] \in \imath(j/kuv)[j/kuv \approx l/mno]$ 7, 9, mp.
12. $5 \supset 10$ cp.
13. $(lmno)12$ ug.
14. $(\exists fxy)(f/xy\ .\approx_{rc}.\ S(hde)(rc)\ :\&:\ N(fxy))$ 13, $*(\text{T}_{6.19})^2$, df(N).
15. $\_\_$ $\imath(s/tw)[s/tw\ .\approx_{rc}.\ S(hde)(rc)] \in \imath(j/kuv)[j/kuv \approx_{qpa} N(qpa)]$ 14, $*(\text{Abs})^2$.
16. $1 \supset 15$ cp.
17. $(hde)16$ ug.

We now proceed to Peano postulate five. Postulate four will be taken up in a later section. We first prove the following lemma:

$(\text{T}_{\text{A}.7})$

$$\vdash (\forall \beta^{(((o)))})(0^{((o))} \in \beta^{(((o)))}\ \&\ (\forall \alpha^{(o)})(\alpha^{((o))} \in \beta^{(((o)))}\ .\supset.\ S\alpha^{((o))} \in \beta^{(((o)))})\ :\supset:\ N^{(((o)))} \subseteq \beta^{(((o)))})$$

$$\vdash (lmno)(\imath(s/tw)[s/tw \approx_{rc} 0(rc)] \in \imath(j/kuv)[j/kuv \approx l/mno]\ .\&.\ (hde)(\imath(stw)[s/tw \approx h/de] \in \imath(j/kuv)[j/kuv \approx l/mno]\ .\supset.\ \imath(s/tw)[s/tw \approx_{rc} S(hde)(rc)] \in \imath(j/kuv)[j/kuv \approx l/mno])\ :\supset:\ \imath(j/kuv)[j/kuv \approx_{qpa} N(qpa)] \subseteq \imath(j/kuv)[j/kuv \approx l/mno)])$$

Proof:

1. $\rightarrow\ \imath(s/tw)[s/tw \approx_{rc} 0(rc)] \in \imath(j/kuv)[j/kuv \approx l/mno]$ $.\&.\ (hde)(\imath(stw)[s/tw \approx h/de] \in \imath(j/kuv)[j/kuv \approx l/mno]\ .\supset.$ $\imath(s/tw)[s/tw \approx_{rc} S(hde)(rc)] \in \imath(j/kuv)[j/kuv \approx l/mno])$
2. $\rightarrow\ \imath(s/tw)[s/tw \approx h/de] \in \imath(j/kuv)[j/kuv \approx_{qpa} N(qpa)]$
3. $(\exists stw)[s/tw \approx h/de\ .\&.\ N(stw))$ 2, $*(\mathrm{Abs})^2$.
4. $N(hde)$ 3, $*(\mathrm{T}_{6.19})^2$.
5. $(lmno)(\imath(s/tw)[s/tw \approx_{rc} 0(rc)] \in \imath(j/kuv)[j/kuv \approx l/mno]$ $.\&.\ (hde)(\imath(s/tw)[s/tw \approx h/de] \in \imath(j/kuv)[j/kuv \approx l/mno]\ .\supset.$ $\imath(s/tw)[s/tw \approx_{rc} S(hde)(rc)] \in \imath(j/kuv)[j/kuv \approx l/mno])$ $:\supset:\ \imath(s/tw)[s/tw \approx h/de] \in \imath(j/kuv)[j/kuv \approx l/mno])$ 4, df(N).
6. $\imath(s/tw)[s/tw \approx_{rc} 0(rc)] \in \imath(j/kuv)[j/kuv \approx l/mno]$ $.\&.\ (hde)(\imath(s/tw)[s/tw \approx h/de] \in \imath(j/kuv)[j/kuv \approx l/mno]\ .\supset.$ $\imath(s/tw)[s/tw \approx_{rc} S(hde)(rc)] \in \imath(j/kuv)[j/kuv \approx l/mno])$ $:\supset:\ \imath(s/tw)[s/tw \approx h/de] \in \imath(j/kuv)[j/kuv \approx l/mno]$ 5, ui.
7. __ $\imath(s/tw)[s/tw \approx h/de] \in \imath(j/kuv)[j/kuv \approx l/mno]$ 1, 6, mp.
8. $2 \supset 7$ cp.
9. $(hde)8$ ug.
10. __ $\imath(j/kuv)[j/kuv \approx_{qpa} N(qpa)] \subseteq \imath(j/kuv)[j/kuv \approx l/mno]$ 9, $\mathrm{df}(\subseteq)^3$.
11. $1 \supset 10$ cp.
12. $(lmno)11$ ug.

The thoerem shows that the following theorem schema is valid:

$*(\mathrm{T}_{A.7})$

$$\vdash A(0^{((o))})\ \&\ (\forall\alpha^{((o))})(A\alpha^{((o))}\ .\supset.\ A(S\alpha^{((o))}))\ .\supset.\ (\forall\delta^{((o))})(\delta^{((o))} \in N^{(((o)))}\ .\supset.\ A\delta^{((o))})$$

$$\vdash (\exists fxy)(f/xy \approx_{rc} 0(rc)\ .\&.\ A(fxy))\ .\&.\ (hde)((\exists fxy)(f/xy \approx h/de\ .\&.\ A(fxy))\ .\supset.\ (\exists fxy)(f/xy \approx_{rc} S(hde)(rc)\ .\&.\ A(fxy)))\ :\supset:\ (hde)(\imath(s/tw)[s/tw \approx h/de] \in \imath(j/kuv)[j/kuv \approx_{qpa} N(qpa)]\ .\supset.\ (\exists fxy)(f/xy \approx h/de\ .\&.\ A(fxy))).$$

The proof is from $*(\mathrm{T}_{6.20})^3$ and $*(\mathrm{Abs})^3$. Of course, every instance of the schema can be proved on the basis of $(\mathrm{T}_{A.7})$ itself.

From this theorem schema of induction we can go on to prove Peano_5 and the schema of mathematical induction. To facilitate our proof, let us use the following abbreviation:

$$\Sigma(xyz) =\mathrm{df}\ \imath(s/tw)[s/tw \approx x/yz] \in \imath(j/kuv)[j/kuv \approx l/mno]\ .\&.\ \imath(s/tw)[s/tw \approx x/yz] \in \imath(j/kuv)[j/kuv \approx_{qpa} N(qpa)]$$

We need to show:

Base step: $(\exists stw)(f/xy \approx_{rc} 0(rc)\ .\&.\ \Sigma(fxy))$
Induction step: $(hde)((\exists fxy)(f/xy \approx h/de\ .\&.\ \Sigma(fxy))\ .\supset.\ (\exists fxy)(f/xy \approx_{rc} S(hde)(rc)\ .\&.\ \Sigma(fxy)))$

Then we put $\Sigma(fxy)$ for $A(fxy)$ in $*(\mathrm{T}_{A.7})$.

$(\mathrm{T}_{A.8})$

$\vdash\ (\forall\beta)(0^{((o))} \in \beta^{(((o)))}\ \&$
$\quad (\forall\alpha^{((o))})(\alpha^{((o))} \in \beta^{(((o)))}\ .\&.\ \alpha^{((o))} \in N^{(((o)))}\ :\supset:\ S\alpha^{((o))} \in \beta^{(((o)))})$
$\qquad .\supset.\ N^{(((o)))} \subseteq \beta^{(((o)))})$

$\vdash\ (lmno)(\iota(s/tw)[s/tw \approx_{rc} 0(rc)] \in \iota(j/kuv)[j/kuv \approx l/mno]\ .\&.$
$\quad (hde)(\iota(s/tw)[s/tw \approx h/de] \in \iota(j/kuv)[j/kuv \approx l/mno]\ .\&.$
$\qquad \iota(s/tw)[s/tw \approx h/de] \in \iota(j/kuv)[j/kuv \approx_{qpa} N(qpa)]\ :\supset:$
$\qquad\quad \iota(s/tw)[s/tw \approx_{rc} S(hde)(rc)] \in \iota(j/kuv)[j/kuv \approx l/mno])$
$\quad :\supset:\ \iota(j/kuv)[j/kuv \approx_{qpa} N(qpa)] \subseteq \iota(j/kuv)[j/kuv \approx l/mno])$

Proof:

1. $\rightarrow\ \iota(s/tw)[s/tw \approx_{rc} 0(rc)] \in \iota(j/kuv)[j/kuv \approx l/mno]\ .\&.$
 $(hde)(\iota(s/tw)[s/tw \approx h/de] \in \iota(j/kuv)[j/kuv \approx l/mno]\ .\&.$
 $\iota(s/tw)[s/tw \approx h/de] \in \iota(j/kuv)[j/kuv \approx_{qpa} N(qpa)]\ :\supset:$
 $\iota(s/tw)[s/tw \approx_{rc} S(hde)(rc)] \in \iota(j/kuv)[j/kuv \approx l/mno])$
2. $\iota(s/tw)[s/tw \approx_{rc} 0(rc)] \in \iota(j/kuv)[j/kuv \approx l/mno]$ 1, simp.
3. $\iota(s/tw)[s/tw \approx_{rc} 0(rc)] \in \iota(k/kuv)[j/kuv \approx_{qpa} N(qpa)]$
 (Peano$_1$).
4. $(\exists fxy)(f/xy \approx_{rc} 0(rc)\ .\&.\ \Sigma(fxy))$ 2, 3, $*(\mathrm{T}_{6.19})^2$.
5. $\rightarrow\ (\exists fxy)(f/xy \approx h/de\ .\&.\ \Sigma(fxy))$
6. $\Sigma(hde)$ 5, $*(\mathrm{T}_{6.19})^2$.
7. $\iota(s/tw)[s/tw \approx h/de] \in \iota(j/kuv)[j/kuv \approx l/mno]\ .\&.$
 $\iota(s/tw)[s/tw \approx h/de] \in \iota(j/kuv)[j/kuv \approx_{qpa} N(qpa)]$ 6, df(Σ).
8. $(hde)(\iota(s/tw)[s/tw \approx h/de] \in \iota(j/kuv)[j/kuv \approx l/mno]\ .\&.$
 $\iota(s/tw)[s/tw \approx h/de] \in \iota(j/kuv)[j/kuv \approx_{qpa} N(qpa)]$
 $:\supset:\ \iota(s/tw)[s/tw \approx_{rc} S(hde)(rc)] \in \iota(j/kuv)[j/kuv \approx l/mno])$
 1, simp.
9. $\iota(stw)[s/tw \approx h/de] \in \iota(j/kuv)[j/kuv \approx l/mno]\ .\&.$
 $\iota(s/tw)[s/tw \approx h/de] \in \iota(j/kuv)[j/kuv \approx_{qpa} N(qpa)]$
 $:\supset:\ \iota(s/tw)[s/tw \approx_{rc} S(hde)(rc)] \in \iota(j/kuv)[j/kuv \approx l/mno]$
 8, ui.
10. $\iota(s/tw)[s/tw \approx_{rc} S(hde)(rc)] \in \iota(j/kuv)[j/kuv \approx l/mno]$ 7, 9, mp.
12. $\iota(stw)[s/tw \approx h/de] \in \iota(j/kuv)[j/kuv \approx_{qpa} N(qpa)]$ 7, simp.
13. $\iota(s/tw)[s/tw \approx_{rc} S(hde)(rc)] \in \iota(j/kuv)[j/kuv \approx_{qpa} N(qpa)]$
 12, (Peano$_3$).
14. __ $(\exists fxy)(f/xy\ .\approx_{rc}.\ S(hde)(rc)\ :\&:\ \Sigma(fxy))$ 10, 13, $*(\mathrm{T}_{6.19})^2$.
15. $5 \supset 14$ cp.
16. $(hde)15$ ug.
17. $(\exists fxy)(f/xy \approx_{rc} 0(rc)\ .\&.\ \Sigma(fxy)\ .\&.$
 $(hde)((\exists fxy)[f/xy \approx h/de\ .\&.\ \Sigma(fxy)]\ .\supset.$
 $(\exists fxy)(f/xy \approx_{rc} S(hde)(rc)\ .\&.\ \Sigma(fxy)))$
 4, 16, conj.
18. $(hde)(\iota(s/tw)[s/tw \approx h/de] \in \iota(j/kuv)[j/kuv\ \approx_{qpa}\ N(qpa)]$
 $.\supset.\ (\exists fxy)(f/xy \approx h/de\ .\&.\ \Sigma(fxy)))$ 17, $*(\mathrm{T}_{A.7})$.
19. $\iota(s/tw)[s/tw \approx h/de] \in \iota(j/kuv)[j/kuv\ \approx_{qpa}\ N(qpa)]$
 $.\supset.\ (\exists fxy)(f/xy \approx h/de\ .\&.\ \Sigma(fxy))$
 18, ui.
20. $\iota(s/tw)[s/tw \approx h/de] \in \iota(j/kuv)[j/kuv\ \approx_{qpa}\ N(qpa)]\ .\supset.\ \Sigma(hde)$
 19, $*(\mathrm{T}_{6.19})^2$.

21 $\imath(s/tw)[s/tw \approx h/de] \in \imath(j/kuv)[j/kuv \approx_{qpa} N(qpa)]$:⊃:
$\imath(s/tw)[s/tw \approx h/de] \in \imath(j/kuv)[j/kuv \approx l/mno]$.&.
$\imath(s/tw)[s/tw \approx h/de] \in \imath(j/kuv)[j/kuv \approx_{qpa} N(qpa)]$
20, df(Σ).

22. $\imath(s/tw)[s/tw \approx h/de] \in \imath(j/kuv)[j/kuv \approx_{qpa} N(qpa)]$.⊃.
$\imath(s/tw)[s/tw \approx h/de] \in \imath(j/kuv)[j/kuv \approx l/mno]$
21, log.

23. $(hde)22$ ug.

24. __ $\imath(j/kuv)[j/kuv \approx_{qpa} N(qpa)] \subseteq \imath(j/kuv)[j/kuv \approx l/mno]$
23, df(⊆).[3]

25. 1 ⊃ 24 cp.

26. $(lmno)25$ ug.

This completes our proof of Peano_5. Of course, from Peano_5 we can prove a theorem schema for mathematical induction $*(\text{Peano}_5)$:

$*(\text{T}_{A.8})$
$\vdash (\exists fxy)(f/xy \approx_{rc} 0(rc)$.&. $A(fxy))$.&.
$(hde)((\exists fxy)(f/xy \approx h/de$.&. $A(fxy))$.&.
$(\exists fxy)(f/xy \approx h/de$.&. $N(fxy))$:⊃:
$(\exists fxy)(f/xy \approx_{rc} S(hde)(rc)$.&. $A(fxy)))$:⊃:
$(hde)(\imath(s/tw)[s/tw \approx h/de] \in \imath(j/kuv)[j/kuv \approx_{qpa} N(qpa)]$
.⊃. $(\exists stw)(s/tw \approx h/de$.&. $A(fxy)))$.

The proof is from $(\text{T}_{A.8})$, $*(\text{Abs})^3$, and $*(\text{T}_{6.20})^3$.

The Infinity Theorem

A.3 Peano Postulate_4 says that if the successors of any two natural numbers are identical then the two numbers are identical. In order to prove this we shall need to first prove that the class of all individuals is infinite (in the "ordinary" or Fregean "inductive" sense). That is, we must show that it is not a member of any natural number ("inductive cardinal"). The reason the infinity of the class of all individuals is needed lies in the nature of the Frege/Russell theory of natural number. If the class of individuals is a member of some natural number, then the natural number which is its successor is the null-class. Also its successor would be null. Consequently, we shall have a violation of Peano Postulate_4.

Frege's definition of "finite" is quite natural given his construction of natural numbers as finite cardinals. Formally he has:

$$\text{fin}(\alpha) = \text{df } (\exists\beta)(\beta \in \mathbb{N} \;\&\; \alpha \in \beta).$$
$$\text{inf}(\alpha) = \text{df} \sim \text{fin}(\alpha)$$

Substitutionally, we put:

$\text{fin}(\imath(p/a)[p/a \approx_z Az]) = \text{df}$
$(\exists hde)([\imath(stw)[s/tw \approx h/de] \in \imath(j/kuv)[j/kuv \approx_{qpa} N(qpa)]$
.&. $\imath(p/a)[p/a \approx_z Az] \in \imath(s/tw)[s/tw \approx h/de])$
$\text{inf}(\imath(p/a)[p/a \approx_z Az]) = \text{df} \sim \text{fin}(\imath(p/a)[p/a \approx_z Az])$

Thus we want to get,

$$\vdash \ \mathrm{inf}(\imath(p/a)[p/a \approx_z Vz]),$$

which is the theorem that the universal class of entities is infinite in Frege's sense.

Different proofs of the infinity of the universal class of entities are possible within the substitutional theory. One technique is to first prove the Dedekind infinity of the universal class (of individuals). But this requires the substitutional proxy of relations-in-extension. A simpler road to the Frege infinity of the universal class (of individuals) is readily available. Russell had suggested it in his 1907 manuscript "On Types." He wrote (*OT*, p. 3):

> The axiom of infinity is
> $(\exists x)(x \text{ ex } p)$ Pp?
> This will allow us to infer *Infin ax*. For we can prove
> $\alpha \in \text{cls ind} \,.\supset\colon (\exists p)\colon x \in \alpha \,.\supset_x.\, x \text{ in } p$
> whence $\alpha \in \text{cls induct} \,.\supset.\, \exists! - \alpha$
> whence $V \sim\in \text{cls induct}$.

The idea of this proof is as follows. Prove by mathematical induction, that for every Frege finite class there is a proposition containing every member of the class as a constituent. It follows that if the universal class of entities were Frege finite, there would be a single proposition q with every entity as its constituent. Now since $\{\sim q\}$ is itself an entity, it is in the universal class of entities; accordingly, $\{\sim q\}$ would be a constituent of q. By Russell's axiom (A9), we arrive at $q = \{\sim q\}$, which yields the contradiction, $q \equiv \sim q$.

($\mathrm{T}_{A.9}$)
$\vdash (\forall\alpha^{((o))})(\alpha^{((o))} \in N^{(((o)))} \,.\supset.\, (\forall\beta^{(o)})(\beta^{(o)} \in \alpha^{((o))} \,.\supset.\, (\exists q^{o})(x^{o})(x^{o} \in \beta^{(o)}$
$.\supset.\, x^{o} \mathit{\ in\ } q^{o})))$
$\vdash (hde)(\imath(s/tw)[s/tw \approx h/de] \in \imath(j/kuv)[j/kuv \approx_{qpa} N(qpa)] \,.\supset.$
$(lm)(\imath(p/a)[p/a \approx l/m] \in \imath(s/tw)[s/tw \approx h/de] \,.\supset.$
$(\exists q)(x)(x \in \imath(p/a)[p/a \approx l/m] \,.\supset.\, x \text{ in } q)))$

Proof:

$I(xyz) \ =\mathrm{df}\ (lm)(\imath(p/a)[p/a \approx l/m] \in \imath(s/tw)[s/tw \approx x/yz] \,.\supset.$
$(\exists q)(x \in \imath(p/a)[p/a \approx l/m] \,.\supset.\, x \text{ in } q))$

1. $\rightarrow\ \imath(p/a)[p/a \approx l/m] \in \imath(s/tw)[s/tw \approx_{rc} 0(rc)]$
2. $\quad \imath(p/a)[p/a \approx l/m] = \imath(p/a)[p/a \approx_z \Lambda z]$ 1, ($\mathrm{T}_{A.2}$).
3. $\quad \rightarrow\ x \in \imath(p/a)[p/a \approx l/m]$
4. $\quad\quad x \in \imath(p/a)[p/a \approx_z \Lambda z]$ 2, 3, $*(\mathrm{Ext})^1$.
5. $\quad\quad F$ 4, $(\mathrm{T}_{6.23})^1$.
6. $\quad$ __ x in q 5, log.
7. $\quad 3 \supset 6$ cp.
8. $\quad (x)7$ ug.
9. __ $(\exists q)8$ eg.
10. $1 \supset 9$ cp.
11. $(lm)10$ ug.
12. $(\exists fxy)(f/xy \approx_{rc} 0(rc) \,.\&.\, I(fxy))$ 11, $*(\mathrm{T}_{6.19})^2$, df(I).

This completes the base step. Now for the inductive step.

13. $\rightarrow$ $(\exists fxy)(f/xy \approx h/de \;.\&.\; I(fxy))$ & $(\exists fxy)(f/xy \approx h/de \;.\&.\; N(fxy))$
show: $(\exists fxy)(f/xy \approx_{rc} S(hde)(rc) \;.\&.\; I(fxy))$
14. $I(hde)$ 13, $*(T_{6.19})^2$
15. $(lm)(\text{℩}(p/a)[p/a \approx l/m] \in \text{℩}(s/tw)[s/tw \approx h/de] \;.\supset.$
$(\exists q)(x)(x \in \text{℩}(p/a)[p/a \approx l/m] \;.\supset.\; x \text{ in } q))$ 14, df(I).
16. $\rightarrow$ $\text{℩}(p/a)[p/a \approx l/m] \in \text{℩}(s/tw)[s/tw \approx S(hde)(rc)]$
17. $(\exists p^\circ a^\circ)(p^\circ/a^\circ \approx l/m \;.\&.\; S(hde)(p^\circ a^\circ))$ 16, $*(\mathrm{Abs})^2$.
18. $S(hde)(lm)$ 17, $*(T_{6.19})^2$.
19. $(\exists y)(y \in \text{℩}(p/a)[p/a \approx l/m] \;.\&.$
$\text{℩}(p/a)[p/a \;.\approx_z.\; (l/m - \iota y)(z)] \in \text{℩}(s/tw)[s/tw \approx h/de])$
18, df(S).
20. $y \in \text{℩}(p/a)[p/a \approx 1/m] \;.\&.$
$\text{℩}(p/a)[p/a \;.\approx_z.\; (l/m - \iota y)(z)] \in \text{℩}(s/tw)[s/tw \approx h/de]$
19, EI.
21. $\text{℩}(p/a)[p/a \;.\approx_z.\; (l/m - \iota y)(z)] \in \text{℩}(s/tw)[s/tw \approx h/de]$
20, simp.
22. $\text{℩}(p/a)[p/a \;.\approx_z.\; (l/m - \iota y)(z)] \in \text{℩}(s/tw)[s/tw \approx h/de] \;.\supset.$
$(\exists q)(x)(x \in \text{℩}(p/a)[p/a \;.\approx_z.\; (l/m - \iota y)(z)] \;.\supset.\; x \text{ in } q))$
15, 21, $*(T_{6.20})^1$.
23. $(\exists q)(x)(x \in \text{℩}(p/a)[p/a \;.\approx_z.\; (l/m - \iota y)(z)] \;.\supset.\; x \text{ in } q)$ 21, 22, mp.
24. $(x)(x \in \text{℩}(p/a)[p/a \;.\approx_z.\; (l/m - \iota y)(z)] \;.\supset.\; x \text{ in } q)$ 23, EI.
25. $\rightarrow$ $n \in \text{℩}(p/a)[p/a \approx l/m]$
26. $\rightarrow$ $n = y$
27. $y \textit{ in } \{q \supset y\}$ $*(A_7)$.
28. __ $n \textit{ in } \{q \supset y\}$ 26, 27, $*(=)$.
29. $n = y \;.\supset.\; n \textit{ in } \{q \supset y\}$ 26–28, cp.
30. $\rightarrow$ $n \neq y$
31. $n \in \text{℩}(p/a)[p/a \;.\approx_z.\; (l/m - \iota y)(z)]$ 25, 30, $*(\mathrm{Abs})^1$, df$(l/m - \iota y)$.
32. $n \textit{ in } q$ 24, 31, uimp.
33. $q \textit{ in } \{q \supset y\}$ $*(A_7)$.
34. __ $n \textit{ in } \{q \supset y\}$ 32, 33, (A_{10}).
35. $n \neq y \;.\supset.\; n \textit{ in } \{q \supset y\}$ 30–34, cp.
36. __ $n \textit{ in } \{q \supset y\}$ 29, 35, log.
37. $25 \supset 36$ cp.
38. $(x)(x \in \text{℩}(p/a)[p/a \approx l/m] \;.\supset.\; x \textit{ in } \{q \supset y\})$ 37, ug.
39. __ $(\exists q)(x)(x \in \text{℩}(p/a)[p/a \approx l/m] \;.\supset.\; x \text{ in } q)$ 38, eg, (24ei), (20ei).
40. $16 \supset 39$ cp.
41. $(lm)40$ ug.
42. __ $(\exists fxy)(f/xy \;.\approx_{rc}.\; S(hde)(rc) \;.\&.\; I(fxy))$ 41, $*(T_{6.19})^2$, df(I).
43. $13 \supset 42$ cp.
44. $(hde)43$ ug.

This completes the induction step. Now applying $*(\mathrm{Peano}_5)$, we get:

45. $(hde)(\text{℩}(s/tw)[s/tw \approx h/de] \in \text{℩}(j/kuv)[j/kuv \approx_{qpa} N(qpa)]$
$.\supset.\; (\exists fxy)(f/xy \approx h/de \;.\&.\; I(fxy)))$
12, 44, $*(\mathrm{Peano}_5)$.
46. $(hde)(\text{℩}(s/tw)[s/tw \approx h/de] \in \text{℩}(j/kuv)[j/kuv \approx_{qpa} N(qpa)] \;.\supset.\; I(hde)$
45, $*(T_{6.19})^2$.
47. $(hde)(\text{℩}(s/tw)[s/tw \approx h/de] \in \text{℩}(j/kuv)[j/kuv \approx_{qpa} N(qpa) \;.\supset.$
$(lm)\,(\text{℩}(p/a)[p/a \approx l/m] \in \text{℩}(s/tw)[s/tw \approx h/de] \;.\supset.$

$$(\exists q)(x)(x \in \iota(p/a)[p/a \approx l/m] \;.\supset.\; x \text{ in } q)))$$
46, df(I).

This completes the proof.

From ($T_{A.9}$), we quickly arrive at the theorem that every Frege finite class is such that there is some proposition containing each member of the class as its constituent.

($T_{A.10}$)
$\vdash (lm)(\text{fin}(\iota(p/a)[p/a \approx l/m]) \;.\supset.$
$(\exists q)(x)(x \in \iota(p/a)[p/a \approx l/m] \;.\supset.\; x \text{ in } q))$

Next we get:

($T_{A.11}$) $\vdash \text{inf}(\iota(p/a)[p/a \approx_z Vz])$

Proof:

1. $\text{fin}(\iota(p/a)[p/a \approx_z Vz]) .\supset.$ $(\exists q)(x)(x \in \iota(p/a)[p/a \approx_z Vz] \;.\supset.\; x \text{ in } q))$ ($T_{A.10}$), $*(T_{6.20})^1$.
2. $\rightarrow$ $\text{fin}(\iota(p/a)[p/a \approx_z Vz])$
3. $(\exists q)(x)(x \in \iota(p/a)[p/a \approx_z Vz] \;.\supset.\; x \text{ in } q))$ 1, 2, mp.
4. $(x)(x \in \iota(p/a)[p/a \approx_z Vz] \;.\supset.\; x \text{ in } q)$ 3, EI.
5. $q \in \iota(p/a)[p/a \approx_z Vz]$ $(T_{6.21})^1$.
6. $\{\sim q\} \in \iota(p/a)[p/a \approx_z Vz]$ $(T_{6.21})^1$.
7. $\{\sim q\}$ in q 4, 6, ui, mp.
8. q *in* $\{\sim q\}$ $*(A_7)$
9. $q = \{\sim q\}$ 7, 8, (A_9).
10. $q \equiv \sim q$ 9, ($T_{4.7}$).
11. $\underline{\quad}$ F 10, log.
12. $2 \supset F$ cp.
13. $\sim \text{fin}(\iota(p/a)[p/a \approx_z Vz])$ 12, reductio.
14. $\text{inf}(\iota(p/a)[p/a \approx_z Vz])$ 13, df(inf).

($T_{A.12}$)
$\vdash (\forall \alpha^{((o))})(\forall \beta^{(o)})(\alpha^{((o))} \in N^{(((o)))} \;\&\; \beta^{(o)} \in \alpha^{((o))} \;.\supset.\; (\exists y^{o}) \sim (y^{o} \in \beta^{(o)}))$
$\vdash (hde)(lm)(\iota(s/tw)[s/tw \approx h/de] \in \iota(j/kuv)[j/kuv \approx_{qpa} N(qpa)]$
$.\&.\; \iota(p/a)[p/a \approx l/m] \in \iota(s/tw)[s/tw \approx h/de] \;:\supset:$
$(\exists y) \sim (y \in \iota(p/a)[p/a \approx l/m]))$

Proof:

1. $\rightarrow$ $\iota(s/tw)[s/tw \approx h/de] \in \iota(j/kuv)[j/kuv \approx_{qpa} N(qpa)]$ $.\&.\; \iota(p/a)[p/a \approx l/m] \in \iota(s/tw)[s/tw \approx h/de]$
2. $\rightarrow \sim (\exists y) \sim (y \in \iota(p/a)[p/a \approx l/m])$
3. $(y)(y \in \iota(p/a)[p/a \approx l/m])$ 2, log.
4. $\iota(p/a)[p/a \approx l/m] = \iota(p/a)[p/a \approx_z Vz]$ 3, $(T_{6.22})^1$.
5. $\iota(p/a)[p/a \approx l/m] \in \iota(s/tw)[s/tw \approx h/de]$ 1, simp.
6. $\iota(p/a)[p/a \approx_z Vz] \in \iota(s/tw)[s/tw \approx h/de]$ 4, 5, $*(T_{6.15})^2$.
7. $\iota(s/tw)[s/tw \approx h/de] \in \iota(j/kuv)[j/kuv \approx_{qpa} N(qpa)]$ $.\&.\; \iota(p/a)[p/a \approx_z Vz] \in \iota, (s/tw)[s/tw \approx h/de]$ 1, 6, log.
8. $(\exists hde)7$ eg.
9. $\text{fin}(\iota(p/a)[p/a \approx_z Vz])$ 10, df(fin).
10. $\underline{\quad}$ F 11, ($T_{A.11}$).

11. $2 \supset F$ cp.
12. __ $(\exists y) \sim (y \in \imath(p/a)[p/a \approx l/m])$ 11, reductio.
13. $1 \supset 12$ cp.
14. $(hde)(lm)13$ ug.

This theorem says that no member of any natural number is empty.

Peano Postulate Four

A.4 Peano postulate four is now at hand. But we need one more lemma. The lemma is,

$$\vdash (\forall\alpha^{((o))})(\alpha^{((o))} \in N^{(((o)))} \;.\supset.\; (\forall\delta^{(o)})(\forall x^{o})(\forall y^{o})(x^{o} \in \delta^{(o)} \;\&\; y^{o} \in \delta^{(o)} \;\&\; \delta^{(o)} - \iota x^{o} \in \alpha^{((o))} \;.\supset.\; \delta^{(o)} - \iota y^{o} \in \alpha^{((o))})$$

$*(T_{A.13})$

$$\vdash (hde)(\imath(s/tw)[s/tw \approx h/de] \in \imath(j/kuv)[j/kuv \approx_{qpa} N(qpa)] \;.\supset.\; (lm)(c)(n)(c \in \imath(p/a)[p/a \approx l/m] \;.\&.\; n \in \imath(p/a)[p/a \approx l/m] \;.\&.\; \imath(p/a)[p/a \;.\approx_z.\; (l/m - \iota c)(z)] \in \imath(s/tw)[s/tw \approx h/de] \;:\supset:\; \imath(p/a)[p/a \;.\approx_z.\; (l/m - \iota n)(z)] \in \imath(s/tw)[s/tw \approx h/de]))$$

$$T(hde) = \text{df}\; (lm)(c)(n)(c \in \imath(p/a)[p/a \approx l/m] \;.\&.\; n \in \imath(p/a)[p/a \approx l/m] \;.\&.\; \imath(p/a)[p/a \;.\approx_z.\; (l/m - \iota c)(z)] \in \imath(s/tw)[s/tw \approx h/de] \;:\supset:\; \imath(p/a)[p/a \;.\approx_z.\; (l/m - \iota n)(z)] \in \imath(s/tw)[s/tw \approx h/de])$$

Proof:

1. $\rightarrow$ $c \in \imath(p/a)[p/a \approx l/m] \;.\&.\; n \in \imath(p/a)[p/a \approx l/m] \;.\&.\; \imath(p/a)[p/a \;.\approx_z.\; (l/m - \iota c)(z)] \in \imath(s/tw)[s/tw \approx_{rc} 0(rc)]$
2. $\imath(p/a)[p/a \;.\approx_z.\; (l/m - \iota c)(z)] \in \imath(s/tw)[s/tw \approx_{rc} 0(rc)]$ 1, simp.
3. $\imath(p/a)[p/a \;.\approx_z.\; (l/m - c)(z)] = \imath(p/a)[p/a \approx_z \Lambda z]$ 2, $(T_{A.2})$.
4. $\rightarrow$ $n \neq c$
5. $n \in \imath(p/a)[p/a \approx l/m] \;\&\; n \neq c$ 1, 4, simp, conj.
6. $n \in \imath(p/a)[p/a \;.\approx_z.\; (l/m - \iota c)(z)]$ 5, df$(-\iota n)$, $*(\text{Abs})^2$.
7. $n \in \imath(p/a)[p/a \approx_z \Lambda z]$ 3, 6, $*(\text{Ext})^1$.
8. __ F 7, $(T_{6.23})^1$.
9. $4 \supset F$ cp.
10. $n = c$ 9, reductio.
11. __ $\imath(p/a)[p/a \;.\approx_z.\; (l/m \sim \iota n)(z)] \in \imath(s/tw)[s/tw \;.\approx_{rc}.\; 0(rc)]$ 1, 10, simp, $*(=)$.
12. $1 \supset 11$ cp.
13. $(lm)(c)(n)12$ ug.
14. $(\exists fxy)(f/xy \approx_{rc} 0(rc) \;.\&.\; T(fxy))$ 13, $*(T_{6.19})^2$, df(T).

This completes the base step. Next, on to the inductive step.

15. $\rightarrow$ $(\exists fxy)(f/xy \approx h/de \;.\&.\; T(fxy)) \;\&\; \imath(s/tw)[s/tw \approx h/de] \in \imath(j/kuv)[j/kuv \approx_{qpa} N(qpa)]$
16. $T(hde)$ 15, $*(T_{6.19})^2$.

17. $(lm)(c)(n)(c \in \text{℩}(p/a)[p/a \approx l/m]$.&. $n \in \text{℩}(p/a)[p/a \approx l/m]$.&.
$\text{℩}(p/a)[p/a .\approx_z. (l/m - \iota c)(z)] \in \text{℩}(s/tw)[s/tw \approx h/de]$
$:\supset: \text{℩}(p/a)[p/a .\approx_z. (l/m - \iota n)(z)] \in \text{℩}(s/tw)[s/tw \approx h/de])$ 16, df(T).

18. $\rightarrow c \in \text{℩}(p/a)[p/a)[p/a \approx l/m]$.&. $n \in \text{℩}(p/a)[p/a \approx l/m]$.&.
$\text{℩}(p/a)[p/a .\approx_z. (l/m - \iota c)(z)] \in \text{℩}(s/tw)[s/tw .\approx_{rc}. S(hde)(rc)]$

19. $\text{℩}(p/a)[p/a .\approx_z. (l/m - \iota c)(z)] \in \text{℩}(s/tw)[s/tw \approx_{rc} S(hde)(rc)]$
18, simp.

20. $\rightarrow c = n$

21. $\_\_$ $\text{℩}(p/a)[p/a .\approx_z. (l/m - \iota n)(z)] \in \text{℩}(s/tw)[s/tw \approx_{rc} S(hde)(rc)]$
19, 20, $*(=)$

22. $20 \supset 21$ cp.

23. $\rightarrow c \neq n$

24. $(\exists qb)(q/b .\approx_z. (l/m - \iota c)(z)$:&: $S(hde)(qb))$ 19, $*(\mathrm{Abs})^2$.

25. $(\exists v)(v \in \text{℩}(p/a)[p/a .\approx_z. (l/m - \iota c)(z)]$.&.
$\text{℩}(p/a)[p/a .\approx_z. (l/m - \iota c - \iota v)(z)] \in \text{℩}(s/tw)[s/tw \approx h/de])$
24, $*(\mathrm{T}_{6.19})^2$, df(S).

26. $v \in \text{℩}(p/a)[p/a .\approx_z. (l/m - \iota c)(z)]$.&.
$\text{℩}(p/a)[p/a .\approx_z. (l/m - \iota c - \iota v)(z)] \in \text{℩}(s/tw)[s/tw \approx h/de])$
25, EI.

27. $\text{℩}(p/a)[p/a .\approx_z. (l/m - \iota c - \iota v)(z)] =$
$\text{℩}(p/a)[p/a .\approx_z. (l/m - \iota v - \iota c)(z)]$ $(\mathrm{T}_{6.29})^1$.

28. $\text{℩}(p/a)[p/a .\approx_z. (l/m - \iota v - \iota c)(z)] \in \text{℩}(s/tw)[s/tw \approx h/de])$
26, 27, $*(\mathrm{T}_{6.15})^2$.

29. $\rightarrow v = n$

30. $\text{℩}(p/a)[p/a .\approx_z. (l/m - \iota n - \iota c)(z)] \in \text{℩}(s/tw)[s/tw \approx h/de])$
28, 29, $*(=)$.

31. $(l/m - \iota n)(c)$ 18, 23, df$(-\iota n)$.

32. $c \in \text{℩}(p/a)[p/a .\approx_z. (l/m - \iota n)(z)]$ 31, $*(\mathrm{Abs})^1$.

33. $(\exists v)(v \in \text{℩}(p/a)[p/a .\approx_z. (l/m - \iota n)(z)]$.&.
$\text{℩}(p/a)[p/a .\approx_z. (l/m - \iota n - \iota v)(z)] \in \text{℩}(s/tw)[s/tw \approx h/de])$
30, 32, conj, eg.

34. $(\exists qb)(q/b .\approx_z. (l/m - \iota n)(z)$:&:
$(\exists v)(v \in \text{℩}(p/a)[p/a .\approx_z. q/b]$.&.
$\text{℩}(p/a)[p/a .\approx_z. (q/b - \iota v)(z)] \in \text{℩}(s/tw)[s/tw \approx h/de])$
33, $*(\mathrm{T}_{6.19})^2$.

35. $(\exists qb)(q/b .\approx_z. (l/m - \iota n)(z)$:&: $S(hde)(qb))$ 34, df(S).

36. $\_\_$ $\text{℩}(p/a)[p/a .\approx_z. (l/m - \iota n)(z)] \in \text{℩}(s/tw)[s/tw \approx_{rc} S(hde)(rc)]$
35, $*(\mathrm{Abs})^2$.

37. $29 \supset 36$ cp.

38. $\rightarrow v \neq n$

39. $(l/m - \iota v)(n)$ 18, 38, df$(-\iota v)$.

40. $n \in \text{℩}(p/a)[p/a .\approx_z. (l/m - \iota v)(z)]$ 39, $*(\mathrm{Abs})^1$.

41. $c \neq v$ 26, $*(\mathrm{Abs})^1$.

42. $(l/m - \iota v)(c)$ 18, 41, df$(-\iota v)$.

43. $c \in \text{℩}(p/a)[p/a .\approx_z. (l/m - \iota v)(z)]$ 42, $*(\mathrm{Abs})^1$.

44. 43 & 40 & 28 conj.

45. $c \in \text{℩}(p/a)[p/a .\approx_z. (l/m - \iota v)(z)]$.&.
$n \in \text{℩}(p/a)[p/a .\approx_z. (l/m - \iota v)(z)]$.&.
$\text{℩}(p/a)[p/a .\approx_z. (l/m - \iota v - \iota c)(z)] \in \text{℩}(s/tw)[s/tw \approx h/de]$
$:\supset: \text{℩}(p/a)[p/a .\approx_z. (l/m - \iota v - \iota n)(z)] \in$
$\text{℩}(s/tw)[s/tw \approx h/de]$ 17, ui, $*(\mathrm{T}_{6.20})^2$.

46. $℩(p/a)[p/a \mathrel{.{\approx_z}.} (l/m - \iota v - \iota n)(z)] \in ℩(s/tw)[s/tw \approx h/de]$ 44, 45, mp.

47. $℩(p/a)[p/a \mathrel{.{\approx_z}.} (l/m - \iota n - \iota v)(z)] =$
$℩(p/a)[p/a \mathrel{.{\approx_z}.} (l/m - \iota v - \iota n)(z)]$ $(\mathrm{T}_{6.29})^1$.

48. $℩(p/a)[p/a \mathrel{.{\approx_z}.} (l/m - \iota n - \iota v)(z)] \in ℩(s/tw)[s/tw \approx h/de]$ 46, 47, $*(\mathrm{T}_{6.15})^1$.

49. $v \in ℩(p/a)[p/a \mathrel{.{\approx_z}.} (l/m - \iota n)(z)]$ 26, 38, $*(\mathrm{Abs})^1$.

50. $(\exists qb)(q/b \mathrel{.{\approx_z}.} (l/m - \iota n)(z) \mathbin{:\&:}$
$(\exists v)(v \in ℩(p/a)[p/a \mathrel{.{\approx_z}.} q/b] \mathbin{.\&.}$
$℩(p/a)[p/a \mathrel{.{\approx_z}.} (q/b - \iota v)(z)] \in ℩(s/tw)[s/tw \approx h/de])$
48, 49, $*(\mathrm{T}_{6.19})^1$.

51. $(\exists qb)(q/b \mathrel{.{\approx_z}.} (l/m - \iota n)(z) \mathbin{:\&:} S(hde)(qb))$ 50, df(S).

52. — $℩(p/a)[p/a \mathrel{.{\approx_z}.} (l/m - \iota n)(z)] \in$
$℩(s/tw)[s/tw \approx_{rc} S(hde)(rc)]$ 51, $*(\mathrm{Abs})^2$.

53. $38 \supset 52$ cp.

54. — $℩(p/a)[p/a \mathrel{.{\approx_z}.} (l/m - \iota n)(z)] \in ℩(s/tw)[s/tw \approx_{rc} S(hde)(rc)]$
37, 53, log, (26ei).

55. $23 \supset 54$ cp.

56. — $℩(p/a)[p/a \mathrel{.{\approx_z}.} (l/m - \iota n)(z)] \in ℩(s/tw)[s/tw \approx_{rc} S(hde)(rc)]$
22, 55, log.

57. $18 \supset 56$ cp.

58. $(lm)(c)(n)57$ ug.

59. $(\exists fxy)(f/xy \mathrel{.{\approx_{rc}}.} S(hde)(rc) \mathbin{:\&:}$
$(lm)(c)(n)(c \in ℩(p/a)[p/a \approx l/m] \mathbin{.\&.} n \in ℩(p/a)[p/a \approx l/m] \mathbin{.\&.}$
$℩(p/a)[p/a \mathrel{.{\approx_z}.} (l/m - \iota c)(z)] \in ℩(s/tw)[s/tw \approx f/xy] \mathrel{:\supset:}$
$℩(p/a)[p/a \mathrel{.{\approx_z}.} (l/m - \iota n)(z)] \in ℩(s/tw)[s/tw \approx f/xy)])$
58, $*(\mathrm{T}_{6.19})^2$.

60. — $(\exists fxy)(f/xy \mathrel{.{\approx_{rc}}.} S(hde)(rc) \mathbin{:\&:} T(fxy))$ 59, df(T).

61. $15 \supset 60$ cp.

62. $(hde)61$ ug.

63. $(hde)(℩(s/tw)[s/tw \approx h/de] \in ℩(j/kuv)[j/kuv \approx_{qpa} N(qpa)] \mathrel{.\supset.}$
$(\exists fxy)(f/xy \approx h/de \mathbin{.\&.} T(fxy)))$ 14, 62, (Peano_5).

64. $(hde)(℩(s/tw)[s/tw \approx h/de] \in ℩(j/kuv)[j/kuv \approx_{qpa} N(qpa)] \mathrel{.\supset.}$
$(lm)(c)(n)(c \in ℩(p/a)[p/a \approx l/m] \mathbin{.\&.} n \in ℩(p/a)[p/a \approx l/m] \mathbin{.\&.}$
$℩(p/a)[p/a \mathrel{.{\approx_z}.} (l/m - \iota c)(z)] \in ℩(s/tw)[s/tw \approx h/de]$
$\mathrel{:\supset:} ℩(p/a)[p/a \mathrel{.{\approx_z}.} (l/m - \iota n)(z)] \in ℩(s/tw)[s/tw \approx h/de])$
63, $*(\mathrm{T}_{6.19})^2$, df(T).

At long last we arrive at Peano postulate four. The proof is now rather straightforward.

$$\vdash (\forall \alpha^{((o))})(\forall \beta^{((o))})(\alpha^{((o))} \in N^{(((o)))} \mathbin{.\&.} \beta^{((o))} \in N^{(((o)))} \mathbin{.\&.} S\alpha^{((o))} = S\beta^{((o))} \mathrel{:\supset:} \alpha^{((o))} = \beta^{((o))})$$

$(\mathrm{T}_{A.14})$

$$\vdash (hde)(fxy)(℩(s/tw)[s/tw \approx h/de] \in ℩(j/kuv)[j/kuv \approx_{qpa} N(qpa)] \mathbin{.\&.} ℩(s/tw)[s/tw \approx fxy] \in ℩(j/kuv)[j/kuv \approx_{qpa} N(qpa)] \mathbin{.\&.} ℩(stw)[s/tw \approx_{rc} S(hde)(rc)] = ℩(s/tw)[s/tw \approx_{rc} S(fxy)(rc)] \mathrel{:\supset:} ℩(s/tw)[s/tw \approx h/de] = ℩(s/tw)[s/tw \approx f/xy]).$$

Proof:

1. $\rightarrow$ $℩(s/tw)[s/tw \approx h/de] \in ℩(j/kuv)[j/kuv \approx_{qpa} N(qpa)] \mathbin{.\&.}$
$℩(s/tw)[s/tw \approx f/xy] \in ℩(j/kuv)[j/kuv \approx_{qpa} N(qpa)]$
$\mathbin{.\&.} ℩(s/tw)[s/tw \approx_{rc} S(hde)(rc)] = ℩(s/tw)[s/tw \approx_{rc} S(fxy)(rc)]$

2. $\rightarrow \iota(p/a)[p/a \approx l/m] \in \iota(s/tw)[s/tw \approx h/de]$

   Show: $\iota(p/a)[p/a \approx l/m] \in \iota(s/tw)[s/tw \approx f/xy]$

3. $(\exists n) \sim (n \in \iota(p/a)[p/a \approx l/m])$ 1, 2, $(\mathrm{T}_{A.12})$.

4. $\sim (n \in \iota(p/a)[p/a \approx l/m])$ 3, EI.

5. $n \in \iota(p/a)[p/a \;.\!\approx_z\!.\; (l/m + \iota n)(z)]$ 4, df$(+\iota n)$, $*(\mathrm{Abs})^1$.

6. $\iota(p/a)[p/a \;.\!\approx_z\!.\; (l/m + \iota n - \iota n)(z)] = \iota(p/a)[p/a \approx l/m]$ 4, $*(V)^1$.

7. $\iota(p/a)[p/a \;.\!\approx_z\!.\; (l/m + \iota n - \iota n)(z)] \in \iota(s/tw)[s/tw \approx h/de]$

   2, 6, $*(\mathrm{T}_{6.15})^1$.

8. 5 & 7 conj.

9. $(\exists v)(v \in \iota(p/a)[p/a \;.\!\approx_z\!.\; (l/m + \iota n)(z)]$.&.

   $\iota(p/a)[p/a \;.\!\approx_z\!.\; (l/m + \iota n - \iota v)(z)] \in \iota(s/tw)[s/tw \approx h/de])$

   8, eg.

10. $(\exists qb)(q/b \;.\!\approx_z\!.\; (l/m + \iota n)(z)$.&.

    $(\exists v)(v \in \iota(p/a)[p/a \approx q/b]$.&.

    $\iota(p/a)[p/a \;.\!\approx_z\!.\; (q/b - \iota v)(z)] \in \iota(s/tw)[s/tw \approx h/de]))$

    9, $*(\mathrm{T}_{6.19})^2$.

11. $(\exists qb)(q/b \;.\!\approx_z\!.\; (l/m + \iota n)(z)$:&: $S(hde)(qb))$ 10, df(S).

12. $\iota(q/b)[q/b \;.\!\approx_z\!.\; (l/m + \iota n)(z)] \in \iota(s/tw)[s/tw \;.\!\approx_{rc}\!.\; S(hde)(rc)]$

    11, $*(\mathrm{Abs})^2$.

13. $\iota(q/b)[q/b \;.\!\approx_z\!.\; (l/m + \iota n)(z)] \in \iota(s/tw)[s/tw \;.\!\approx_{rc}\!.\; S(fxy)(rc)]$

    1, 12, $*(\mathrm{Ext})^2$.

14. $(\exists qb)(q/b \;.\!\approx_z\!.\; (l/m + \iota n)(z)$:&: $S(fxy)(qb))$ 13, $*(\mathrm{Abs})^2$.

15. $(\exists qb)(q/b \;.\!\approx_z\!.\; (l/m + \iota n)(z)$:&:

    $(\exists v)(v \in \iota(p/a)[p/a \approx q/b]$.&.

    $\iota(p/a)[p/a \;.\!\approx_z\!.\; (q/b - \iota v)(z)] \in \iota(s/tw)[s/tw \approx f/xy]))$

    14, df(S).

16. $(\exists v)(v \in \iota(p/a)[p/a \;.\!\approx_z\!.\; (l/m + \iota n)(z)]$.&.

    $\iota(p/a)[p/a \;.\!\approx_z\!.\; (l/m + \iota n - \iota v)(z)] \in \iota(s/tw)[s/tw \approx f/xy])$

    15, $*(\mathrm{T}_{6.19})^2$.

17. $v \in \iota(p/a)[p/a \;.\!\approx_z\!.\; (l/m + \iota n)(z)]$.&.

    $\iota(p/a)[p/a \;.\!\approx_z\!.\; (l/m + \iota n - \iota v)(z)] \in \iota(s/tw)[s/tw \approx f/xy]$

    16, EI.

18. $\iota(p/a)[p/a \;.\!\approx_z\!.\; (l/m + \iota n - \iota v)(z)] \in \iota(s/tw)[s/tw \approx f/xy]$

    17, simp.

19. $n \in \iota(p/a)[p/a \;.\!\approx_z\!.\; (l/m + \iota n)(z)]$ &

    $v \in \iota(p/a)[p/a \;.\!\approx_z\!.\; (l/m + \iota n)(z)]$ &

    $\iota(p/a)[p/a \;.\!\approx_z\!.\; (l/m + \iota n - \iota v)(z)] \in \iota(s/tw)[s/tw \approx f/xy]$

    & $\iota(s/tw)[s/tw \approx f/xy] \in \iota(j/kuv)[j/kuv \approx_{qpa} N(qpa)]$

    1, 5, 17, conj.

20. $\iota(p/a)[p/a \;.\!\approx_z\!.\; (l/m + \iota n - \iota n)(z)] \in \iota(s/tw)[s/tw \approx f/xy]$

    19, $(\mathrm{T}_{A.13})$.

21. __ $\iota(p/a)[p/a \approx l/m] \in \iota(s/tw)[s/tw \approx f/xy]$ 6, 20, $*(\mathrm{T}_{6.15})^1$,

    (17ei), (4ei).

22. $2 \supset 21$ cp.

The reader may easily complete the proof from here. This completes the Peano postulates.

Appendix B: Axioms, Theorems, and Definitions

Principle Theses of Chapter 4

The asterisk "$*$" tags axiom and theorem schemata.

$*(A_1)$ $\alpha\ .\supset.\ \beta \supset \alpha$
$*(A_2)$ $\alpha\ .\supset.\ \beta \supset \delta\ :\supset:\ \beta\ .\supset.\ \alpha \supset \delta$
$*(A_3)$ $\alpha \supset \beta\ :\supset:\ \beta \supset \delta\ .\supset.\ \alpha \supset \delta$
$*(A_4)$ $\alpha \supset \beta\ .\supset.\ \alpha\ :\supset:\ \alpha$
$*(A_5)$ $(\mu)(A\mu \supset A\alpha|\mu,$
where α is free for μ in A.
$*(A_6)$ $(\mu)(\alpha \supset A\mu)\ .\supset.\ \alpha \supset (\mu)A\mu,$
where μ does not occur free in α and u is not α.

Modus ponens:
From A and $\{A\} \supset \{B\}$, infer B.
Universal generalization
From A, infer $(\mu)A$,
where μ is an individual variable free in A.
Replacement rule for defined signs
Definiendum may replace and be replaced by definiens wherever it occurs.

df($\exists$) $(\exists\mu)A = \text{df} \sim (\mu) \sim A$
df(A) $(\mu_1, \ldots, \mu_n)A\ = \text{df}\ (\mu_1), \ldots, (\mu_n)A$
df(E) $(\exists\mu_1, \ldots, \mu_n)A\ = \text{df}\ (\exists\mu_1), \ldots, (\exists\mu_n)A$
df($\sim$) $\sim \alpha\ = \text{df}\ (\mu)(\alpha \supset \mu)$
df($\vee$) $\alpha \vee \beta = \text{df} \sim \alpha \supset \beta$
df(&) $\alpha\ \&\ \beta\ = \text{df} \sim (\alpha \supset \sim \beta)$
df($\equiv$) $\alpha \equiv \beta\ = \text{df}\ \alpha \supset \beta\ .\&.\ \beta \supset \alpha$

df(*out*) $\alpha\ \mathit{out}\ \beta\ = \text{df}\ (\mu)(\beta/\alpha;\mu!\beta)$
df(*in*) $\alpha\ \mathit{in}\ \beta\ = \text{df} \sim (\alpha\ \mathit{out}\ \beta)$

df(ex) $\alpha\ ex\ \beta = \text{df} \sim (\exists\mu)(\mu\ in\ \alpha\ .\&.\ \mu\ in\ \beta)$
df(ind) $\alpha\ ind\ \beta = \text{df}\ \alpha\ out\ \beta\ .\&.\ \beta\ out\ \alpha$

df(=) $\alpha = \beta = \text{df}\ (p, a, q, r)(p/a;\alpha!q\ \&\ p/a;\beta!r\ .\supset.\ q \supset r)$
df(≠) $\alpha \neq \beta = \text{df} \sim (\alpha = \beta)$
$@ = \text{df}\ (x)(x \supset x)$
$f = \text{df}\ (x)(x \supset x\ .\supset.\ x)$
$*(A_7)$ $\alpha\ in\ \{A\alpha\}$,
$*(A_8)$ $\alpha\ in\ \{A\beta_1, \ldots, \beta_n\}\ :\supset:$
$\alpha = \{A\beta_1, \ldots, \beta_n\}\ .\vee.\ \alpha\ in\ \beta_1\ .\vee., \ldots, .\vee.\ \alpha\ in\ \beta_n,$
where A is any wff all of whose free terms are $\beta_1, \ldots, \beta_n$.
(A_9) $(x, y)(x\ in\ y\ .\&.\ y\ in\ x\ :\supset:\ x = y)$
(A_{10}) $(x, y, z)(x \text{ in } y\ .\&.\ y \text{ in } z\ :\supset:\ x \text{ in } z)$
(A_{11}) $(p, a)(q)(x, y)(p/a;x!q\ .\&.\ p/a;y!q\ .\&.\ a\ in\ p\ :\supset:\ x = y)$
(A_{12}) $(p, a)(z)(q)(p/a;z!q\ .\&.\ a\ in\ p\ .\&.\ a \neq p\ :\supset:\ z\ in\ q\ .\&.\ z \neq q)$
(A_{13}) $(x, y)(x/x;y!y)$
(A_{14}) $(x, y)(x/y;y!x)$
$(p)/(A_{15})$ $(p, a)(x)(\exists q)(p/a;x!q\ \&\ (r)(p/a;x!r\ .\supset.\ r = q))$
$(q)/(A_{16})$ $(p)(\exists q)(q \text{ ex } p)$

$*(r)/*(A_{17})$ $(\exists\mu)(\alpha \text{ out } \{A\mu|\nu\})\ .\supset.\ (\mu)(\{A\alpha|\nu\}/\alpha;\mu!\{A\mu|\nu),$
where α and μ are free for ν in A.

$*(s)/*(A_{18})$
$(\exists\mu_1, \ldots, \mu_n)(\alpha\ out\ \{A\mu|\nu_1, \ldots, \mu|\nu_n\})\ :\&:$
$\alpha \neq \{A\sigma_1|\nu_1, \ldots, \sigma_n|\nu_n\}\ .\&.\ \alpha \neq \beta_1\ .\&., \ldots, .\&.\ \alpha \neq \beta_m\ .:\supset:.$
$(\mu)(\exists\tau_1, \ldots, \tau_n)(\{A\sigma_1|\nu_1, \ldots, \sigma_n|\nu_n\}\alpha;\mu!\{A\tau_1|\nu_1, \ldots, \tau_n|\nu_n\}$
$:\&:\ \sigma_1/\alpha;\mu!\tau_1\ .\&., \ldots, .\&.\ \sigma_n/\alpha;\mu!\tau_n),$
where each μ_i, σ_i, τ_i, $1 \geq i \geq n$, $n \in \mathbb{N} - \{0\}$, is free for its respective ν_i in the wff $A\nu_1, \ldots, \nu_n$, and $\beta_1, \ldots, \beta_m$, $m \in \mathbb{N}$, are all the distinct singular terms occurring free in the wff $A\sigma_1|\nu_1, \ldots, \sigma_n|\nu_n$ excepting $\sigma_1, \ldots, \sigma_n$.
$*(A_{19})$ $\{(\mu)A\mu\} = \{(\nu)A\nu|\mu\}$, where ν is free for μ in A.
$*(A_{20})$ $\{(\mu)A\mu\} = \{(\mu)B\mu\}\ .\supset.\ (\mu)(\{A\mu\} = \{B\mu\})$
$*(A_{21})$ $\{(\mu)A\mu\} \neq \{\alpha \supset \beta\}$
$*(A_{22})$ $\alpha\ in\ \{(\mu)A\mu\}\ .\&.\ \alpha \neq \{(\mu)A\mu\}\ :\supset:\ (\mu)(\alpha\ in\ \{A\mu\}),$
where μ is not α and is not free in α
$*(A_{23})$ $\supset \neq \{\alpha \supset \beta\}\ .\&.\ \supset \neq \{(\mu)A\mu\}$

Theorems

$(T_{4.0})$ $\vdash (x)(x\ in\ x)$
$(T_{4.1})$ $\vdash (x, y)(x \neq y\ :\supset:\ x \text{ in } y\ .\supset.\ y \text{ out } x)$
$(T_{4.2})$a $\vdash (p, a)(z)(q)(p/a;z!q\ :\supset:\ a\ in\ p\ .\supset.\ z\ in\ q)$
$(T_{4.2})$b $\vdash (p, a, b)(p/a;b!b\ :\supset:\ a \text{ out } p\ .\vee.\ a = p)$
$(T_{4.3})$ $\vdash (p, a)(x)(q, r)(p/a;x!q\ \&\ p/a;x!r\ .\supset.\ q = r)$
$(T_{4.4})$ $\vdash (p, a)(z, q)(p/a;z!q\ :\supset:\ a \text{ out } p\ .\supset.\ p = q)$
$(T_{4.5})$a $\vdash (p, q)(q \text{ ex } p\ .\supset.\ p \text{ ex } q)$
$(T_{4.5})$b $\vdash (p, q)(q \text{ ex } p\ :\supset:\ q \text{ ind } p\ .\&.\ q \neq p)$
$(T_{4.5})$c $\vdash (p, q)(q \text{ ex } p\ .\supset.\ (z)(z\ in\ p\ .\supset.\ q \text{ ex } z))$

$*(T_{4.6})/$
$*(CP_{sub})$[1]

$\vdash (\exists\mu^\circ, \nu^\circ)(\nu^\circ \text{ in } \mu^\circ .\&. (\mu)(\mu^\circ/\nu^\circ;\mu!\{A\}))$,
where μ° and ν° do not occur free in A and μ is free in A.

$(T_{4.7}) \vdash (x, y)(x = y .\supset. x \equiv y)$
$(T_{4.8})a \vdash (z)(z = z)$
$(T_{4.8})b \vdash (x, y)(x = y .\supset. y = x)$
$(T_{4.8})c \vdash (x, y, z)(x = y .\&. y = z :\supset: x = z)$
$*(T_{4.8})d \vdash \alpha = \beta .\supset. A(\alpha, \alpha) \supset A(\beta|\alpha, \alpha)$,
where β is free for free occurrences of α in A.

$*(=)$

$\alpha = \beta, A\alpha \vdash A\beta$,
where $A\beta$ results from replacing one or more free occurrences of α by β and β is free for α in A.

$\sim \{A\alpha\}, A\beta \vdash \sim \{\alpha = \beta\}$

$\alpha = \beta \vdash \beta = \alpha$

$(T_{4.9}) \vdash (\exists p, q)(p \equiv q .\&. p \neq q)$
$(T_{4.10})a \vdash @ \neq f$
$(T_{4.10})b \vdash (z)(z \neq @ .\supset. z \text{ out } @)$
$(T_{4.10})c \vdash (z)(z \neq f .\supset. z \text{ out } f)$
$(T_{4.11})a \vdash (p, a)(p \neq \{p \supset a\})$
$(T_{4.11})b \vdash (p, a)(p \neq \{a \supset p\})$
$(T_{4.11})c \vdash (p, a, b, q)(p \neq \{p/a;b!q\})$
$(T_{4.11})d \vdash (p, a, b, q)(a \neq \{p/a;b!q\})$
$(T_{4.11})e \vdash (p, a, b, q)(q \neq \{p/a;b!q\})$
$(T_{4.12})a \vdash (p, a, r)(\{p \supset a\} = \{r \supset a\} .\supset. p = r)$
$(T_{4.12})b \vdash (p, a, s)(\{p \supset a\} = \{p \supset c\} .\supset. a = c)$
$(T_{4.13})a \vdash (p)(\exists y)(p \text{ out } \{\sim y\})$
$(T_{4.13})b \vdash (p)(\exists y, z)(p \text{ out } \{y \supset z\})$.
$(T_{4.13})c \vdash (p)(\exists x, y, z, w)(p \text{ out } \{x/y;z!w\})$.
$(T_{4.14}) \vdash (p, q)(\{\sim p\} = \{\sim q\} .\supset. p = q)$
$(T_{4.15}) \vdash (p)(\{\sim p\} \neq \{r \supset s\})$
$(T_{4.16})a \vdash (p, r, c)(\{p \supset p\} = \{r \supset c\} .\supset. p = r)$
$(T_{4.16})b \vdash (p, a)(\{p \supset a\} = \{a \supset p\} .\supset. p = a)$
$(T_{4.17})a \vdash (p, a, r, c)(\{p \supset a\} = \{r \supset c\} .\supset. p = r)$
$(T_{4.17})b \vdash (p, a, r, c)(\{p \supset a\} = \{r \supset c\} :\supset: p = r .\&. a = c)$
$(T_{4.18})a \vdash (b, q)(p, a, r)(\{p/a;b!q\} = \{r/a;b!q\} :\supset: p = r)$
$(T_{4.18})b \vdash (b, q)(p, a, c)(\{p/a;b!q\} = \{p/c;b!q\} .\supset. a = c)$
$(T_{4.18})c \vdash (b, q)(p, r)(\{p/p;b!q\} = \{r/r;b!q\} .\supset. p = r)$
$(T_{4.18})d \vdash (b, q)(p, r, c)(\{p/p;b!q\} = \{r/c;b!q\} .\supset. p = r)$
$(T_{4.18})e \vdash (p, a, q)(m)(m \text{ ex } p .\&. p/a;m!q :\supset: q = m .\vee. q \text{ out } m)$
$(T_{4.18})f \vdash (b)(p, a)(\{p/a;b!a\} = \{a/p;b!a\} .\supset. p = a)$
$(T_{4.18})g \vdash (b, q)(p, a)(\{p/a;b!q\} = \{a/p;b!q\} .\supset. p = a)$
$(T_{4.19})a \vdash (b, q)(p, a, r, c)(\{p/a;b!q)\} = \{r/c;b!q\} .\supset. p = r)$
$(T_{4.19})b \vdash (b, q)(p, a, r, c)(\{p/a;b!q\} = \{r/c;b!q\} :\supset: p = r .\&. a = c)$
$(T_{4.20}) \vdash (a)(a \neq \{(\exists p)(p = a .\&. \sim p)\})$

Principle Theses of Chapter 5

Definitions

df(desc)

$[\imath zAz]\ B(\imath zAz|v)\ = \text{df}\ (\exists x)(Ax|z\ .\&.\ (y)(Ay|z\ .\supset.\ y = x)\ \&\ Bx|v)$,
where x and y are free for z in Az and x is free for v in Bv.

df(desc)^1

$[(\imath q)(p/a;b!q)]\ B((\imath q)(p/a;b!q)|v) = \text{df}$
$(\exists q)(p/a;b!q\ \&\ (r)(p/a;b!r\ .\supset.\ r = q)\ \&\ Bq|v)$

df(desc.a)^1

$[p/a;b]\ B(p/a;b) = \text{df}\ [(\imath q)(p/a;b!q)]\ B((\imath q)(p/a;b!q))$.

df(desc)^2

$[(\imath q)(p/a,b;x,y!q)]\ A((\imath q)(p/a,b;x,y!q)|v) = \text{df}$
$(\exists q)((r)(p/a,b;x,y!r\ .\equiv.\ r = q)\ \&\ Aq|v)$,
where q is free for v in Av.

df(desc.a)^2 $[p/a,b;x,y]\ A(p/a,b;x,y) = \text{df}$
$[(\imath q)(p/a,b;x,y!q)]\ A((\imath q)(p/a,b;x,y!q))$.

$[\imath xAx, \imath zBz]\ C(\imath xAx, \imath zBz)\ = \text{df}\ [\imath xAx]\ ([\imath zBz]\ C(\imath xAx, \imath zBz))$.

$[\imath zBz]\ C(\imath xAx, \imath zBz)\ = \text{df}\ [\imath zBz, \imath xAx]\ C(\imath xAx, \imath zBz)$.

df($\exists^1$)

$(\exists^1 q)(p/a;b!q\ .\&.\ Aq) = \text{df}$
$(\exists q)(p/a;b!q\ .\&.\ (r)(p/a;b!r\ .\supset.\ r = q)\ .\&.\ Aq)$

df(dual)

$p/a,b;x,y!q = \text{df}$
$(\exists h, e, t)(p/a;e!h\ \&\ h/b;y!t\ \&\ t/e;x!q)$.

df(triple)

$p/a,b,c;x,y,z!q\ = \text{df}\ (\exists h, e, t, m, s, l)(p/a;e!h\ .\&.\ h/b;m!t$
$.\&.\ t/c;z!s\ .\&.\ s/m;y!l\ .\&.\ l/e;x!q)$

df($\equiv$.a)^1 $p/a \equiv r/c = \text{df}\ (x)(p/a;x\ .\equiv.\ r/c;x)$.

df($\equiv$.a)^2 $q/p,a \equiv s/t,w = \text{df}\ (r,c)(q/p,a;r,c\ .\equiv.\ s/t,w;r,c)$

df($=$.a)^1

$p/a = r/c = \text{df}\ (s,t,w)(s/t,w;r,c\ .\equiv.\ s/t,w;p,a)$

df($=$.a)^2

$s/t,w = q/p,a = \text{df}\ (\exists j, k, u, v)(j/k,u,v;s,t,w\ .\equiv.\ j/k,u,v;q,p,a)$,

Theorems

$*(\text{T}_{5.1})\ \vdash E!(\imath xAx)\ :\supset:\ (x)Cx \supset [\imath xAx]C(\imath xAx)$

$*(\text{T}_{5.1})\text{a}\ \vdash\ (x)Cx \supset [p/a;b]C(p/a;b)$

$*(\text{T}_{5.2})$

$\vdash\ (p)(q)(p \equiv q)\ .\supset.\ Cp \equiv Cq)\ \&\ E!(\imath zAz)\ :\supset:$
$[\imath zAz]C(B(\imath zAz))\ .\equiv.\ C([\imath zAz]B(\imath zAz))$.

$*(\text{T}_{5.3})^1/$

$*(\text{Scope})^1$

$\vdash\ (p, q)(p \equiv q\ .\supset.\ Cp \equiv Cq)\ .\supset.$
$(\exists q)(p/a;b!q\ :\&:\ (r)(p/a;b!r\ .\supset.\ r = q)\ \&\ Cq) \equiv$
$C((\exists q)(p/a;b!q\ :\&:\ (r)(p/a;b!r\ .\supset.\ r = q)\ \&\ q))$.

$*(\text{T}_{5.3})^2/$

$*(\text{scope})^2$

$(p, q)(p \equiv q \ .\supset.\ Cp \equiv Cq)\ :\supset:$
$(\exists q)((r)(p/a,b;x,y!r \ .\equiv.\ r = q)\ \&\ Cq) \equiv$
$C(\exists q)(p/a,b;x,y!q\ \&\ q).$

$*(\mathrm{T}_{5.4})^2/$
$*(\mathrm{CP}_{\mathrm{sub}})^2$
$\vdash\ (\exists p, a, b)(a\ in\ p\ .\&.\ b\ in\ p\ .\&.\ (x, y)(p/a,b;x,y!\{A\}),$
where p, a, b are not free in A, and x, y are free in A.

$(\mathrm{T}_{5.5})/$
(p^2)
$\vdash\ (p, a, b)(x, y)E!(\iota q)(p/a,b;x, y!q)$
$*(\mathrm{CP}_{\mathrm{sub}})^3$
$(\exists j, k, u, v)(k\ in\ j\ .\&.\ u\ in\ j\ .\&.\ v\ in\ j\ .\&.$
$(q, p, a)(j/k,u,v;q,p,a!\{A\}),$
where j, k, u, v are not free in A.

$*(\mathrm{T}_{5.6})/$
$*(\mathrm{CP})^1$
$\vdash\ (\exists p, a)(z)(p/a;z\ .\equiv.\ A),$
where p, a are not free in A, and z is free in A.

$*(\mathrm{T}_{5.6})^2/$
$*(\mathrm{CP})^2$
$\vdash\ (\exists p, a, b)(x, y)(p/a,b;x,y\ .\equiv.\ A)$
where p, a, b are not free in A, and x, y are free in A.

$*(\mathrm{T}_{5.6})^3/$
$*(\mathrm{CP})^3$
$\vdash\ (\exists j, k, u, v)(q, p, a)(j/k,u,v;q,p,a\ .\equiv.\ A),$
where j, k, u, v are not in A, and q, p, a are free in A.

$(\mathrm{T}_{5.7})^1\quad \vdash\ (p, a)(p/a \equiv p/a)$
$(\mathrm{T}_{5.8})^1\quad \vdash (p, a)(l, m)(p/a \equiv l/m\ .\supset.\ l/m \equiv p/a)$
$(\mathrm{T}_{5.9})^1\quad \vdash\ (p, a)(l, m)(r, c)(p/a \equiv l/m\ .\&.$
$l/m \equiv r/c\ :\supset:\ p/a \equiv r/c)$
$(\mathrm{T}_{5.10})^1\quad \vdash\ (p, a)(x)(\sim (p/a;x)\ .\equiv.\ \sim (\exists^1 q)(p/a;x!q\ \&\ q))$
$(\mathrm{T}_{5.11})^1\quad \vdash\ (p, a)(l, m)(p/a \equiv l/m\ .\equiv.$
$(x)[(\exists^1 q)(p/a;x!q\ \&\ q) \equiv (\exists^1 q)(l/m;x!q\ \&\ q)]$
$(\mathrm{T}_{5.7})^2\quad \vdash\ (q, p, a)(q/p,a \equiv q/p, a)$
$(\mathrm{T}_{5.8})^2\quad \vdash\ (q, p, a)(s, t, w)(q/p,a \equiv s/t,w\ .\supset.\ s/t,w \equiv q/p,a)$
$(\mathrm{T}_{5.9})^2\quad \vdash\ (q, p, a)(s, t, w)(h, d, e)(q/p,a \equiv s/t,w\ .\&.\ s/t,w \equiv h/d,e$
$\supset:\ q/p,a \equiv h/d,e)$
$(\mathrm{T}_{5.10})^2\quad \vdash\ (q, p, a)(r, c)(\sim (q/p,a;r,c)\ .\equiv.\ \sim (\exists^1 t)(q/p,a;r,c!t\ \&\ t))$
$(\mathrm{T}_{5.11})^2\quad \vdash\ (q, p, a)(s, t, w)(q/p,a \equiv s/t,w\ .\equiv.$
$(r, c)[(\exists^1 h)(q/p,a;r,c!h\ \&\ h)\ .\equiv.\ (\exists^1 h)(s/t,w;r,c!h\ \&\ h)])$

$(\mathrm{T}_{5.12})^1\quad \vdash\ (l, m)(l/m = l/m)$
$(\mathrm{T}_{5.13})^1\quad \vdash\ (l, m)(r, c)(l/m = r/c\ .\supset.\ r/c = l/m)$
$(\mathrm{T}_{5.14})^1\quad \vdash\ (l, m)(r, c)(t, w)(l/m = r/c\ .\&.\ r/c = t/w\ .\supset.\ l/m = t/w)$
$(\mathrm{T}_{5.12})^2\quad \vdash\ (l, m, n)(1/m,n = l/m,n)$
$\mathrm{T}_{5.13})^2\quad \vdash\ (l, m, n)(s, t, w)(l/m,n = s/t,w\ .\supset.\ s/t,w = l/m,n)$
$(\mathrm{T}_{5.14})^2\quad \vdash\ (l, m, n)(s, t, w)(q, p, a)(l/m,n = s/t,w\ .\&.$
$s/t,w = q/p,a\ :\supset:\ l/m,n = q/p,a)$

$(T_{5.15})^1$ $\vdash$ $(l, m)(r, c)(l/m = r/c\ .\supset.\ l/m \equiv r/c)$
$(T_{5.15})^2$ $\vdash$ $(l, m, n)(s, t, w)(l/m,n = s/t,w\ .\supset.\ l/m,n \equiv s/t,w)$,
$(T_{5.16})^1$ $\vdash$ $(p, a)(r, c)(p/a = r/c :\equiv: p = r\ .\&.\ a = c)$.
$(T_{5.16})^2$ $\vdash$ $(s, t, w)(q, p, a)(s/t,w = q/p,a :\equiv: s = q\ .\&.\ t = p\ .\&.\ w = a)$.

Principle Theses of Chapter 6

Definitions

df(C)[1]
$x \in \imath(p/a)[p/a \approx_z Az] = \text{df } (\exists pa)[p/a \approx_z Az\ .\&.\ x \in p/a]$
df(C.a)[1]
$x \in p/a = \text{df } (\exists q)(p/a;x!q\ \&\ q)$
df(C)[2]
$\imath(p/a)[p/a \approx_z Az] \in \imath(s/tw)[s/tw \approx_{rc} B(rc)] = \text{df}$
$(\exists stw)(s/tw \approx_{rc} B(rc)\ .\&.\ (\exists pa)(p/a \approx_z Az\ .\&.\ p/a \in s/tw))$
df(C.a)[2]
$p/a \in s/tw = \text{df } (\exists q)(s/tw;pa!q\ \&\ q)$
df(C)[3]
$\imath(s/tw)[s/tw \approx_{rc} A(rc)] \in \imath(j/kuv)[j/kuv \approx_{qpa} B(qpa)] = \text{df}$
$(\exists jkuv)(j/kuv \approx_{qpa} B(qpa)\ .\&.$
$(\exists stw)(s/tw \approx_{rc} A(rc)\ .\&.\ s/tw \in j/kuv))$
df(C.a)[3]
$s/tw \in j/kuv = \text{df } (\exists q)(j/kuv;stw!q\ \&\ q)$
df(C =)[1]
$\imath(p/a)[p/a \approx_z Az] = \imath(p/a)[p/a \approx_z Bz] = \text{df}$
$(\exists pa)(p/a \approx_z Az\ .\&.\ (\exists qb)(q/b \approx_z Bz\ .\&.\ p/a = q/b))$
df(C =)[2]
$\imath(s/tw)[s/tw \approx_{rc} A(rc)] = \imath(s/tw)[s/tw \approx_{rc} B(rc)] = \text{df}$
$(\exists stw)(s/tw \approx_{rc} A(rc)\ .\&.$
$(\exists lmn)(l/mn \approx_{rc} B(rc)\ .\&.\ s/tw = l/mn))$
df(A)[1]
$[r/c]\ A(rc) = \text{df } (\exists pa)(p/a \approx r/c\ .\&.\ A(pa))$
df(A)[2]
$[q/pa]\ A(qpa) = \text{df } (\exists stw)(s/tw \approx q/pa\ .\&.\ A(qpa))$
df(≈)[1]
$p/a \approx r/c = \text{df } (z)(z \in p/a\ .\equiv.\ z \in r/c)$
df(≈.a)[1]
$p/a \approx_z Az = \text{df } (z)(z \in p/a\ .\equiv.\ Az)$
df(≈.b)[1]
$Az \approx_z Bz = \text{df } (z)(Az \equiv Bz)$
df(≈)[2]
$s/tw \approx l/mn = \text{df } (rc)([r/c][r/c \in s/tw]\ .\equiv.\ [r/c][r/c \in l/mn])$
df(≈.a)[2]
$s/tw \approx_{rc} A(rc) = \text{df } (rc)([r/c][r/c \in s/tw]\ .\equiv.\ [r/c]A(rc))$
df(≈.b)[2]
$A(rc) \approx_{rc} B(rc) = \text{df } (rc)([r/c]A(rc)\ .\equiv.\ [r/c]B(rc))$
df(≈)[3]
$j/kuv \approx l/mno = \text{df}$

$(qpa)([q/pa][q/pa \in j/kuv] .\equiv. [q/pa][q/pa \in l/mno])$

df$(\approx.\text{a})^3$

$j/kuv \approx_{qpa} A(qpa) = \text{df } (qpa)([q/pa][q/pa \in j/kuv] .\equiv. [q/pa]A(qpa))$

df$(\approx.\text{b})^3$

$A(qpa) \approx_{qpa} B(qpa) = \text{df } (qpa)([q/pa]A(qpa) .\equiv. [q/pa]B(qpa))$

df$(-\iota y)$ $(r/c - \iota y)(z) = \text{df } z \in \imath(p/a)[p/a \approx r/c] .\&. z \neq y$

df$(+\iota y)$ $(r/c + \iota y)(z) = \text{df } z \in \imath(p/a)[p/a \approx r/c] .\vee. z = y$

df$(+\iota y - \iota x)$ $(r/c + \iota y - \iota x)(z) = \text{df}$

$z \in \imath(p/a)[p/a \approx r/c] .\vee. z = y \;:\&:\; z \neq x$

df$(-\iota y + \iota x)$ $(r/c - \iota y + \iota x)(z) = \text{df}$

$z \in \imath(p/a)[p/a \approx r/c] .\&. z \neq y :\vee: z = x$

df$(-\iota y - \iota x)$ $(r/c - \iota y - \iota x)(z) = \text{df}$

$z \in \imath(p/a)[p/a \approx r/c] .\&. z \neq y :\&: z \neq x$

df$(V)^1$ $Vz = \text{df } z = z.$

df$(\Lambda)^1$ $\Lambda z = \text{df } z \neq z.$

df$(\subseteq)^1$

$\imath(p/a)[p/a \approx_z Az] \subseteq \imath(p/a)[p/a \approx_z Bz] = \text{df}$

$(y)(y \in \imath(p/a)[p/a \approx_z Az] .\supset. y \in \imath(p/a)[p/a \approx_z Bz])$

df$(\subset)^1$

$\imath(p/a)[p/a \approx_z Az] \subset \imath(p/a)[p/a \approx_z Bz] = \text{df}$

$\imath(p/a)[p/a \approx_z Az] \subseteq \imath(p/a)[p/a \approx_z Bz] .\&.$

$\sim (\imath(p/a)[p/a \approx_z Az] = \imath(p/a)[p/a \approx_z Bz])$

df$(\subseteq)^2$

$\imath(s/tw)[s/tw \approx_{rc} A(rc)] \subseteq \imath(s/tw)[s/tw \approx_{rc} B(rc)] = \text{df}$

$(lm)(\imath(p/a)[p/a \approx lm] \in \imath(s/tw)[s/tw \approx_{rc} A(rc)] .\supset.$

$\imath(p/a)[p/a \approx l/m] \in \imath(s/tw)[s/tw \approx_{rc} B(rc)])$

df$(\subset)^2$

$\imath(s/tw)[s/tw \approx_{rc} A(rc)] \subset \imath(s/tw)[s/tw \approx_{rc} B(rc)] = \text{df}$

$\imath(s/tw)[s/tw \approx_{rc} A(rc)] \subseteq \imath(s/tw)[s/tw \approx_{rc} B(rc)] .\&.$

$\sim (\imath(s/tw)[s/tw \approx_{rc} A(rc)] = \imath(s/tw)[s/tw \approx_{rc} B(rc)])$

$E!\imath(p/a)(p/a \approx_z Az) = \text{df } (\exists pa)(p/a \approx_z Az)$

$E!\imath(s/tw)(s/tw .\approx_{rc}. A(rc) = \text{df } (\exists stw)(s/tw \approx_{rc} A(rc))$

$E!\imath(j/kuv)(j/kuv \approx_{qpa} A(qpa) = \text{df } (\exists jkuv)(j/kuv \approx_{qpa} A(qpa))$

Theorems

$(\mathrm{T}_{6.1})^1$ $\vdash (pa)(p/a \approx p/a)$

$(\mathrm{T}_{6.2})^1$ $\vdash (pa)(lm)(p/a \approx l/m .\supset. l/m \approx p/a)$

$(\mathrm{T}_{6.3})^1$ $\vdash (pa)(lm)(rc)(p/a \approx l/m .\&. l/m \approx r/c :\supset: p/a \approx r/c)$

$(\mathrm{T}_{6.4})^1$ $\vdash (pa)(rc)(p/a \equiv r/c .\supset. p/a \approx r/c)$

$(\mathrm{T}_{6.4})^2$ $\vdash (stw)(lmn)(s/tw \equiv l/mn .\supset. s/tw \approx l/mn)$

$(\mathrm{T}_{6.4})^2$ $\vdash (jkuv)(lmno)(j/kuv \equiv l/mno .\supset. j/kuv \approx l/mno),$

$(\mathrm{T}_{6.5})^1$ $\vdash (pa)(rc)(p/a = r/c .\supset. p/a \approx r/c)$

$(\mathrm{T}_{6.5})^2$ $\vdash (stw)(lmn)(s/tw = l/mn .\supset. s/tw \approx l/mn)$

$(\mathrm{T}_{6.5})^3$ $\vdash (jkuv)(lmno)(j/kuv = l/mno .\supset. j/kuv \approx l/mno),$

$*(\mathrm{T}_{6.6})^1$ $\vdash Az \approx_z Bz :\supset:$

$(\exists pa)(p/a \approx_z Az .\&. C(pa)) .\equiv. (\exists pa)(p/a \approx_z Bz .\&. C(pa)))$

$*(\mathrm{T}_{6.6})^2$ $\vdash A(rc) \approx_{rc} B(rc) :\supset:$

$(\exists stw)(s/tw \approx_{rc} A(rc) .\&. C(stw)) .\equiv.$
$(\exists stw)(s/tw \approx_{rc} B(rc) .\&. C(stw))$

$*(T_{6.7})^1 \quad \vdash (lm)(l/m \approx_z Bz :\supset:$
$(\exists pa)(p/a \approx l/m .\&. C(pa)) .\equiv. (\exists pa)(p/a \approx_z Bz .\&. C(pa)))$

$*(T_{6.7})^2 \quad \vdash (hde)(h/de \approx_{rc} B(rc) :\supset:$
$(\exists stw)(s/tw \approx h/de .\&. C(stw)) .\equiv.$
$(\exists stw)(s/tw \approx_{rc} B(rc) .\&. C(stw))$

$*(T_{6.8})^1 \quad \vdash (\exists pa)(p/a \approx_z A)$,
where p, a are not free in A.

$*(T_{6.8})^2 \quad \vdash (\exists stw)(s/tw \approx_{rc} A)$, where s, t, w are not free in A.

$*(T_{6.8})^3 \quad \vdash (\exists jkuv)(j/kuv \approx_{qpa} A)$, where j, k, u, v are not free in A.

These number $*(T_{6.9})$–$*(T_{6.14})$:

$*20.3 \quad \vdash x \in \hat{z}Az .\equiv. Az$

$*(\mathrm{Abs})^1 \quad \vdash x \in \imath(p/a)[p/a \approx_z Az] .\equiv. Ax$

$*(\mathrm{Abs})^2 \quad \vdash \imath(p/a)[p/a \approx_z Az] \in \imath(s/tw)[s/tw \approx_{rc} B(rc)] .\equiv.$
$(\exists pa)(p/a \approx_z Az .\&. B(pa))$

$*20.15 \quad \vdash (x)(Ax \equiv Bx) .\equiv. \hat{z}Az = \hat{z}Bz$

$*(\mathrm{V})^1 \quad \vdash (z)(Az .\equiv. Bz) .\equiv.$
$\imath(p/a)[p/a \approx_z Az] = \imath(p/a)[p/a \approx_z Bz]$

$*(\mathrm{V})^2 \quad \vdash A(rc) \approx_{rc} B(rc) .\equiv.$
$\imath(s/tw)[s/tw \approx_{rc} A(rc)] = \imath(s/tw)[s/tw \approx_{rc} B(rc)]$

$*(\mathrm{Ext.a})^1 \quad \vdash \imath(p/a)[p/a \approx_z Az] = \imath(p/a)[p/a \approx_z Bz] :\supset:$
$y \in \imath(p/a)[p/a \approx_z Az] .\equiv. y \in \imath(p/a)[p/a \approx_z Bz]$

$*(\mathrm{Ext.a})^2 \quad \vdash \imath(s/tw)[s/tw \approx_{rc} A(rc)] = \imath(s/tw)[s/tw \approx_{rc} B(rc)] :\supset:$
$\imath(p/a)[p/a \approx_z Dz] \in \imath(s/tw)[s/tw \approx_{rc} A(rc)] .\equiv.$
$\imath(p/a)[p/a \approx_z Dz] \in \imath(s/tw)[s/tw \approx_{rc} B(rc)]$

$*20.31 \quad \vdash \hat{z}Az = \hat{z}Bz .\equiv. (y)(y \in \hat{z}Az .\equiv. y \in \hat{z}Bz)$

$*(\mathrm{Ext})^1 \quad \vdash \imath(p/a)[p/a \approx_z Az] = \imath(p/a)[p/a \approx_z Bz] .\equiv.$
$(y)(y \in \imath(p/a)[p/a \approx_z Az] .\equiv. y \in \imath(p/a)[p/a \approx_z Bz])$

$*(\mathrm{Ext})^2 \quad \vdash \imath(s/tw)[s/tw \approx_{rc} A(rc)] = \imath(s/tw)[s/tw \approx_{rc} B(rc)] .\equiv.$
$(rc)(\imath(p/a)[p/a \approx r/c] \in \imath(s/tw)[s/tw \approx_{rc} A(rc)] .\equiv.$
$\imath(p/a)[p/a \approx r/c] \in \imath(s/tw)[s/tw \approx_{rc} B(rc)])$

$*20.18 \quad \vdash \hat{z}Az = \hat{z}Bz .\supset. f(\hat{z}Az) .\equiv. f(\hat{z}Bz)$

$*(\mathrm{E.a})^1 \quad \vdash \imath(p/a)[p/a \approx_z Az] = \imath(p/a)[p/a \approx_z Bz] .\supset.$
$(\exists pa)(p/a \approx_z Az .\&. C(pa)) .\equiv.$
$(\exists pa)(p/a \approx_z Bz .\&. C(pa))$

$*(\mathrm{E.a})^2 \quad \vdash \imath(s/tw)[s/tw \approx_{rc} A(rc)] = \imath(s/tw)[s/tw \approx_{rc} B(rc)] .\supset.$
$(\exists stw)(s/tw \approx_{rc} A(rc) .\&. C(stw)) .\equiv.$
$(\exists stw)(s/tw \approx_{rc} B(rc) .\&. C(stw))$

$*20.19 \quad \vdash \hat{z}Az = \hat{z}Bz .\equiv. (f)(f(\hat{z}Az) .\equiv. f(\hat{z}Bz))$

$*(\mathrm{E})^1 \quad \vdash \imath(p/a)[p/a \approx_z Az] = \imath(p/a)[p/a \approx_z Bz] .\equiv.$
$(stw)((\exists pa)(p/a \approx_z Az .\&. p/a \in s/tw)$
$.\equiv. (\exists pa)(p/a \approx_z Bz .\&. p/a \in s/tw))$

$*(\mathrm{E})^2 \quad \vdash \imath(s/tw)[s/tw \approx_{rc} A(rc)] = \imath(s/tw)[s/tw \approx_{rc} B(rc)] .\equiv.$
$(lmno)((\exists stw)(s/tw \approx_{rc} A(rc) .\&. s/tw \in l/mno) .\equiv.$
$(\exists stw)(s/tw \approx_{rc} B(rc) .\&. s/tw \in l/mno))$

$*(T_{6.15})^1 \quad \vdash \imath(p/a)[p/a \approx_z Az] = \imath(p/a)[p/a \approx_z Bz] :\supset:$
$\imath(p/a)[p/a \approx_z Az] \in \imath(s/tw)[s/tw \approx_{rc} C(rc)] .\equiv.$

$\imath(p/a)[p/a \approx_z Bz] \in \imath(s/tw)[s/tw \approx_{rc} C(rc)]$

$*20.2 \quad \vdash \hat{z}Az = \hat{z}Az$
$*(T_{6.16})^1 \quad \vdash \imath(p/a)[p/a \approx_z Az] = \imath(p/a)[p/a \approx_z Az]$
$*20.21 \quad \vdash \hat{z}Az = \hat{z}Bz \ .\supset.\ \hat{z}Bz = \hat{z}Az$
$*(T_{6.17})^1 \quad \vdash \imath(p/a)[p/a \approx_z Az] = \imath(p/a)[p/a \approx_z Bz] \ .\supset.$
$\imath(p/a)[p/a \approx_z Bz] = \imath(p/a)[p/a \approx_z Az]$
$*20.22 \quad \vdash \hat{z}Az = \hat{z}Bz \ .\&.\ \hat{z}Bz = \hat{z}Cz \ :\supset:\ \hat{z}Az = \hat{z}Cz$
$*(T_{6.18})^1 \quad \vdash \imath(p/a)[p/a \approx_z Az] = \imath(p/a)[p/a \approx_z Bz] \ .\&.$
$\imath(p/a)[p/a \approx_z Bz] = \imath(p/a)[p/a \approx_z Cz] \ :\supset:$
$\imath(p/a)[p/a \approx_z Az] = \imath(p/a)[p/a \approx_z Cz]$
$*(T_{6.19})^1 \quad \vdash C(B(\imath(p/a)[p/a \approx_z Az])) \equiv$
$(\exists qb)(q/b \approx_z Az \ .\&.\ C(B(\imath(p/a)[p/a \approx q/b]))),$
where C is truth-functional, and B is a wff with
$\ulcorner\imath(p/a)[p/a \approx_z Az]\urcorner$ flanking a membership sign or an identity sign.

$*(T_{6.19})^2 \quad \vdash C(B(\imath(s/tw)[s/tw \approx_{rc} A(rc)])) \equiv$
$(\exists hde)(h/de \approx_{rc} Arc \ .\&.\ C(B(\imath(s/tw)[s/tw \approx h/de])),$
where C is truth-functional, and B is a wff with
$\ulcorner\imath(s/tw)\ [s/tw \approx_{rc} Arc]\urcorner$ flanking a membership sign or an identity sign.

$*(T_{6.20})^1 \quad \vdash (lm)(C(B(\imath(p/a)[p/a \approx l/m]))) \ .\supset.\ C(B(\imath(p/a)[p/a \approx_z Az])),$
where $C$is truth-functional, and B is a wff with $\ulcorner\imath(p/a)[p/a \approx_z Az]\urcorner$ flanking a membership sign or an identity sign.
$*(T_{6.20})^2 \quad \vdash (hde)(C(B(\imath(s/tw)[s/tw \approx h/de])) \ .\supset.\ C(B(\imath(s/tw)[s/tw \approx_{rc} A(rc)])),$
where C is truth-functional, and B is a wff with
$\ulcorner\imath(s/tw)\ [s/tw \approx_{rc} Arc]\urcorner$ flanking a membership sign or an identity sign.

$(T_{6.21})^1$
$\vdash (\forall x^o)(x^o \in V^{(o)})$
$\vdash (x)(x \in \imath(p/a)[p/a \approx_z Vz])$
$(T_{6.22})^1$
$\vdash (\forall\alpha^{(o)})((\forall x^o)(x^o \in \alpha^{(o)}) \ .\supset.\ \alpha^{(o)} = V^{(o)})$
$\vdash (lm)((x)(x \in \imath(p/a)[p/a \approx l/m]) \ .\supset.$
$\imath(p/a)[p/a \approx l/m] = \imath(p/a)[p/a \approx_z Vz])$

$(T_{6.23})^1$
$\vdash (\forall x^o) \sim (x^o \in \Lambda^{(o)}),$
$\vdash (x) \sim (x \in \imath(p/a)[p/a \approx_z \Lambda z])$

$(T_{6.24})^1$
$\vdash (\forall\alpha^{(o)})((\forall y^o) \sim (y^o \in \alpha^{(o)}) \ .\equiv.\ \alpha^{(o)} = \Lambda^{(o)})$
$\vdash (lm)((y) \sim (y \in \imath(p/a)[p/a \approx l/m)]) \ .\equiv.$
$\imath(p/a)[p/a \approx l/m] = \imath(p/a)[p/a \approx_z \Lambda z])$

$(T_{6.25})^1$
$\vdash (\forall\alpha^{(o)})(\alpha^{(o)} \subseteq V^{(o)})$
$\vdash (lm)(\imath(p/a)[p/a \approx l/m] \subseteq \imath(p/a)[p/a \approx_z Vz]$
$(T_{6.26})^1$
$\vdash (\forall x^o)(x^o \in \iota x^{(o)})$
$\vdash (x)(x \in \imath(p/a)[p/a \ .\approx_z.\ z = x])$

$(\mathrm{T}_{6.27})^1$
$\vdash\ (\forall y^{\mathrm{o}})(\forall\alpha^{(\mathrm{o})})(y^{\mathrm{o}} \in \alpha^{(\mathrm{o})} \cup \iota y^{(\mathrm{o})})$
$\vdash\ (y)(lm)(y \in \imath(p/a)[p/a \ .\!\approx_z\!. \ (l/m + \iota y)(z)])$

$*(\mathrm{T}_{6.28})^1$
$\vdash\ (\forall y^{\mathrm{o}})(\forall\alpha^{(\mathrm{o})})(y^{\mathrm{o}} \in \alpha^{(\mathrm{o})} \ :\supset:\ \alpha^{(\mathrm{o})} \ .\!=\!. \ \alpha^{(\mathrm{o})} - \iota y^{(\mathrm{o})} \cup \iota y^{(\mathrm{o})})$
$\vdash\ (y)(lm)(y \in \imath(p/a)[p/a \approx l/m] \ :\supset:$
$\quad \imath(p/a)[p/a \approx l/m] \ .\!=\!. \ \imath(p/a)[p/a \ .\!\approx_z\!. \ (l/m - \iota y + \iota y)(z)])$

$*(\mathrm{T}_{6.29})^1$
$\vdash\ (\forall\alpha^{(\mathrm{o})}(\forall x^{\mathrm{o}})(\forall y^{\mathrm{o}})(\alpha^{(\mathrm{o})} - \iota x^{\mathrm{o}}) - \iota y^{\mathrm{o}} \ .\!=\!. \ (\alpha^{(\mathrm{o})} - \iota y^{\mathrm{o}}) - \iota x^{\mathrm{o}})$
$\vdash\ (lm)(x)(y)(\imath(p/a)[p/a \ .\!\approx_z\!. \ (l/m - \iota x - \iota y)(z)] =$
$\quad \imath(p/a)[p/a \ .\!\approx_z\!. \ (l/m - \iota y - \iota x)(z)])$

Bibliography

Works by Bertrand Russell

1900a. *A Critical Examination of the Philosophy of Leibniz*, George Allen and Unwin, Second edition; 1949 (first edition 1900).

1900b. "On the Logic of Relations," dated Oct. 1900, manuscript in the Russell Archives (McMaster University, Ontario, Canada).

1901a. "Sur la logique des relations avec des applications à la théorie des séries," undated, manuscript in the Russell Archives (McMaster University, Ontario, Canada).

1901b. (*MM*) "Recent Work in the Philosophy of Mathematics," *The International Montly*, 1901. Reprinted in *MyL* with the title "Mathematics and the Metaphysicians," pp. 59–74.

1901c. (*Sur*). "Sur la logique des relations avec des applications à la théorie des séries," *Revue de mathématiques* 7, pp. 115–48.

1902. "Théorie générale des séries bien-ordonnées," Revue de mathématiques, 8, pp. 12–43.

1903a. "Frege: Sinn and Bedeutung," undated, manuscript in the Russell Archives (McMaster University, Ontario, Canada).

1903b. "General Theory of Classes,' undated, manuscript numbered 230.030940 in the Russell Archives (McMaster University, Ontario, Canada).

1903c. "General Theory of Classes," undated, manuscript in the Russell Archives (McMaster University, Ontario, Canada).

1903d. (*P*) "Points About Denoting," undated manuscript; published in *CP*, vol. 4, pp. 305–13.

1903g. (*PoM*) *The Principles of Mathematics* (Cambridge University Press: London, 1937).

1904a. (*N&P*) "Necessity and Possibility," manuscript published in *CP*, vol. 4, pp. 507–20.

1904a. "On Functions," manuscript dated Oct. 1904; published in *CP*, vol. 4, pp. 96–110.
1904b. (*M*) "On Meaning and Denotation," undated manuscript; published in *CP*, vol. 4, pp. 314–58.
1904c. (*M&D*) "On the Meaning and Denotation of Phrases," manuscript published in *CP*, vol. 4, pp. 283–96.
1904d. (*P*) "Points About Denoting," undated manuscript; published in *CP*, vol. 4, pp. 305–13.
1904e. "The Axiom of Infinity," *Hibbert Journal* 2, pp. 809–812. Reprinted in *EA*, pp. 256–9.
1905a. (*OD*) "On Denoting," *Mind* 14, pp. 479–93. Reprinted in *EA*, pp. 103–19.
1905b. (*OF*) "On Fundamentals," dated 7 June, ms. 230.030710. Published in *CP*, vol. 4, pp. 359–413.
1905c. (*S*) "On Substitution," dated 22 Dec. 1905, manuscript in the Russell Archives (McMaster University, Ontario, Canada).
1905d. (*EI*) "On the Existential Import of Propositions," *Mind* 14, pp. 398–401.
1905e. (*RML*) "Sur la relation des mathématiques à la logistique," in the *Revue de métaphysique et de morale*, 13, pp. 906–917. English translation in EA, pp. 260–71.
1906a. (*InS*) "Les paradoxes de la logique," *Revue de métaphysique et de morale* 14, pp. 627–50. Reprinted in *EA* with the original English title "On 'Insolubilia' and Their Solution by Symbolic Logic," pp. 190–214.
1906b. (*L*) "Logic in which Propositions are not Entities," dated April–May 1906, ms. 230.030720.
1906c. (*TN*) "On Some Difficulties in the Theory of Transfinite Numbers and Order Types," *Proceedings of the London Mathematical Society* 4, pp. 29–53. Reprinted in *EA*, pp. 135–64. (Received by the Society 24 November 1905).
1906d. (*OS*) "On Substitution," dated April–May 1906, manuscript in the Russell Archives (McMaster University, Ontario, Canada).
1906e. (*F*) "On the Functional Theory of Propositions, Classes, and Relations," dated May 1906, manuscript in the Russell Archives (McMaster University, Ontario, Canada).
1906f. (*SCR*) "On the Substitutional Theory of Classes and Relations," manuscript received by the London Mathematical Society on 24 April 1906. Published in *EA*, pp. 165–89.
1906g. (*PL*) "The Paradox of the Liar," dated September 1906, manuscript in the Russell Archives (McMaster University, Ontario, Canada).
1906h. (*TofI*) "The Theory of Implication," *American Journal of Mathematics* 28, pp. 159–202. (Written in 1905).
1907a. "Fundamentals," dated June 1907, manuscript in the Russell Archives (McMaster University, Ontario, Canada).
1907b. Letter to Hawtrey, dated 22 January 1907, manuscript in the Russell Archives (McMaster University, Ontario, Canada).
1907c. (*OT*) "On the Nature of Truth," *Proceedings of the Aristotelian Society* 7, pp. 28–49.
1907d. (*T*) "On Types," undated, manuscript in the Russell Archives (McMaster University, Ontario, Canada).
1908a. (*ML*) "Mathematical Logic as Based on the Theory of Types," *American Journal of Mathematics* 30, pp. 222–62. Reprinted in van Heijenoort 1967, pp. 150–82. (Written in 1907.)

1910a. (*KAKD*) "Knowledge by Acquaintance and Knowledge by Description," *Proceedings of the Aristotelian Society* (1910–11), pp. 108–28. Reprinted in *MyL*, pp. 152–67.

1910b. (*TT*) "La théorie des types logiques," *Revue de métaphysique et de morale* 18, pp. 263–301. Reprinted in *EA* with original English title "The Theory of Logical Types," pp. 215–52.

1910c. *Philosophical Essays* (London: Longman).

1910d. (*PM*) *Principia Mathematica*, with A. N. Whitehead, vols. 1–3 1910, 1912, 1913 (Cambridge: Cambridge University Press).

1912a. (*ONC*) "On the Notion of Cause," *Proceedings of the Aristotelian Society* 13 (1912–13), pp. 1–26. Reprinted in *MyL*, pp. 132–51.

1912b. (*Pr*) *The Problems of Philosophy* (London: Home University Library).

1913. (*TK*) *Theory of Knowledge: The 1913 Manuscript*, vol. 7 of *The Collected Papers of Bertrand Russell* ed. Elizabeth Ramsden Eames in collaboration with Kenneth Blackwell (London: Routledge, 1984).

1914a. *Our Knowledge of the External World* (Chicago: Open Court). Second edition, with revisions (London: Allen & Unwin, 1926); with new preface and slightly new revisions (New York: Norton, 1929).

1914b. "The Relation of Sense-data to Physics," *Scientia*. Reprinted in *MyL*, pp. 108–31.

1917. (*MyL*) *Mysticism and Logic* (New Jersey: Barnes & Nobel, 1981).

1919. (*IMP*) *Introduction to Mathematical Philosophy* (London: Allen & Unwin).

1925. (*PM2*) *Principia Mathematica*, second edition with new Introduction and Appendices (Cambridge: Cambridge University Press).

1944. (*MMD*) "My Mental Development," in Schilpp, pp. 3–20.

1956. *Logic and Knowledge: Essays 1901–1950* (London: Allen & Unwin).

1959. (*MPD*) *My Philosophical Development* (London: Allen & Unwin).

1967. (*A*) *Autobiography of Bertrand Russell*, 3 vols. (London: Allen & Unwin, 1967–9).

1973. (*EA*) *Essays in Analysis by Bertrand Russell*, ed. D. Lackey (New York: Braziller).

1994. (*CP*) *The Collected Papers of Bertrand Russell 1903–1905*, vol 4, ed. Alasdair Urquhart (London: Routledge).

The Collected Papers of Bertrand Russell, vols 1–28 (London, Routledge).

Works by Other Authors

Anderson, C. Anthony. 1986. "Some Difficulties Concerning Russellian Intensional Logic," *Noûs* 20, pp. 35–43.

Anscombe, G. E. M. 1959. *An Introduction to Wittgenstein's Tractatus* (New York: Harper Torchbooks).

Bernays, P. 1926. "Aussagen-Kalkukldér PM," *Mathematische Zeitschrift* 25, pp. 305–30.

Blackburn and Code. 1978. "The Power of Russell's Criticism of Frege," *Analysis* 38, pp. 65–77.

Blackwell, Kenneth. 1984. "Part I of *The Principles of Mathematics*," *Russell* 4, pp. 271–88.

Böer, Steven. 1973. "Russell on Classes as Logical Fictions," *Analysis* 33, pp. 206–8.

Boyer, Carl. 1949. *The History of the Calculus and Its Conceptual Development* (New York: Dover).

Burali-Forti, Cesare. 1896. "Le classi finite," *Atti della R. Accademia delle scienze, di Torino*, vol. 32, pp. 34–52.

—1897. "Una questione sui numeri transfiniti," *Rediconti de Palmero* 11, pp. 154–64.

—1899. "Les propriétiés formales des opérations algébriques," *Revue de mathématiques* 6, pp. 141–77.

—1900. "Sur les différentes méthodes logiques pour la définition du nombre réel," Ier Congrès Internationale de Philosophie, Tome 3, pp. 289–307.

Butler, R. J. 1954. "The Scaffolding of Russell's Theory of Descriptions," *Philosophical Review* 63, pp. 350–64.

Byrd, Michael. 1987. "Part II of *The Principles of Mathematics*," *Russell* 7, pp. 60–70.

—1989. "Russell, Logicism and the Choice of The Logical Constants," *Notre Dame Journal of Formal Logic* 30, pp. 343–61.

Carnap, Rudolf. 1947. *Meaning and Necessity* (Chicago: University of Chicago Press).

Cassin, Chrystine. 1970. "Russell's Discussion of Meaning and Denoting: A Reexamination," in *Essays on Bertrand Russell*, ed. E. Klempke (Urbana: University of Illinois Press), pp. 256–72.

Chihara, Charles. 1973. *Ontology and the Vicious Circle Principle* (Ithaca: Cornell University Press).

Church, Alonzo. 1956. *Introduction to Mathematical Logic* (Princeton: Princeton University Press).

—1974. "Russellian Simple Type Theory," *Proceedings and Addresses of the American Philosophical Association*, pp. 21–33.

—1978. "A Comparison of Russell's Resolution of the Semantical Antinomies with that of Tarski," *Journal of Symbolic Logic* 41, pp. 747–60.

—1984. "Russell's Theory of the Identity of Propositions," *Philosophia Naturalis* 21, pp. 513–22.

Cocchiarella, Nino. 1980. "The Development of the Theory of Logical Types and the Notion of a Logical Subject in Russell's Early Philosophy," *Synthese* 45, pp. 71–115.

—1986. *Logical Investigations of Predication Theory and the Problem of Universals* (Naples: Bibliopolis Press).

—1989. "Russell's Theory of Logical Types and the Atomistic Hierarchy of Sentences," in *Rereading Russell: Essays on Bertrand Russell's Metaphysics and Epistemology*, vol. XI of *Minnesota Studies in Philosophy of Science*, eds. C. Wade Savage and C. Anthony Anderson (Minneapolis: University of Minnesota Press), pp. 41–62.

Coffa, Alberto. 1979. "The Humble Origins of Russell's Paradox," *Russell* 33, pp. 31–7.

—1981. "Russell and Kant," *Synthese* 46, pp. 247–63.

Copi, Irving. 1950. "The Inconsistency or Redundancy of *Principia Mathematica*," *Philosophy and Phenomenological Research* 11, pp. 190–9.

—1971. *The Theory of Types* (London: Routledge & Kegan Paul).

— 1979. *Introduction to Symbolic Logic* (New York: Macmillan Publishing Co.).

Dau, Paulo. 1985. "The Complex Matter of Denoting," *Analysis* 45, pp. 190–9.

—1986. "Russell's First Theory of Denoting and Quantification," *Notre Dame Journal of Formal Logic* 27, pp. 133–66.

Dixon, A. C. 1906. "On 'Well-Ordered' Aggregates," *Proceedings of the London Mathematical Society*, 4, pp. 18–20.

Dummett, Michael. 1981. *The Interpretation of Frege's Philosophy* (Cambridge, Mass.: Harvard University Press).

Frege, Gottlob. 1879. *Begriffsschrift, eine der Arithmetischen Nachgebildete Formelsprache des Reinen Denkens* (Halle).

— 1884. *Die Grundlagen der Arithmetik* (Breslau).

— 1891. *Funktion und Begriff* (Jena). Translated in *Translations from the Philosophical Writings of Gottlob Frege*, eds. P. T. Geach and M. Black (Oxford: Basil Blackwell, 1980), pp. 21–41.

— 1892a. "Über Funktion und Gegenstand," in Vierteljachrsschrift für wissenschaftliche Philosophie, vol. xiv, pp. 192–205. Translated in *Translations of the Philosophical Writings of Gottlob Frege*, eds. P. T. Geach and M. Black (Oxford: Basil Blackwell, 1980), pp. 21–41.

— 1892b. "Über Sinn und Bedeutung," *Zeitschrift für Philosophie und Philosophische Kritik* vol. c., pp. 25–50. Translated in *Translations from the Philosophical Writings of Gottlob Frege*, eds. P. T. Geach and M. Black (Oxford: Basil Blackwell, 1980), pp. 21–41.

— 1893. *Grundgesetze Der Arithmetik*, vol. 1 (Jena).

— 1969. "On Herr Peano's *Begriffsschrift* and My Own," *Australasian Journal of Philosophy*, 47, pp. 1–14.

— 1980. *Gottlob Frege: Philosophical and Mathematical Correspondence*, eds. G Gabriel, H. Hermes, F. Kambartel, C. Thiel, and A. Veraart. Abridged from the German edition by B. McGuinness and translated by Hans Kaal (Chicago: University of Chicago Press).

Geach, P. T. 1950. "Russell's Theory of Descriptions," *Analysis* 10, pp. 84–8.

— 1959. "Russell on Meaning and Denotation," *Analysis* 19, pp. 69–72.

— 1962. *Reference and Generality*, 3rd edition (Ithaca: Cornell University Press).

— 1969. "On Frege's Way Out," in *Logic Matters* (Berkeley: University of California Press), pp. 235–7.

Goldfarb, Warren. 1989. "Russell's Reasons for Ramification," in *Rereading Russell: Essays on Bertrand Russell's Metaphysics and Epistemology*, vol. XI of *Minnesota Studies in Philosophy of Science*, eds. C. Wade Savage and C. Anthony Anderson (University of Minnesota Press), pp. 24–40.

Grattan-Guinness, Ivor. 1972. "Bertrand Russell on His Paradox and the Multiplicative Axiom: An Unpublished Letter to Philip Jourdain," *Journal of Philosophical Logic* 1, pp. 103–10.

— 1974. "The Russell Archives: Some New Light on Russell's Logicism," *Annals of Science* 31, pp. 387–406.

— 1977a. "Bertrand Russell's Manuscripts: An Apprehensive Brief," *History and Philosophy of Logic* 6, pp. 53–74.

— 1977b. *Dear Russell – Dear Jourdain* (London: Duckworth).

— 1978. "How Bertrand Russell Discovered his Paradox," *Historica Mathematica* 5, pp. 127–37.

— 1980. "Georg Cantor's Influence on Bertrand Russell," *History and Philosophy of Logic*, pp. 61–93.

— 1984. "Notes on the Fate of Logicism from *Principia Mathematica* to Gödel's Incompletability Theorem," *History and Philosophy of Logic* 5, pp. 67–78.

— 1987. "The Tenability of Russell's Early Philosophy," (Panel discussion: I. Grattan-Guinness, Nicholas Griffin, Robert Tully, and W. V. O. Quine, with A. J. Ayer moderating), *Russell* 8, pp. 232–46.

Griffin, Nicholas. 1980. "Russell on the Nature of Logic," *Synthese* 45, pp. 117–88.

— 1985. "Russell's Multiple Relation Theory Judgment," *Philosophical Studies* 47, pp. 213–47.

— 1986. "Wittgenstein's Criticism of Russell's Theory of Judgment," *Russell* 4, pp. 132–45.

—1992. *The Selected Letters of Bertrand Russell* vol 1, 1884–1914 (Boston: Houghton Mifflin).

Grelling, K. and Nelson, L. 1907. "Bemerkungen zu den Paradoxien van Russell and Burali-Forti," *Abhandlung der Friesschen Schule*, 2, pp. 300–24.

Hatcher, William. 1968. *The Logical Foundations of Mathematics* (Oxford: Pergamon Press).

Hintikka, Jaakko. 1969. "On Kant's Notion of Intuition," in *The First Critique*, eds. Terence Penelhum and J. J. MacIntosh (Belmont, Calif.: Wadsworth Publishing Co.), pp. 38–53.

Hochberg, Herbert. 1980. "Descriptions, Scope, and Identity," *Analysis* 18, pp. 205–8.

—1987. "Russell's Early Analysis of Relational Predication and the Asymmetry of the Predication Relation," *Philosophica* 17, pp. 439–59.

Hylton, Peter. 1980. "Russell's Substitutional Theory," *Synthese* 45, pp. 1–31.

—1990. *Russell, Idealism and the Emergence of Analytic Philosophy* (Oxford: Oxford University Press).

Johnson, W. E. 1922. *Logic*, vol. 2 (Cambridge: Cambridge University Press).

Kennedy, H. C. 1973. "What Russell Learned from Peano," *Notre Dame Journal of Formal Logic* 14, pp. 367–72.

—1974. "Peano's Conception of Number," *Historica Mathematica* 1, pp. 387–408.

—1975. "New Letters from Guiseppe Peano to Bertrand Russell," *Journal of the History of Philosophy* 13, pp. 203–20.

Kilmister, C. W. 1984. *Russell* (New York: St. Martin's Press).

Kripke, Saul. 1976. "Is There a Problem about Substitutional Quantification?" in *Truth and Meaning: Essays in Semantics*, ed. Gareth Evans and John McDowell (Oxford: University Press), pp. 375–419.

Lackey, Douglas. 1974. "Russell's Anticipation of Quine's Criterion," *Russell* 16, pp. 27–31.

—1976. "Russell's Unknown Theory of Classes: The Substitutional System of 1906," *Journal of the History of Philosophy* 14, pp. 69–78.

Lambert, Karel. 1990. "Russell's Theory of Definite Descriptions," *Dialectica* 44, pp. 137–52.

Landini, Gregory. 1989. "New Evidence Concerning Russell's Substitutional Theory of Classes," *Russell* 9, pp. 26–42.

—1991. "A New Interpretation of Russell's Multiple-Relation Theory of Judgment," *History and Philosophy of Logic* 12, pp. 37–69.

—1992. "Russell to Frege 24 May 1903: 'I Believe I Have Discovered That Classes are Entirely Superfluous,'" *Russell* 12, pp. 160–85.

—1993. "Reconciling *PM's* Ramified Type Theory with the Doctrine of the Unrestricted Variable of the *Principles*," in eds. A. D. Irving and G. A. Weber, *Russell and Analytic Philosophy* (Toronto: Toronto University Press), pp. 361–94.

—1996a. "The *Definability* of the Set of Natural Numbers in the 1925 *Principia Mathematica*," *Journal of Philosophical Logic*, 25, pp. 597–615.

—1996b. "Logic in Russell's *Principles of Mathematics*," *Notre Dame Journal of Formal Logic*, 37, pp. 554–84.

—1966c. "Decomposition and Analysis in Frege's *Grundgesetze*," *History and Philosophy of Logic*, 17, pp. 121–39.

—1998. "Principia's *9," *History and Philosophyy of Logic*, forthcoming.

Linsky, Bernard. 1988. "Propositional Functions and Universals in *Principia Mathematica*," *Australasian Journal of Philosophy*, 66, pp. 447–60.

—1990. "Was the Axiom of Reducibility a Principle of Logic?" *Russell* 10, pp. 125–40.

Linsky, Leonard. 1983. *Oblique Contexts* (Chicago: University of Chicago Press).

Linsky, Leonard and Schumm, George. 1971. "Frege's Way Out: A Footnote," *Analysis* 32, pp. 5–7.

Lukasiewicz, Jan. 1963. *Elements of Mathematical Logic* (Oxford: Pergamon Press).

McGuinness, Brian, ed. 1980. *Gottlob Frege: Philosophical and Mathematical Correspondence* (Chicago: University of Chicago Press).

Myhill, John. 1974. "The Undefinability of the Set of Natural Numbers in the Ramified Principia in *Bertrand Russell's Philosophy*, ed. George Nakhnikian (New York: Barnes & Noble), pp. 19–27.

— 1979. "A Refutation of an Unjustified Attack on the Axiom of Reducibility," in *Bertrand Russell Memorial Volume*, ed. George Roberts (London: Allen & Unwin), pp. 81–90.

Padoa, Alessandro. 1886. "Note di logica mathematica," *Revue de mathématiques* 6, pp. 105–21.

— 1938. "Ce que la logique doit à Peano," *Actes du Congrès International de Philosophie Scientifique*, vol 8, 1938, pp. 282–36.

Peano, Guiseppe. 1889. *Arithmetices principia, novo methodo exposita* (Turin). Translation in *From Frege to Gödel: A Source Book on Mathematical Logic, 1879–1931*, ed. Jean van Heijenoort (Cambridge Mass.: Harvard University Press, 1967), pp. 83–97.

— 1889–1901. *Formulaire de mathématiques*, vols. 1–5 (Turin).

Pelham, Judy and Urquhart, Alasdair. 1991. "Russellian Propositions," *Logic, Methodology and the Philosophy of Science IX*, Elsevier.

Peirce, Charles Sanders. 1970. "Descriptions of a Notation for the Logic of Relations," *Memoirs of the American Academy of Arts and Sciences*, **9**, pp. 313–87.

Poincaré Henri. 1905–6. "Les mathématiques et la logique." *Revue de métaphysique et de morale* 13, pp. 815–35; 14, pp. 17–34, 294–317.

Quine, W. V. O. 1940. *Mathematical Logic* (Cambridge Mass.: Harvard University Press; revised edition 1951; paperback, New York: Harper, 1962).

— 1941. "Whitehead and the Rise of Modern Logic," in *The Philosophy of A. N. Whitehead*, ed. P. Schilpp (La Salle: Open Court), pp. 125–63.

— 1955. "On Frege's Way Out," *Mind 64*, pp. 145–9.

— 1963. *Set Theory and Its Logic* (Cambridge Mass.: Harvard University Press).

— 1966. "Russell's Ontological Development," *Journal of Philosophy* 63, pp. 657–67. Reprinted in *Bertrand Russell, Philosopher of the Century*, ed. R. Schoenman (London: Allen & Unwin, 1967), pp. 304–14.

— 1967. "Introduction to Russell's 'Mathematical Logic as Based on the Theory of Types,'" in *From Frege to Gödel: A Source Book in Mathematical Logic, 1879–1931*, ed. Jean van Heijenoort (Cambridge, Mass.: Harvard University Press), pp. 150–2.

Robinson, R. M. 1949. "Definability and Decision Problems in Arithmetic," *Journal of Symbolic Logic* 14, pp. 98–114.

Rodriguez-Consuegra, Francisco. 1987. "Russell's Logicist Definitions of Numbers 1899–1913: Chronology and Significance," *History and Philosophy of Logic* 8, pp. 141–69.

— 1991. *The Mathematical Philosophy of Bertrand Russell: Origins and Development* (Boston: Birkhäuser Press).

— 1992. "A Comparison of the Theories of Descriptions by Frege, Peano and Russell," manuscript.

Sainsbury, Mark. 1979. *Russell* (London: Routledge and Kegan Paul).

Schilpp, Paul Arthur, ed. 1946. *The Philosophy of Bertrand Russell* (Evanston, Ill.: The Library of Living Philosophers).

Schröder, F. W. 1890–1905. *Vorlesungen über die Algebra der Logik* (3 vols.) Leipzig.

Schutte, Kurt. 1960. *Beweistheorie* (Berlin: Springer).

Tichý, Pavel. 1988. *The Foundations of Frege's Logic* (New York: de Gruyter).

Urquhart, Alasdair. 1988. "Russell's Zig-Zag Path to the Ramified Theory of Types," *Russell* 8, pp. 82–91.

Turnau, Paweł. 1991. "Russell's Argument against Frege's Sense–Reference Distinction," *Russell* 11, pp. 52–66.

van Heijenoort, Jean. 1967a. *From Frege to Gödel: A Source Book in Mathematical Logic, 1879–1931* ed. Jean van Heijenoort (Cambridge Mass.: Harvard University Press), pp. 150–2.

— 1967b. "Logic as Calculus and Logic as Language," *Synthese* 17, pp. 324–30.

— 1987. "Systeme et Mètasystéme ches Russell," in Paris Logic Group (eds), *Logic Colloquium* 1985 (Amsterdam: North Holland).

Wahl, Russell, 1983. "Russell's Theory of Meaning and Denotation," *Journal of the History of Philosophy* 31, pp. 71–94.

Vuillemin, Jules. 1968. *Leçons sur la premère philosophie de Russell* (Paris: Colin).

— 1972. "Platonism in Russell's Early Philosophy and the Principle of Abstraction," in *Bertrand Russell*, ed. D. Pears (New York: Doubleday), pp. 305–24.

Wittgenstein, Ludwig. 1921. *Tractatus Logico–Philosophicus in Annalen der Naturphilosophie*. Trans. by D. F. Pears and B. F. McGuinness (New York: Routledge & Kegan Paul).

Index